Biophysical Chemistry
Membranes and Proteins

Edited by

Richard H. Templer and Robin Leatherbarrow
Imperial College of Science and Technology, London, UK

RS•C
ROYAL SOCIETY OF CHEMISTRY

The Proceedings of the first annual international conference, Biophysical Chemistry 2001, held at Imperial College, London, UK on 19–21 September 2001

Special Publication No. 283

ISBN 0-85404-851-0

A catalogue record for this book is available from the British Library

Published by The Royal Society of Chemistry,
Thomas Graham House, Science Park, Milton Road,
Cambridge CB4 0WF, UK
Registered Charity No. 207890

For further information see our web site at www.rsc.org

Printed by George Over Ltd., London and Rugby

Preface

On the 19th of September 2001 the first annual meeting organised by the Biophysical Chemistry Group of the Royal Society of Chemistry was held in London. Called Biophysical Chemistry 2001 the conference focussed on four areas: modelling of biological systems; membrane structures and interactions; methods for probing biomolecules and channels and receptors. The aim of the conference was to mix together scientists from the physical and life sciences who were bound only by their shared interest in understanding the molecular details of biological processes. The areas we chose as a focus for this sociological experiment were designed to have the maximum potential for cross-fertilisation of scientific thought.

Despite the tragic events that occurred in the week preceding the conference over 170 scientists from around the world were able to attend what was a lively and invigorating three days of scientific discussion. This book consists of a selection of the contributions that were made to the conference. The authors were asked to write articles, which although they would be subjected to refereeing were expected to be opinionated and that the authors' opinions would not be altered. The articles should be read in the same spirit. Indeed the opinions and views expressed during the conference and we hope replicated to some extent in these papers is in some part a reflection of the different ways in which our scientific training prepares us to view the world around us. We hope that this diversity in viewpoint will be evident from the collection of papers we present here and that you will find them as stimulating as we did.

Richard Templer and Robin Leatherbarrow, Imperial College, London, May 2002

Contents

I Probing Biological Molecules: Theory and Experiment

Flow Oriented Linear Dichroism to Probe Protein Orientation in Membrane
Environments 3
*A. Rodger, J. Rajendra, R. Mortimer, T. Andrews, J.D. Hirst, A.T.B. Gilbert,
R. Marrington, T.R. Dafforn, D.J. Halsall, M. Ardhammar, B. Nordén,
C.A. Woolhead, C. Robinson, T.J.T. Pinheiro, J. Kazlauskaite, M. Seymour,
N. Perez and M.J. Hannon*

Quantitative Protein Circular Dichroism Calculations 20
N.A. Besley and J.D. Hirst

Probing Cellular Structure and Function by Atomic Force Microscopy 31
M.A. Horton, P.P. Lehenkari and G.T. Charras

Physical Characterization of Wild Type and *mnn9* Mutant Cells of
Saccharomyces cerevisiae by Atomic Force Microscopy (AFM) 50
A. Méndez-Vilas, I. Corbacho, M.L. González-Martín and M.J. Nuevo

Probing Supramolecular Organisation at Immune Synapses 58
*F.E. McCann, K. Suhling, L.M. Carlin, K. Eleme, K. Yanagi, P.M.W. French,
D. Phillips and D.M. Davis*

Probing the Structure of Viral Ion Channel Proteins: A Computational
Approach 72
W.B. Fischer and M.S.P. Sansom

The Impact of H_2O_2 on the Structure of Catalases by Molecular
Modelling Methods 78
S.G. Kalko, J.Ll. Gelpí and M. Orozco

Entropy in the Alignment and Dimerization of Class C G-Protein Coupled
Receptors 85
*M.K. Dean, C. Higgs, R.E. Smith, P.D. Scott, R.P. Bywater, T.J. Howe and
C.A. Reynolds*

Electrostatic Stability of Wild Type and Mutant Transthyretin Oligomers 94
S. Skoulakis and J.M. Goodfellow

Simulations of Human Lysozyme: Conformations Triggering Amyloidosis in
156T Mutant 103
G. Moraitakis and J.M. Goodfellow

Collective Excitation Dynamics in Molecular Aggregates: Exciton Relaxation,
Self-Trapping and Polaron Formation   118
M. Dahlbom, W. Beenken, V. Sundström and T. Pullerits

Surprising Electro-magnetic Properties of Close Packed Organized Organic
Layers – Magnetization of Chiral Monolayers of Polypeptide   136
I. Carmeli, V. Shakalova, R. Naaman and Z. Vager

Barrier Crossing by a Flexible Long Chain Molecule – The Kink Mechanism   147
K.L. Sebastian

II Proteins, Lipids and Their Interactions

Lipid Interaction with Cytidylyltransferase Regulates Membrane Synthesis   163
S. Jackowski and I. Baburina

Models and Measurements on the Monolayer Bending Energy of Inverse
Lyotropic Mesophases   177
A.M. Squires, J.M. Seddon and R.H. Templer

Hemolytic and Antibacterial Activities of LK Peptides of Various Topologies:
A Monolayer and PM-IRRAS Approach   191
S. Castano, B. Desbat, H. Wróblewski and J. Dufourcq

A Novel Approach for Probing Protein–Lipid Interactions of MscL,
a Membrane-Tension-Gated Channel   199
P.C. Moe and P. Blount

Folding of The α-Helical Membrane Proteins DsbB and NhaA   208
D.E. Otzen

FhuA, an *Escherichia coli* Transporter and Phage Receptor   215
P. Boulanger, L. Plançon, M. Bonhivers and L. Letellier

Morpholgical Aspects of *in cubo* Protein Crystallisation   221
*C. Sennoga, B. Hankamer, A. Heron, J.M. Seddon, J. Barber and
R.H. Templer*

Mobility of Proteins and Lipids in the Photosynthetic Membranes of
Cyanobacteria   237
C.W. Mullineaux and M. Sarcina

Partitioning and Thermodynamics of Chlordiazepoxide in *n*-Octanol/Buffer
and Liposome System   243
C. Rodrigues, P. Gameiro, S. Reis, J.L.F.C. Lima and B. De Castro

Distribution of Vitamin E in Model Membranes 248
P.J. Quinn

Theory on Opening-up of Liposomal Membranes by Adsorption of Talin 254
Y. Suezaki

Differential Scanning Calorimetry and X-Ray Diffraction Studies of
Glycolipid Membranes 267
O. Ces, J.M. Seddon, R.H. Templer, D.A. Mannock and R.N. McElhaney

Subject Index 277

I Probing Biological Molecules: Theory and Experiment

FLOW ORIENTED LINEAR DICHROISM TO PROBE PROTEIN ORIENTATION IN MEMBRANE ENVIRONMENTS

Alison Rodger,[a]* Jascindra Rajendra,[a] Rhoderick Mortimer,[b] Terrence Andrews,[b] Jonathan D. Hirst,[c] Andrew T.B. Gilbert,[c] Rachel Marrington,[a] Timothy R. Dafforn,[d] David J. Halsall,[d] Malin Ardhammar,[e] Bengt Nordén,[e] Cheryl A. Woolhead,[a] Colin Robinson,[f] Teresa J.T. Pinheiro,[f] Jurate Kazlauskaite,[f] Mark Seymour,[g] Niuvis Perez,[h] Michael J. Hannon[a]

a   Department of Chemistry, University of Warwick, Coventry, CV4 7AL, UK
b   Crystal Precision Ltd, Little Farm, Faraday Road, Rugby, Warwickshire, CV22 5NB, UK
c   School of Chemistry, University of Nottingham, University Park, Nottingham, NG7 2RD, UK
d   Cambridge Institute of Medical Research, Wellcome Trust/MRC Building, Hills Rd, Cambridge, CB2 2XY, UK
e   Physical Chemistry, Chalmers University of Technology, S-41296, Gothenburg, Sweden
f   Biological Sciences, University of Warwick, Coventry, CV4 7AL, UK
g   Syngenta, Jealott's Hill, Bracknell, Berkshire, RG42 6EY, UK
h   Center of Molecular Immunology, 216 St. and 15, Atabey, Playa, PO Box 16040, Habana 11600 Cuba

* email: a.rodger@warwick.ac.uk   Fax (+44) 24 76524112

1   INTRODUCTION

Processes occurring on or in membranes are essential in most biological systems, and the study of these processes has been engendering an increasing interest for a long time, as has the creation of artificial lipid membrane systems. Studies of, for example, membrane transport and membrane protein function call for a thorough knowledge of molecular interactions within the membrane, between the lipids themselves and between lipids and other species (proteins, drugs, and ions). To this end, the locations and orientations of molecules bound to the membrane can give important information. However, to date no simple experimental method has been established to achieve this for membrane bound proteins. In this work we report the first flow linear dichroism (LD) [1,2,3] study of proteins bound to liposomes. Flow LD of molecules bound to the bilayer of shear-deformed liposomes is one of the few direct methods potentially available for the study of the orientation of membrane guest molecules, provided that the molecules of interest have significant absorption in the visible and near-UV regions.

Linear dichroism is the difference in absorption of light polarised parallel to an orientation direction and light polarised perpendicular to that direction. The LD signal is related to the oscillator strength of a transition (its absorbance intensity) and the polarisation of the transition relative to the orientation axis. It is thus the ideal technique to use to probe the orientation of an analyte and it is widely used, for example, for determining the orientation of drugs bound to flow oriented DNA. The most effective method for achieving flow orientation has proved to be Couette flow [1,2,3] where two concentric cylinders with a small annular gap (usually 500 μm) are aligned and one of them rotates. The light is incident radially on the cell so the stationary cylinder needs to have two windows and the rotating cylinder needs to be transparent to the intended radiation.

Liposomes can be considered models of cell membranes, and be used for studying transport and signal mechanisms of membrane proteins in situ. They are also used for drug delivery and as transfecting agents in gene therapy. Ardhammar, Mikati, Lincoln and Norden [4] have shown that aromatic moieties or the aromatic 'arm' of ruthenium dipyridophenazine can be oriented in liposomes and their orientation detected when the liposomes are subjected to shear flow. In this work we probed the orientation achievable by flow with different size model membranes and then the application of *LD* to determine the orientation of proteins bound on or in liposomes. The proteins studied include gramicidin, cytochrome *c*, pre-PsbW (a thylakoid membrane protein precursor) and a monoclonal antibody.

2   LINEAR DICHROISM OF ANALYTES IN SHEAR DISTORTED LIPOSOMES

As noted above there are two literature precedents [4] for flow orienting liposomes to assess the orientation of solutes in their bilayer. Ardhammar *et al.* found that pyrene and anthracene in soybean liposomes (produced by extrusion) had a negative *LD* for their long axes, in accord with their expectations that the molecules would be oriented with their long axis parallel to the lipid hydrocarbon chains and thus on average perpendicular to the elongation and orientation axis of the liposome. With perylene, however, both the long axis and the short axis had negative *LD* signals, which the authors took to mean that the diagonal of the molecule was oriented parallel to the lipid chains. For the later ruthenium work the authors modelled a deformed liposome as a cylinder with hemi-spherical caps, assumed that the lipids redistributed evenly in the deformed liposome, and determined that the reduced *LD*, $LD^r = LD$/isotropic absorbance, to be:

$$LD^r = \frac{3S}{4}(1 - 3\cos^2\alpha_i) \tag{1}$$

where α_i is the angle the transition moment of interest makes with the normal to the cylinder surface (*i.e.* to the lipid long axis), *S* is the orientation factor that denotes the fraction of the liposome that is oriented as a cylinder perfectly parallel to the flow direction.

This equation can be formally derived as follows. Consider an ellipsoidal liposome under the flow conditions to be approximated by a cylindrical bilayer, whose lipids are perpendicular to the long axis of the cylinder, capped by two hemispheres. Analytes oriented within the hemispherical caps, assuming an average uniform distribution of analytes, will have no net *LD*, and can therefore be ignored.

The *LD* experiment defines one axis system $\{x, y, z\}$, where *z* is the orientation axis (the long axis of the cylinder) and the *x/y* plane is perpendicular to this (Figure 1). In order to determine the *LD*, it is useful to use another axis system $\{X, Y, Z\}$ defined by the bilayer and an analyte. Z is the long axis of the cylinder (so $z = Z$) and X is the normal to the cylinder surface that passes through the origin of the analyte.

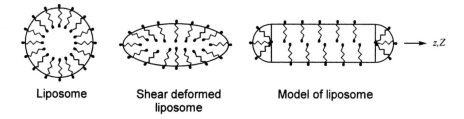

| Liposome | Shear deformed liposome | Model of liposome |

Figure 1   Schematic illustration of a shear deformed liposome.

The analyte orientation will not be affected by the shear flow (the forces are too small), so on average any analyte transition moment, μ_i, will be uniformly distributed about the X axis. Let α_i be defined as above. Let β_i be the angle between the projection of μ_i onto the Y/Z plane and the Z axis. Thus in the $\{X, Y, Z\}$ coordinate system

$$\mu_i = \mu(\cos a_i, \sin \beta_i \sin \alpha_i, \cos \beta_i \sin \alpha_i)_{XYZ} \qquad (2)$$

where μ is the magnitude of the transition moment and β is the angle the YZ projection of μ makes with the Z axis.

The LD is by definition:

$$LD = (A_z - A_y)$$
$$= S(\mu_z^{\ 2} - \mu_y^{\ 2}) \qquad (3)$$

where μ_z is the z component of the transition dipole in the $\{x, y, z\}$ coordinate system and, in this case, also in the $\{X, Y, Z\}$ coordinate system. μ_y may be written as the dot product of the transition moment vector and the vector for the y axis in the $\{X, Y, Z\}$ coordinate system. Thus

$$\mu_y = \mu \cdot y$$
$$= \mu_i(\cos a_i, \sin \beta_i \sin \alpha_i, \cos \beta_i \sin \alpha_i)_{XYZ} \cdot (\sin \gamma, \cos \gamma, 0)_{XYZ} \qquad (4)$$
$$= \mu_i(\cos a_i \sin \gamma + \sin \beta_i \sin \alpha_i \cos \gamma)$$

where γ can take any value from 0 to 2π. Since the isotropic absorbance is $\mu^2/3$, we can write

$$\frac{LD^r}{3S} = (\cos \beta_i \sin \alpha_i)^2 - (\cos a_i \sin \gamma + \sin \beta_i \sin \alpha_i \cos \gamma)^2$$
$$= (\cos^2 \beta_i \sin^2 \alpha_i) - (\cos^2 a_i \sin^2 \gamma + \sin^2 \beta_i \sin^2 \alpha_i \cos^2 \gamma + 2\cos a_i \sin \gamma \sin \beta_i \sin \alpha_i \cos \gamma) \qquad (5)$$

Both β_i and γ_i can take any value from 0 to 2π, so upon averaging over them:

$$\frac{LD^r}{3S} = \frac{2\sin^2 \alpha_i - 2\cos^2 \alpha_i - \sin^2 \alpha_i}{4}$$
$$= \frac{2 - 2\cos^2 \alpha_i - 2\cos^2 \alpha_i - 1 + \cos^2 \alpha_i}{4} \qquad (6)$$
$$= \frac{1 - 3\cos^2 \alpha_i}{4}$$

as given by Ardhammar *et al.*[4] Two special cases of interest are when a transition moment is parallel to the cylinder normal (the lipids), for which $\alpha_i = 0$ and $LD^r/3S = -1/2$, and when it is perpendicular to that, *e.g.* on the surface of the cylinder, for which $\alpha_i = 90°$ and $LD^r/3S = +1/4$.

3   MATERIALS AND METHODS

The solvents were all analytical grade obtained from BDH laboratory supplies and were dried where necessary. Other chemicals including soybean lipid (L-α-Lecithin, Type IV-S: from soybean approx. 40% (TLC)), Gramicidin D from Bacillus Brevis and cytochrome *c* (Horse heart muscle) were obtained from Sigma Chemical Company and were used as received. Lyso-1-myristoyl-sn-glycero-3-phosphocholine (LMPC) and 1-palmitoyl-2-oleoyl-phosphatidylcholine were obtained from Avanti Polar Lipids. Pre-PsbW was made according to the methods in reference [5]. The humanised monoclonal antibody was prepared as described in reference [6].

The methods were used to create the membrane mimicking environments for these experiments depended on the sample to be studied and the size of membrane structure required. The smallest structures were produced by sonication as the final stage; the largest by vortexing the lipid preparations (this results in large multi-lamellar liposomes as shown by microscopy; unilamellar liposomes were produced from multi-lamellar liposomes by putting them through a LiposoFast Basic Extruder (with a polycarbonate membrane) from 12 and 25 times following reference[3]). In general lipids and analytes were in molar ratio approximately 50:1 for pyrene and ~ 5 mg/mL lipid to ~ 1 mg/mL protein for the proteins, though in some cases further dilution was required to avoid excessive absorbance. If the analyte was not water soluble, it was generally simplest to mix lipid and analyte in chloroform. The solvent was then evaporated and kept under vacuum for 24 hours. To the dry lipid mixture was added 5 mM phosphate buffer (pH 7) to a final concentration of lipid ~ 5 mg/mL. Then the sample was sonicated or vortexed or extruded. If the analytes were water soluble they were added with the buffer or to the liposome solution after this had been prepared. Although Gramicidin is not very water soluble, in this case it was introduced to the membrane by treating as if it were water soluble and extruding the sample which seemed to facilitate its solubilisation by the lipids. Alternatively, Gramicidin was added to a liposome preparation from an ethanol solution. The LD experiments were performed on newly prepared samples.

UV-visible absorbance spectra were recorded using a Cary 1E spectrophotometer. Flow linear dichroism spectra were recorded using a Jasco J-715 circular dichroism spectropolarimeter, which was adapted for flow *LD* measurements. The photomultiplier tube was moved into the sample compartment next to the *LD* cell to reduce artefacts due to scattered light using a housing fabricated for the instrument by European Chirality Services. The cell used was developed for this work and is described below. The rotation speed used in the experiments was ~1000 rpm. In each case the speed chosen was the maximum value that avoided significant bubble formation in the cell. An *LD* baseline was measured on the same sample but without any rotation.

With light scattering samples, this method gave a slightly better baseline correction than a water baseline. For non-scattering samples the same baseline was recorded for the *LD* sample in a non-spinning cell and for a water or other non-orientatable solution in a spinning cell.

The geometries for the determination of protein structural motif transition moments were constructed using the CHARMM program.[7] Two α-helices were considered: the ideal geometry of Pauling and Corey [8] with (φ=−48°, ψ=−57°) and the mean values observed in experimentally determined structures of (φ=−63°, ψ=−41°). For the β-strand a single geometry with (φ=−135°, ψ=135°) was chosen. Two different geometries of gramicidin were also considered.[9] For the α-helices and β-strand, constrained minimisations of the blocked peptides Amn-Ala$_{200}$ cbx were performed in the absence of solvent effects. A length of 200 amino acids was chosen to minimise end effects, and thus mimic an infinite helix or strand.

The polarisation calculations were performed using the Matrix Method [10] in which the protein is considered to be a set of *n* chromophores, each with a characteristic set of excitation bands. These effectively form a basis for the transitions for the entire protein, and are allowed to couple through diagonalisation of the Frenkel Hamiltonian. This yields a set of energies (eigenvalues) and transition couplings (eigenvectors) for the protein transitions. In our calculations we consider only the $n{\rightarrow}\pi^*$ and $\pi{\rightarrow}\pi^*$ transitions for each peptide link in the protein backbone. To date, this approach is the most accurate for the calculation of protein circular dichroism from first principles.[11] The polarisation tensor can be calculated using a sum over states

$$\alpha_{\lambda v} = \sum_b \frac{2\mu^{\lambda}_{ab}\mu^{v}_{ab}}{\Delta E(a \rightarrow b)} \qquad (7)$$

where μ^{λ}_{ab} is the transition dipole moment from state *a* to state *b* in the λ direction. This can be obtained from the eigenvectors. This approach allows the contribution to the polarisation from each individual transition to be considered and, in particular, allows the polarisation due to the $n{\rightarrow}\pi^*$ bands to be isolated.

4   DESIGN AND CONSTRUCTION OF *LD* CELL

Preliminary experiments on the lipids with a 500 μm annular gap *LD* cell [3] were unsatisfactory, at least in part due to excessive bubble formation reducing the signal to noise ratio. For this and other work it was therefore decided to design a new *LD* cell with a significantly smaller annular gap of 50 μm. Although this reduces the pathlength of the sample, it was hoped that the increase in average orientation would compensate for this. It was also decided to construct the optics out of CaF_2 to extend the measurement range into the infra red region if required. This led to problems relating to the effect of heat shock on CaF_2 (it shatters). The final cell consisted of the following components as illustrated in Figure 2:

Main Body: Stainless steel construction.
Rotating Bearings: Stainless steel/water and dust resistant.
Stainless steel crystal supporting spindles.
Stainless steel window housing.

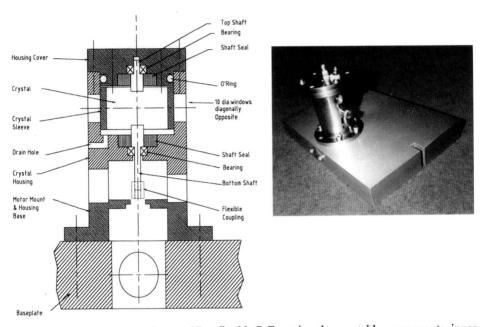

Figure 2   50 μm annular gap *LD* cell with CaF_2 optics, demountable components, inner rotating cylinder.

To enable flexibility in applications and later design modifications it was decided to have the optical components as demountable units. An inner rotating cylinder design was adopted with the motor housed under the unit. Practical considerations of sample loading, cell cleaning *etc.* are in conflict with the optimal smooth running of the cell and a robust design. In the end a compromise was adopted whereby the whole cell can be disassembled when required but can be loaded and cleaned for normal operation with a syringe and compressed air. The cell top, which needs to be in place for smooth running, can readily be removed to aid cleaning, but the small annular gap means that if anything precipitates out of solution or sticks to the cylinder walls disassembly is the only option

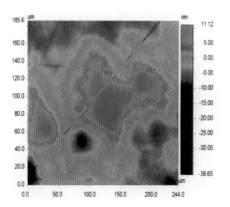 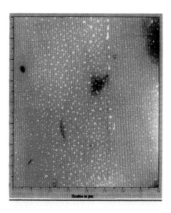

Figure 3 Surface roughness of the polished CaF_2 crystal determined using a Wyko Optical surface measuring instrument (left, Ra=3.21 nm, Rq= 4.11 nm, Rz=43.57 nm, Rt=49.77 nm) and a Nano surf stylus-measuring instrument (right, surface characteristics: average height:10.35 nm, mean deviation 1.92 nm).

To design and fabricate an instrument that rotates in a sealed environment where water and solvents are in contact with the inner cell means that the materials should be immune to water and solvent fatigue. Investigative studies of several types of bearing that could perform within the working environment of the instrument were undertaken. Nylon and glass bearings designed to run in a liquid environment were found to have a high noise level and variable rotation. PTFE bearings which were of a solid construction were found to break down in a short working time span. Stainless steel bearings were found not to be 100% water and dust resistant and also to have a high friction level, which put strain on the motor drive interface. Though not ideal, the latter were incorporated into the design. The finished polished CaF_2 core crystal was supported between two precision ground spindles 180° apart. The crystal after being cut/ground/lapped and polished was scanned for surface damage using a Wyko Optical surface measuring instrument and a Nano surf stylus-measuring instrument (Figure 3). The machining of the crystal created no problems as the hardness factor of CaF_2 is between 5 and 6 on the Mohs scale. However, it was found advisable to have a coolant in constant contact with the crystal during the polishing process. After assembly the maximum alignment error was 6 μm. The drive for the instrument is provided by a 6 V DC motor interfaced with a variable amps/volts control box.

The new *LD* cell (couette *LD* cell (CaF_2, 50 μm)) was initially tested with calf thymus DNA (280 μM solutions in water) and the signal intensity compared with that of our existing quartz cell with 500 μM annular gap. When rotating at approximately the same speed (as determined by a μs kinetics run detecting variation in absorbance as the cells rotate) of 1000 rpm, the *LD* signal for the two cells was almost identical (data not shown). This suggests that the average orientation of the DNA is 10 times greater with the 50 μm pathlength. The improved mechanical efficiency of the new cell may also be playing a role. The smaller pathlength cell is of course attractive as it uses less sample (in our cases: less than 400 μL versus 1800 μL in the longer pathlength cell). Both the cell and also the sample (due to reduced path length) have a lower absorbance signal for the same *LD* signal and thus a wider wavelength range. The lower intrinsic cell absorbance may be in part a feature of the CaF_2 versus quartz optics, though it may simply be due to better polishing.

5   PYRENE

Before attempting to incorporate proteins into lipid structures, we considered different lipids and different preparation methods with the model compound pyrene that had previously been inserted into lipid bilayers using an extrusion method by Ardhammar *et al.* to give an *LD* spectrum.[4] As noted above, if an analyte is incorporated with its long axis parallel to the lipids then the long axis transitions will have an $LD'/3S$ of $-1/2$ and the short axis transitions a value of $+1/4$. Since the 337 nm band has approximately 1/3 of the absorbance signal of the 274 nm and 242 nm bands, this *LD* spectrum should thus have the magnitude pattern $-6:+3:-2$ for 240 nm:275 nm:340 nm. If the pyrene is on the surface then the *LD* spectrum magnitude pattern would be: $+3:+3:+1$. If the short axis is parallel to the lipids then one would expect: $+3:-6:+1$. If the pyrene lies parallel to the lipids but with no distinction between its long and short axis, then we expect: $-3:-3:-1$.

We incorporated pyrene into lipid structures using the sonication method and a range of lipids. The results for the different lipids are similar but not identical. With LMPC our data are approximately: $-6:-1:-3$; with POPC: $-12:-3:-6$ (data shown in Figure 4a); and with soybean: $-2:-1:-2$. These ratios are not consistent with each other nor with any of the limiting cases given above suggesting that either the pyrene adopts a non-symmetry defined orientation as previously concluded by Ardhammar *et al.* [4] for perylene, or there is an orientation distribution with a varying preference for the pyrene long axis being parallel to the lipids.

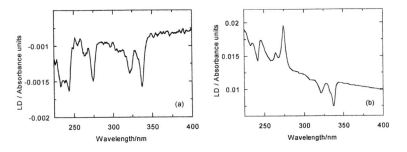

Figure 4   *LD* spectra of pyrene in (a) sonicated POPC micelles and (b) vortexed soybean liposomes.

Soybean lipid structures prepared by vortexing (Figure 4b) gave *LD* ratios of approximately $-2:+2:-1$ (though it is hard to estimate exactly where the baseline is at the lower wavelengths). Using the extruder (data not shown) $-2:+1:-1$ was obtained in accord with Ardhammar *et al.* [4] who found a ratio of $+1:-1$ for the two longer wavelength bands (and could not measure the lower wavelength one). These data suggest that the pyrene adopts a distribution of positions with some surface bound (~1/4 for the vortexed sample) and some with the plane of the pyrene parallel to the lipids and a preference for the long axis being parallel to the lipids (~1/2). The proportion in each position depends on the lipid environment.

The smaller structures are attractive as the baseline scatter was able to be removed by simply subtracting the baseline *LD* recorded using a non-spinning sample. This was not possible with the larger liposomes. Despite this, soybean lipid, which forms multi- or uni-lamellar liposomes, was adopted for later experiments due to its significantly higher signal to noise ratio. Although the multi-lamellar soybean experiment gave a slightly better pyrene *LD* spectrum (presumably as the larger multi-lamellar liposomes were shear distorted to a greater extent so oriented better), we later found that the extrusion process facilitated protein incorporation into the liposomes in some cases. It also produces a more reproducible liposome structure. This method was therefore used for all the protein samples.

6 PROTEIN MOTIF TRANSITION POLARISATIONS AND EXPECTED *LD*

To analyse any protein data we obtained it is necessary to know the transition polarisations of the protein, which means the transition polarisations of its secondary structural motifs. A somewhat simplistic model for determining the backbone $\pi \rightarrow \pi^*$ transition polarisations of α-helical peptides is to assume that we may represent the peptide by a right-handed helix of transition moments oriented along each C=O bond.[2] The lowest energy excited state (~208 nm from experiment) will result from the head-to-tail couplings of neighbouring transition moments. The net transition moment of this state is parallel to the helix axis. The head-to-head, tail-to-tail coupling will give the highest energy component (at ~190 nm) whose polarization is perpendicular to the helix axis. These deductions are consistent with the work of Brahms *et al.* who oriented a film of poly-gamma-ethyl-L-glutamate on glass plates and were able to resolve three peptide *LD* bands: a weak mainly perpendicular polarised one around 225 nm; one parallel polarised band centered around 210 nm; and a strong $\pi \rightarrow \pi^*$ transition polarized perpendicular to helix axis at around 190 nm.[12]

β-sheets are in many ways simpler than α-helices since the $\pi \rightarrow \pi^*$ transition does not have the 208 nm component and its polarisation can be assumed to be along the C=O bond axis, thus (assuming there is more 'sheet' than 'turn' in the structure) the net polarisation is perpendicular to the backbone run. No simplistic modelling can give the transition polarisation for the $n \rightarrow \pi^*$ transitions and so it was necessary to calculate these for α-helices and β-sheets as well as gramicidin.

In the case of the α-helices and β-sheet, as discussed above, we have 201 peptide bonds, so consideration of the $n \rightarrow \pi^*$ and $\pi \rightarrow \pi^*$ transitions will yield a total of 402 transitions for the protein. Similarly the 32 peptide bonds of the gramicidin structures give rise to 64 transitions. Not all of these will contribute significantly, and in the results given below we consider only $n \rightarrow \pi^*$ bands with significant oscillator strengths. In all cases the axis of the helix (for the β-sheet this means the long axis of the structure illustrated in Figure 5) is taken as z'. The results are summarised in Table 1.

For the $n \rightarrow \pi^*$ band in an infinite α-helix, three main transitions are predicted, one polarised parallel to the helix, and two perpendicular. For finite helices some splitting may occur. In both α-helices and the β-strand considered here (Figure 5) the splitting is very small, justifying our choice of 200 amino acids for the length of our peptide. For the ($\varphi = -48°$, $\psi = -57°$) helix, two transitions contribute to the parallel transition, both at 220.83 nm. The third and fifth transitions in Table 1a contribute to one perpendicular transition, while the fourth constitutes the other. The three bands are all perpendicular to each other to within $\pm 2°$. The total strengths of these bands should be the same, however, some discrepancy can be seen here due to the neglect of other transitions with small oscillator strengths (<0.4). The ($\varphi = -63°$, $\psi = -41°$) helix shows similar results, although the parallel band (transitions one and two in Table 1b) is a lot stronger.

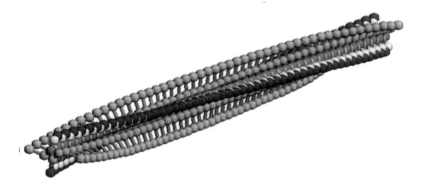

Figure 5   Schematic illustration of a β-strand showing the characteristic twist. The *z*-axis
is along the long axis of the twisted sheet.

For the β-strand, the parallel band at 219.38 nm is not split. The intensity of the
perpendicular bands, however, is scattered over the five remaining transitions in Table 1c.
With angles of up to 37° from perpendicular to each other, it is difficult to separate these
into the two transitions. The perpendicular bands of the β-strand occur at a lower energy
than the parallel band. This contrasts with the results for the α-helices.

For the first gramicidin structure (1MAG, Figure 6, the more likely one in our
experiments [13]) the three transitions in Table 1d are quite clear, and show the parallel
transition lying between the two perpendicular ones with the perpendicular polarisation
having greater intensity. Again there is some discrepancy between the strengths of the
perpendicular bands, as discussed above. The geometry of the 1MIC structure (Figure 6) is
not as regular, and the direction of the axis is not as well defined. The transitions at 220.55
nm and 220.40 nm are only 4° from being parallel to each other, and form the 'parallel'
band. The remaining two transitions in Table 1e make up the perpendicular bands,
although these are 26° from perpendicular to each other, indicating a strong mixing.

Table 1: Wavelengths in nm; short axes and long axis components; and oscillator strengths
of the n((* transitions of a (a) (=(48(, (=(57(200 alanine residue (-helix; (b) (=(63(, (=(41(
200 alanine residue (-helix, Z is the helix axis; (c) (=(135(, (=(135(200 alanine residue (-
strand; (d) Gramicidin (1MAG); and (e) Gramicidin (1MIC).

(a) Wavelength/nm	μ_{ab}/au			Oscillator strength
220.83	0.000	0.001	0.501	0.810
220.83	0.000	0.000	0.589	1.120
219.94	0.220	0.339	−0.001	0.527
219.94	−0.897	0.531	0.001	3.495
219.93	−0.516	−0.827	−0.002	3.057

(b)

Wavelength/nm	μ_{ab}/au			Oscillator strength
221.27	0.000	0.012	−0.504	0.822
220.84	0.000	0.000	0.842	2.290
219.93	0.315	−0.805	−0.009	2.403
219.92	0.987	0.286	0.007	3.401
219.92	0.088	-0.521	0.008	0.897
219.89	0.070	0.361	−0.003	0.434

(c)

Wavelength/nm	μ_{ab}/au			Oscillator strength
220.76	0.306	0.187	0.000	0.415
220.76	−0.331	0.147	0.001	0.425
220.72	−0.241	0.296	−0.002	0.470
220.72	0.219	0.332	0.001	0.511
220.72	0.034	−0.700	0.000	1.587
219.38	0.001	0.001	−0.767	1.886

(d)

Wavelength/nm	μ_{ab}/au			Oscillator strength
220.57	−0.354	0.028	0.000	0.406
220.40	0.000	0.000	0.314	0.317
219.41	0.012	0.241	0.000	0.187

(e)

Wavelength/nm	μ_{ab}/au			Oscillator strength
220.55	0.153	−0.1408	0.031	0.141
220.51	−0.109	−0.153	−0.025	0.115
220.40	0.222	−0.179	0.047	0.269
220.17	0.033	0.181	0.048	0.117

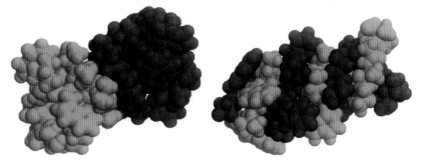

Figure 6 Two gramicidin structures used for the transition moment polarisation calculations. 1MAG (left), 1MIC (right).

Thus for an α-helix, the net effect is a ~221 nm band polarised parallel to the helix axis and a larger 220 nm band polarised perpendicular to the helix axis. The $\pi \to \pi^*$ band (calculation data not shown) has a band polarised parallel to the helix axis at 202 nm and another perpendicular to it at 189 nm. The ratio of calculated transition intensities (for 190 nm:202 nm:220 nm:221 nm) is approximately 230:80:3:1. From these values and Equation (6) it follows that the *LD* expected for a helical peptide whose long axis lies parallel (Figure 7, bottom) to the lipid molecules of a liposome is very approximately: +115:−80:+1.5:−1 (from low wavelength to higher wavelengths) with the last two components and the first two components showing significant cancellation. By way of contrast, if an α-helix lies on the bilayer surface (Figure 7, top) so that the short axis polarised transitions are half polarised parallel and half perpendicular to the lipids, we expect: −230:+80:−3:+1, again with cancellation occurring.

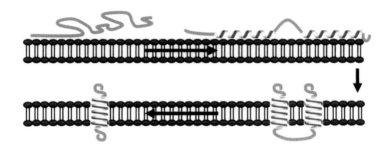

Figure 7   Schematic illustration of the proposed thylakoid membrane insertion mechanism for pre-PsbW indicating a fairly randomly oriented surface bound protein and α-helices in different orientations on a lipid bilayer.

The net $n \to \pi^*$ transition intensity is short axis polarised perpendicular to the axis of the twisted β-strand of Figure 5. Thus if a twisted β-strand is oriented so that the peptide backbone runs parallel to the lipids (Figure 8, middle), then for both the $n \to \pi^*$ and $\pi \to \pi^*$ transitions $LD^r/3S = +1/4$; if, however, the peptide backbone runs perpendicular to the lipids (Figure 8, left and right), then $LD^r/3S = -1/4$ since half the transitions will be parallel to the lipids and half perpendicular.

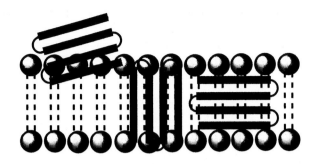

Figure 8   Schematic illustration of β-sheets (from left to right) on the surface of, oriented
parallel to the lipids and perpendicular to the lipids of the bilayer.

7   PROTEIN *CD* AND *LD* SPECTRA

The proteins we chose to study were: gramicidin, cytochrome *c*, a thylakoid membrane
protein precursor pre-PsbW, and a monoclonal antibody. Gramicidin is a well known
membrane protein with an unusual helix dimer spanning the bilayer;[14] Cytochrome *c* is a
membrane associated protein with a mixture of helical and other motifs and also a strong
Soret band absorbance;[13] Pre-PsbW has two transmembrane α-helices in the thylakoid
membrane prior to being processed to having the one (Figure 7);[5] and the monoclonal
antibody [6] has no α-helical character in aqueous solution as shown by *CD* spectroscopy.
Thus the proteins have a range of structural motifs. Our hope was that we would see
evidence of the orientation on the liposomes of these proteins.

Gramicidin is not water soluble so the samples were prepared in two different
ways. In the first the protein was extruded with the liposomes and gradually incorporated.
In the second method it was dissolved in ethanol and added to an aqueous solution of
liposomes. The latter method produced much better spectra which were consistent with
those obtained by the other method but showed more details.

The gramicidin *LD* absorbance spectrum (Figure 9) has an unusually large
absorbance band in the aromatic region due to the high concentration of tryptophan
residues in this protein. The corresponding *CD* signal is a structured band   (also of
unusually large intensity) with alternating positive and negative components, presumably
from opposite signed contributions from the vibronic components of the overlapping L_a
and L_b (starts at longer wavelength) bands whose polarisations are approximately
perpendicular to one another and across the diagonals of the tryptophan.[15] The aromatic *LD*
also has alternating signed bands. Assuming these are due to the components of the two
bands having opposite signs, then we must conclude that the tryptophans are on average
oriented so that one of these two bands is more parallel and one more perpendicular to the
lipids.

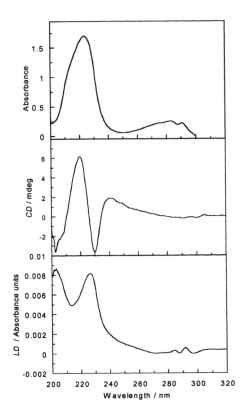

Figure 9   Absorbance, *CD* (averaged over 64 scans), and *LD* spectra of gramicidin (0.5 mg in 2.5 mg/mL lipid in phosphate buffer) in soybean liposomes prepared by adding ethanol solubilised gramicidin to an aqueous extruded liposome solution. The sample has held at 4° for 2 hours prior to analysis. Absorbance and *CD* baselines were collected with the same sample preparation method but without gramicidin.

Following reference [13] the *CD* confirms that the head to tail dimer of Figure 6a is the dominant species in our experiments. As this hydrogen bonding pattern resembles that of a β-sheet rather than that of an α-helix, there is no 208 nm $\pi\rightarrow\pi^*$ transition apparent in the spectra. The tryptophan *LD* from 270 –300 nm requires the tryptophans not to lie parallel to the liposome surface. The *LD* spectrum below 250 nm has positive maxima at 227 nm and ~200 nm and a negative maxima at 213 nm and below 180 nm (from data for a more dilute sample which is not shown). Assuming the longer wavelength bands are the result of cancellation of the $n\rightarrow\pi^*$ components (see Tables 1d and 1e), the longest wavelength component would be perpendicular to the helix axis and the shorter wavelength component parallel to it. The 200 nm maximum is then a short axis polarised $\pi\rightarrow\pi^*$ transition. This requires the gramicidin helices to lie parallel to the lipids, presumably inserted into the membrane.

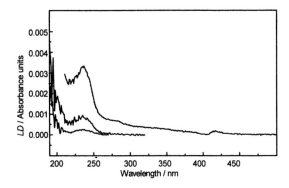

Figure 10   *LD* spectrum of cytochrome *c* (1 mg in 1 mL of 2.5 mg/mL lipid in phosphate buffer and serial dilutions 1:2 and 1:4, with spectra decreasing in magnitude) in extruded soybean liposomes.

The *CD* spectrum of cytochrome *c* (not shown) indicated that the protein was in its usual folded conformation in the presence of the liposomes. The 1 mg/mL cytochrome *c* *LD* spectrum (Figure 10) shows the presence of a positive tryptophan signal at ~ 280 nm and the in-plane polarised heme chromophore Soret band with a bisignate signal at ~ 425 nm. The plane of the heme group is not parallel to the surface (otherwise both components would be positive), however, the positive band is not half the magnitude of the negative component, indicating the heme is not perpendicular to the surface either. With further calculations or experiments to determine the polarisations of the Soret band, coupled with information about the tryptophan groups, it would be possible to determine the orientation factor *S* and the orientations of these chromophores in the membrane. A positive maximum, presumably due to the $n{\rightarrow}\pi^*$ transition occurs at ~240 nm, a negative maximum at ~220 nm followed by a large positive signal below 200 nm. These data all indicate that the average α-helix orientation is more parallel than perpendicular to the lipids.

Pre-PsbW is a thylakoid membrane protein that can exist in tris-buffer soluble multimeric form that is folded into a significantly α-helical structure.[5] There is still significant debate about how it inserts into thylakoid membranes. The helix of the mature protein and a hydrophobic region in the presequence are all α-helical in SDS micelles.[5] Our attempts to orient the SDS samples gave signals of the order of 10^{-5} absorption units and terrible signal to noise ratios. When added to soybean lipids, pre-PsbW gave a *CD* spectrum indicative of either no secondary structure or some β-sheet structure. However, due to uncertainties in concentrations it was not possible to fit these data.[16] The *LD* spectrum of this system is given in Figure 11. It shows a very large positive signal at 204 nm (no signal above 220 nm) and an even larger negative signal below 190 nm that overlaps with the 204 nm band. It is thus likely that the 204 nm band is the $\pi{\rightarrow}\pi^*$ band (shifted to longer wavelength due to cancellation), which indicates that the protein is unfolded on the surface of the liposome, perhaps retaining some β-sheet like hydrogen bonding. Assuming similar behaviour on the thylakoid membrane prior to insertion, this supports the mechanism of pre-PsbW insertion schematically illustrated in Figure 7.

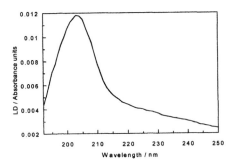

Figure 11 *LD* spectrum of pre-PsbW (1 mg in 2.5 mg/mL lipid in phosphate buffer) in soybean liposomes.

The *CD* spectrum of the monoclonal antibody (data not shown) fitted using CDsstr [16] indicates about 35% β-sheet, 12% turns, and 10% poly-proline type II helix. The *LD* spectrum shows a broad positive band whose maximum is at ~ 230 nm. Below 215 nm a large negative signal occurs. Assuming these two bands are the components of the $n{\rightarrow}\pi^*$ transition, we have the short axis polarised band being positive and the long axis polarised being negative suggesting the strand axis is parallel to the lipids. Such a picture is consistent with the schematic pictures of antibodies binding to bilayers.

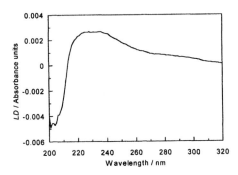

Figure 12 *LD* spectrum of a monoclonal antibody (1 mg in 2.5 mg/mL lipid in phosphate buffer) in soybean liposomes.

8 CONCLUSIONS

The work reported here is the first time flow linear dichroism has been used to probe the orientation of proteins in lipid bilayers. In order to accomplish this, we have designed and constructed a new flow *LD* cell with 50 μm annular gap and calcium fluoride optics that has enabled us to measure flow oriented spectra of micellar and liposomal systems down to 185 nm in some cases.

Although the wavelength range is potentially wider with the micellar systems the orientation is less so the *LD* signal to noise was generally worse than with the liposomes. The orientation of multi-lamellar liposomes was higher but the background light scattering was worse than with unilamellar liposomes. So in each case a choice has to be made that is optimal for a given system.

A set of model protein systems, encompassing helical and sheet proteins, as well as an unfolded protein have been placed in membrane mimicking environments and their secondary structure motif orientations probed using linear dichroism. It has been possible to deduce qualitatively the orientations of the proteins with respect to the normal to the lipid bilayers using transition moments for different secondary structure motifs calculated by the Matrix Method.

The technique of flow *LD* of proteins has potential for monitoring the kinetics of insertion of proteins into membrane mimicking environments as well as determining steady state orientations. In order to use the data more quantitatively to determine absolute orientations of proteins in membranes, a more effective way of removing the background scatter is needed since adding the proteins to the liposomes increases the scatter as does flowing the samples. Work is in progress on this problem.

ACKNOWLEDGEMENTS

Professor Malcolm Mackley and Sirilak Wannaborworn, Cambridge University, are thanked for use of their shear flow microscope system to visualise the liposomes. Support of the Engineering and Physical Sciences Research Council is gratefully acknowledged (GR/M91105(AR), GR/R40869/01(AR), GR/N66971(JDH)

REFERENCES:

1. B. Nordén, M. Kubista, T. Kurucsev, T. *Q. Rev. Biophysics* **1992**, *25*, 51–170
2. A. Rodger, B. Nordén *Circular dichroism and linear dichroism, Oxford University Press*, Oxford, **1997**
3. A. Rodger, A. *Methods in Enzymology*, **1993**, *226*, 232–258
4. M. Ardhammar, N. Mikati, B. Nordén *J. Amer. Chem. Soc.* **1998**, *120*, 9957–9958; M. Ardhammar, P. Lincoln, B. Nordén *J. Phys. Chem B.* **2001**, *105*, 11363-11368
5. C. Woolhead, A. Mant, S.J. Kim, C. Robinson, A. Rodger *J. Biol. Chem.*, **2001**, *276*, 14607–14613
6. A.A.M. Morales, G. Nunez-Gandolff, N.P. Perez, B.C. Veliz, I. Caballero-Torres, J. Duconge, E. Fernandez, F.Z. Crespo, A. Veloso, N. Iznaga-Escobar, *Nuclear Medicine and Biology*, **1999**, *26*, 717-723
7. B. R. Brooks, R. E. Bruccoleri, B. D. Olafson, D. J. States, S. Swaminathan, M. Karplus, *J. Comp. Chem.* **1983**, *4*, 187–217
8. L. Pauling, R. B. Corey, *Proc. Nat. Acad. Sci. USA* **1951**, *37*, 235–240 & D. J. Barlow, J. M. Thornton *J. Mol. Biol.* **1988**, *201*, 601–619
9. R. R. Ketchem, K. C. Lee, S. Huo, T. A. Cross, *J. Biomol. NMR*, **1996**, *8*, 1–14; Y. Chen, A. Tucker, B. A. Wallace, *J. Mol. Biol.* **1996**, *264*, 757–769
10. P. M. Bayley, E. B. Nielsen, J. A. Schellman, *J. Phys. Chem.*, **1969**, *73*, 228–243; W. J. Goux, T. M. Hooker Jr., *J. Am. Chem. Soc.* **1980**, *102*, 7080–7087
11. N. A. Besley, J. D. Hirst, *J. Am. Chem. Soc.* **1999**, *121*, 9636–9644
12. J. Brahms, J. Pilet, H. Damany, V. Chandrasekharan, *P. N.A.S.* **1968**, *60*, 1130–1137
13. T.P. Galbraith, B.A. Wallace, *Faraday Discuss.* **1998**, *111*, 159–164
14. D. Voet, J.G. Voet, *Biochemistry* 2nd edn, John Wiley and Sons **1995**
15. B. Albinsson, M. Kubista, B. Nordén, E.W. Thulstrup *J. Phys. Chem.* **1989**, *93*, 6646–6654
16. W.C. Johnson, *Proteins: Structure, Function and Genetics* **1999**, *35*, 307–312

QUANTITATIVE PROTEIN CIRCULAR DICHROISM CALCULATIONS

N. A. Besley and J. D. Hirst*

School of Chemistry, University of Nottingham, University Park,
Nottingham NG7 2RD, UK

1 INTRODUCTION

Despite inherent complexities and challenges, the determination of protein structure is an active area of research. Much of the motivation for this interest arises from the potential benefit to areas such as drug design. Furthermore, the recent fruition of genome sequencing efforts have heightened the demand for protein structure determination. Membrane proteins are an important class of proteins and constitute a significant proportion of pharmaceutical targets. The structures of these proteins are very difficult to determine through NMR or X-ray crystallography. Consequently, alternative techniques need to be explored.

The measurement of protein electronic circular dichroism (CD) spectra in the far-UV has become routine, and provides a probe of protein secondary structure.[1,2] The characteristic CD spectrum of an α-helix consists of a positive band at 190 nm and two negative bands at 208 nm and 220 nm.[3,4] The CD spectra of β-sheet proteins can be divided into two classes.[5] The first class has a maximum at 195 nm and a minimum in the region 210-220 nm. Alternatively, other β-sheet proteins have a CD spectrum similar to that of a random coil with a minimum at about 200 nm.

In recent years, a number of developments have catalysed renewed interest in protein CD. Synchrotron radiation CD measurements can determine protein CD down to wavelengths as low as 160 nm,[6] compared to ~180 nm for traditional CD measurements. This additional information allows secondary structural components to be resolved in greater detail. In addition, time-resolved far-UV protein CD measurements on the microsecond[7] and nanosecond[8] time scales coupled with the increasing length of molecular dynamics simulations[9] means that detailed modelling and understanding of the early events in protein folding is tantalisingly close. However, to exploit this convergence of experiment and theory requires quantitative calculations of protein CD spectra. Herein, we will describe our recent work toward the development of quantitative protein CD calculations and highlight some current applications of these methods.

2 THEORY

Circular dichroism is the differential absorption of left and right circularly polarised light, and can be expressed as the difference between the left and right extinction coefficients.

$$\Delta\mathcal{E} = \mathcal{E}_L - \mathcal{E}_R \tag{1}$$

The rotational strength represents the integrated intensity beneath a band in the CD spectrum. Rotational strengths can be evaluated through the Rosenfeld equation.[10] For an electronic transition $A \leftarrow 0$, the rotational strength R_{0A} is given by

$$R_{0A} = Im\left(\left\langle\psi_0|\vec{\mu}_e|\psi_A\right\rangle \cdot \left\langle\psi_A|\vec{\mu}_m|\psi_0\right\rangle\right) \tag{2}$$

where $\vec{\mu}_e$ and $\vec{\mu}_m$ represent the electronic and magnetic transition dipole moments. In contrast, the intensity of a band in traditional electronic absorption spectra is given by

$$D_{0A} = \left\langle\psi_0|\vec{\mu}_e|\psi_A\right\rangle \cdot \left\langle\psi_A|\vec{\mu}_e|\psi_0\right\rangle \tag{3}$$

Consequently, the intensity of bands in CD spectroscopy depends on both electronic and magnetic dipole moments while in electronic absorption spectroscopy only the electronic dipole moment determines the intensity. Physically, CD involves both linear and rotational displacement of charge.

In seminal theoretical work on the CD of biopolymers, Moffitt[11-13] applied exciton theory to the study of α-helices. The 190 nm amide $\pi\pi^*$ transition was shown to split into components, polarized parallel and perpendicular to the helix axis. Furthermore, the right-handedness of α-helical polypeptides was predicted. These predictions were later confirmed by experiments.[3,4] Currently, the majority of calculations of the CD of large systems adopt the matrix method[14-16] formalism. In this approach, the biopolymer is assumed to comprise M non-interacting chromophoric groups. Electronic excitations can occur only within a chromophoric group and not between groups. The wave function of the whole molecule is expressed as a linear superposition of basis functions Φ_{ia}

$$\Psi_T = \sum_i^M \sum_a^{n_i} c_{ia} \Phi_{ia} \tag{4}$$

These basis functions are a product of wave functions describing the individual chromophoric groups, which can be in an electronically excited state

$$\Phi_{ia} = \varphi_{10}\cdots\varphi_{ia}\cdots\varphi_{j0}\cdots\varphi_{M0} \tag{5}$$

where φ_{ia} represents the wave function of the chromophore i, which has undergone an electronic excitation $a \leftarrow 0$. A Hamiltonian matrix is constructed by considering the electronic Hamiltonian of the whole system. The diagonal elements of this matrix are the energies of the electronic excited states of the chromophoric groups, while the off-diagonal elements describe the interactions of a chromophore with the rest of the molecule. The off-diagonal elements typically have the form

$$V^{ij}_{0a;0b} = \int_i \int_j \varphi_{i0}\varphi_{ia}\hat{V}_{ij}\varphi_{j0}\varphi_{jb}d\tau_i d\tau_j \qquad (6)$$

If only electrostatic interactions are considered then

$$V^{ij}_{0a;0b} = \int_{r_{i1}} \int_{r_{j1}} \frac{\rho_{i0a}(r_i)\rho_{j0b}(r_j)}{4\pi\varepsilon_0 r_{i,j}}d\tau_i d\tau_j \qquad (7)$$

where ρ_{i0a} and ρ_{j0b} are the respective permanent or transition densities. These densities are usually represented by a set of point charges.[17] Then eqn. (7) becomes

$$V^{ij}_{0a;0b} = \sum_{s=1}^{N_s}\sum_{t=1}^{N_t} \frac{q_s q_t}{r_{st}} \qquad (8)$$

If only $n\pi^*$ and $\pi\pi^*$ transitions are considered, then in the simplest case of a diamide the Hamiltonian matrix takes the form

$$\mathbf{H} = \begin{pmatrix} E^1_{n\pi^*} & V^{11}_{n\pi^*\pi\pi^*} & 0 & V^{12}_{n\pi^*\pi\pi^*} \\ V^{11}_{n\pi^*\pi\pi^*} & E^1_{\pi\pi^*} & V^{21}_{n\pi^*\pi\pi^*} & V^{12}_{\pi\pi^*\pi\pi^*} \\ 0 & V^{21}_{n\pi^*\pi\pi^*} & E^2_{n\pi^*} & V^{22}_{n\pi^*\pi\pi^*} \\ V^{12}_{n\pi^*\pi\pi^*} & V^{12}_{\pi\pi^*\pi\pi^*} & V^{22}_{n\pi^*\pi\pi^*} & E^2_{\pi\pi^*} \end{pmatrix} \qquad (9)$$

This matrix is then diagonalised by a unitary transformation. The transition energies of the interacting system are given by the diagonal elements. The electronic and magnetic moments are obtained by transforming the local moments by the same unitary transformation. Subsequently, the CD can be evaluated through eqn. (2).

Despite over 40 years since the publication of Moffitt's work, quantitative calculations of protein CD from first principles have not been achieved. If the basic assumptions of the matrix method are valid (we will return to this later) then the accuracy of the calculations depends on the construction of the Hamiltonian matrix. In most matrix method calculations only the protein backbone is considered. *N*-methylacetamide (NMA) is used as a model of the protein backbone chromophore. Consequently, in order to construct the matrix we require the energies of the excited states of NMA, in addition to accurate representations of the permanent and transition densities associated with the electronic excitations. In recent years there has been a flurry of investigations both experimental and theoretical toward characterising the electronic excited states of amides

3 ELECTRONIC EXCITED STATES OF AMIDES

3.1 Electronic Structure

A key feature of the electronic structure of amides is the π electron system. The C, N and O atoms each contribute one-electron to the π electron system. The π bonding orbital, π_b, is the in-phase combination of three $2p$ atomic orbitals with no nodes perpendicular to the plane defined by the amide group. The non-bonding π orbital, π_{nb}, has a perpendicular node at the C atom. The anti-bonding π^* orbital has two perpendicular nodes. Another important feature of amide electronic structure is the two lone pairs on the oxygen atom, n and n′, which lie in the plane defined by the amide group and are orientated along the C-O bond and perpendicular to it, respectively.

3.2 Experimental Studies

A number of early experimental studies,[18-22] measured the electronic spectrum of a variety of mono-amides in both gas-phase and solvent. In these studies the characteristic electronic spectrum of amides in gas-phase was established.[22] The spectra are dominated by an intense band, the V band (V for valence), that was assigned to the $\pi_{nb}\pi^*$ transition. In addition, a weak band, the W band (W for weak), corresponding to the $n\pi^*$ transition appears as a shoulder on the high wavelength side of this band. Superimposed on these bands are a number of sharp bands that are attributed to transitions to Rydberg states, involving excitations to diffuse atomic-like Rydberg orbitals. A further broad band appears in the spectrum at higher energy. This band is often termed the Q band and was originally assigned to the $\pi_b\pi^*$ transition. Furthermore, these studies show that the position of the $\pi_{nb}\pi^*$ band is sensitive to substitution of the amide hydrogens. More recently, the electronic spectrum of formamide was measured.[23] The $n\pi^*$ and $\pi_{nb}\pi^*$ bands were found at 5.82 eV and 7.36 eV respectively. It was also proposed that the Q band arises from a superposition of transitions to several Rydberg states and the $\pi_b\pi^*$ transition lies at higher energy.

While there has been considerable effort in measuring the electronic absorption spectra of amides, there has been relatively few studies that determine the orientation of the transition dipole moment of the $\pi_{nb}\pi^*$ band. This is of interest, since this angle is critical for calculations of protein CD. In 1957 Peterson and Simpson[19] determined a value of –41° for myristamide, where the angle is characterised by the angle between it and the carbonyl bond with the OCN angle taken to be positive. In later work, Clark[24] reported values of –35° and –55° for propanamide and N-actetylglycine respectively. For hydrated crystals of glycine-glycine and angle of –46° was found.[25]

In solution the amide absorption spectra show large changes.[22] The $\pi_{nb}\pi^*$ band undergoes a large red-shift of up to 0.5 eV and its intensity is reduced. In polar solvents the $n\pi^*$ band is subject to a blue-shift. In addition, there is no evidence of Rydberg bands.

3.3 Theoretical Studies

Modern computers have made it possible to study electronic excited states of amides using state-of-the-art *ab initio* methods. Multi-reference configuration interaction (MRCI) calculations have been reported for a number of amides, including formamide[26] and NMA.[27] Amides have also been studied using complete-active-space self-consistent-field with multi-configurational perturbation theory (CASSCF/CASPT2)[28] and equation of motion coupled cluster theory.[29] These studies illustrate the problems associated with the study of amide excited states. The large number of low lying Rydberg states gives rise to artificial Rydberg-valence mixing which results in an overestimation of the $\pi_{nb}\pi^*$ transition energy. Within CASSCF/CASPT2 procedures this can be overcome by a second calculation to determine the valence state properties in which the Rydberg states have been 'deleted'. Using this procedure the excitation energies of a number of amides were determined within a few tenths of an electron volt.[28] In a recent study,[30] the thermally broadened VUV absorption spectrum of formamide was determined using an implementation of time-dependent density functional theory (TDDFT) in a plane wave basis set formalism. Excellent agreement with experiment was found, and importantly the Q band was shown to arise predominantly from n→3d transitions.

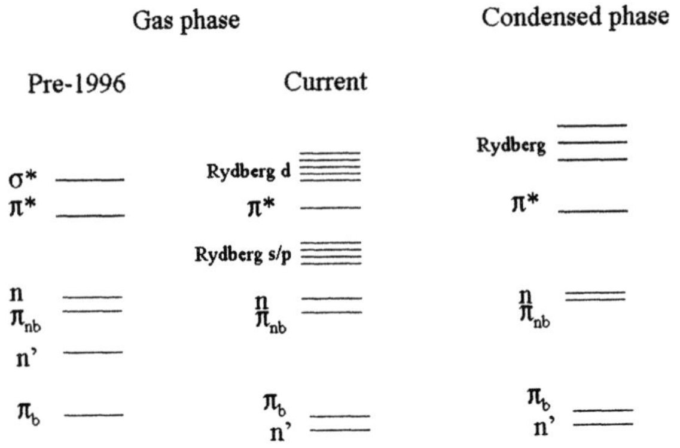

Figure 1 Schematic mono-amide molecular orbital diagram

A number of studies have addressed the problem of modelling the effect of solvent on amide electronic spectra. The description of solvent within an *ab initio* framework is a difficult problem since a large number of solvent molecules and their numerous configurations need to be considered. Generally, three approximate approaches are adopted. In the continuum approach, the solvent is modelled by a continuum characterised by a dielectric constant in which there is a cavity that contains the solute. This provides a gross description of the solvent in which short-range solute-solvent interactions, such as hydrogen bonding, are neglected. Alternatively, explicit solvent molecules can be included within the *ab initio* treatment in a discrete model. The computational cost of this approach rises rapidly and only a small number of solvent molecules can be considered. These two approaches can be combined in a semi-continuum model.

The additional computational cost arising from the introduction of solvent has meant most studies of solvated amide electronic spectra have concerned formamide. Sobolewski[31] reported CASSCF/CASPT2 calculations on the formamide-H_2O complex using a double-ζ plus polarisation basis set. They calculated transition energies and oscillator strengths (in parentheses) of 6.03 eV (0.001) and 7.52 eV (0.337) for the $n\pi^*$ and $\pi_{nb}\pi^*$ transitions respectively. These showed a blue-shift for the $n\pi^*$ band and small red-shift for the $\pi_{nb}\pi^*$ band when compared to the corresponding values of 5.85 eV (0.001) and 7.67 eV (0.317) for isolated formamide. The formamide-H_2O complex was also studied using multi-configurational self-consistent field, a $n\pi^*$ transition energy of 6.14 eV was found.[32] Within a continuum approach, we have examined the electronic spectrum of formamide and NMA.[33] In this study, the continuum model was supplemented with a repulsive potential[34] outside the cavity that represents the Pauli repulsion between the solute and solvent. This repulsion has the effect of destabilising the larger Rydberg states relative to the valence states. A by-product of this is a decrease in the Rydberg-valence mixing problem experienced in gas-phase studies. Transition energies for the $\pi_{nb}\pi^*$ band in water of 6.95 eV (0.256) and 6.60 eV (0.277) were found for formamide and NMA respectively. These are in good agreement with experimental values of 6.81 eV[22] and 6.67 eV.[21] However, the computed $n\pi^*$ transition energies of approximately 5.50 eV did not show a significant blue-shift. In a subsequent study,[35] the electronic spectrum of formamide was investigated using a discrete solvent model. A motivation for this study was to confirm the destabilisation of the Rydberg states by the solvent. The introduction of a small number of water molecules was to destabilise of the Rydberg states by up to 0.5 eV. When combined with a continuum all the features of the electronic spectrum of formamide in solution were reproduced, including the blue-shift in the $n\pi^*$ transition energy. A similar approach was applied to the study of NMA.[36] Following these studies the amide molecular orbital picture has been revised, as depicted in Figure 1.

3.4 Diamide Electronic Structure

Calculations of electronic structure of diamides represent the first step towards protein electronic structure. Hirst and Persson[37] presented CASSCF/CASPT2 calculations of the planar form of diketopiperazine (DKP). The high symmetry of DKP precludes optical activity. However, the coupling between the two $\pi_{nb}\pi^*$ transitions can be examined. These transitions were found to lie at 5.98 eV (0.580) and 7.09 eV (0.076), representing a splitting of 1.11 eV with the intensity appearing in the band of lower energy. Another feature of the study of diamides are charge transfer transitions. These are characterised by a large displacement of electron density from one amide group to the other. In DKP these are forbidden. In another CASSCF/CASPT2 study,[38] a linear dipeptide was studied and the electronic spectrum computed for a number of configurations. This work identified charge transfer bands that were computed to lie at ~7.1-7.5 eV, matching experimentally observed bands. In a recent study,[39] experimental and theoretical infra-red, UV and CD spectra were reported for the bridged cyclic diamide diazabicyclo[2.2.2]octane-3,6-dione I.

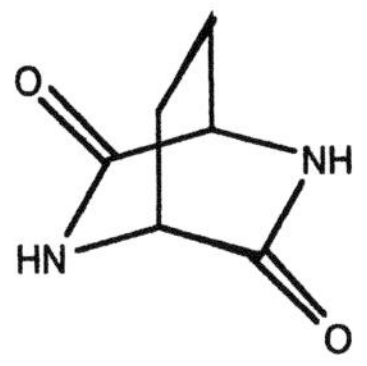

Figure 2 Diazabicyclo[2.2.2]octane-3,6,dione, I

The bridged cyclic diamide is an attractive molecule in which to study the interactions between the amide groups because it is conformational rigid and its structure is known by X-ray crystallography.[40] Consequently, ambiguity regarding its conformation in solution is removed. Furthermore, its C_2 symmetry provides computational advantages. Table 1 shows the electronic and CD spectra of I in gas-phase and solution.

Similarly to mono-amides, the spectra are dominated by an intense $\pi_{nb}\pi^*$ transition which is calculated to lie at 6.28 eV in gas-phase and 6.07 eV in solution. The symmetry of I means that it is not possible to classify the electronic transitions rigorously as local or charge transfer within a delocalised molecular orbital picture. It is possible to determine transitions present in the diamide that have no analogues in mono-amides. These transitions can be associated with charge transfer transitions found in dipeptides of lower symmetry. However, some caution must be taken since these studies compute vertical transitions and this may be a poorer approximation for these bands. Furthermore, I is chiral and may be studied using CD spectroscopy. The computed CD spectra are also shown. The results showed that the CD spectra arise predominately from transitions that can be considered local in nature. Both $n\pi^*$ and $\pi_{nb}\pi^*$ transitions give rise to two bands in the CD spectrum of opposite sign. However, the splitting of the $n\pi^*$ bands is small and only one band is evident, while the larger splitting of the $\pi_{nb}\pi^*$ transitions leads to two resolved bands. Overall, the agreement with experiment can be regarded as qualitative with results in solution closer to experiment. Larger basis sets and non-vertical transition energies are likely to lead to improved agreement with experiment.

State	Gas Phase		Solution		Experiment	
	ΔE (eV)	R (DBM)	ΔE (eV)	R (DBM)	ΔE (eV)	R (DBM)
$2^1A(n\pi^*)$	5.48	-0.04	5.70	-0.03		
$3^1A(\pi_{nb}\pi^*)$	6.28	-0.31	6.67	-0.27	5.69	-0.17
$1^1B(n\pi^*)$	5.47	+0.13	5.77	+0.10	5.30	+0.11
$2^1B(\pi_{nb}\pi^*)$	6.61	+0.23	6.48	+0.29	6.23	+0.11

Table 1 Experimental and Calculated Electronic and CD Spectra of **I**

4 CALCULATIONS OF PROTEIN CD

While *ab initio* calculations of amides become prohibitive beyond diamides, they can prove important in the calculation of protein CD. The matrix method requires energies of the excited states and parameters that describe the permanent and transition electron densities. These are problems well suited to *ab initio* methods.

Recently, we have used *ab initio* studies of NMA as a basis for reparametrisation of the amide chromophore. This has resulted in better agreement with experiment.[41-43] This improved agreement arises for three main reasons. Firstly, *ab initio* methods should provide a more accurate description of electron densities than the semi-empirical methods used previously. Secondly, using parameters based on condensed phase calculations provides a more suitable description of the amide chromophore since the environment of a protein is usually considered to be condensed phase. A third source of improvement comes from a more accurate representation of the electron densities in eqn. (8). Traditionally, a small number of point charges, typically 6-12, located around the atom centres are fitted to reproduce the computed dipole and quadrupole moments. We have adopted an alternative approach in which a larger number of charges, 20 or 32 depending on symmetry, are fitted to reproduce the electrostatic potential. To assess the accuracy of the different parameter sets the computed CD spectra for a set of 29 proteins are compared with experiment. This set includes (PDB codes shown in parentheses) cytochrome c (3cyt), hemoglobin (1hco), myoglobin (1mbn), bacteriorhodopsin (2brd), alcohol dehydrogenase (5adh), glutathione reductase (3grs), lactate dehydrogenase (6ldh), lysozyme (7lyz), papain (9pap), rhodanese (1rhd), subtilisin (1sbt), thermolysin (4tln), triose phosphate isomerase (1tim), flavodoxin (2fx2), carbonic anhydrase (1ca2), concanavalin A (3cna), λ-immunoglobulin (1rei), ribonuclease A (3rn3), ribonuclease S (2rns), erabutoxin (3ebx), plastocyanin (1plc), porin (3por), prealbumin (2pab), α-chymotrypsinogen A (2cga), α-chymotrypsin II (5cha), elastase (3est), superoxide dismutase (2sod), trypsin inhibitor (4pti) and trypsin (3ptn). This set of proteins is representative of all types of protein CD. Table 2 shows the

Spearman rank correlation coefficients are computed at three important wavelengths of 190 nm, 208 nm and 220 nm for a number of different parameter sets. Figures in bold represent a significant correlation at the 1% level.

Theory	Model	Number of Charges	$r_{[\lambda=190nm]}$	$r_{[\lambda=208nm]}$	$r_{[\lambda=220nm]}$
CNDO/S	NMA (g)	μ, Q, 6-12	0.44	-0.14	**0.48**
CNDO/S	acetamide (g)	μ, Q, 6-12	0.08	-0.16	**0.65**
MRCI	NMA (g)	μ, Q, 6	0.12	-0.24	**0.74**
CASSCF	NMA (aq)	μ, Q, 6	**0.50**	**0.51**	**0.67**
CASSCF	NMA	ESP, 20-32	**0.84**	**0.73**	**0.90**

Table 2 Comparison with experiment (data set comprising 29 proteins)

The correlation between experiment and theory is poor at all three wavelengths for the parameter sets based on semi-empirical calculations. Gas phase MRCI parameters, comprising 6 charges fitted to reproduce the molecular dipole and quadrupole, show an improved correlation at 220 nm. The CASSCF parameters derived from condensed phase studies are the first to show moderate correlation at 190 nm and 208 nm. This is initially surprising since the CASSCF wave function is inferior to the MRCI wave function, containing no description of dynamic correlation. However, this improved agreement arises from an improved orientation of the $\pi_{nb}\pi^*$ transition dipole moment. Physically this may be because there is a significant difference between this angle in gas-phase and solution. More likely is that the removal of Rydberg-valence mixing in the condensed phase calculations results in a better description of the $\pi_{nb}\pi^*$ state. Using this wave function and improving the description of the charge densities leads to a significant improvement in the correlation at all three wavelengths. Spearman rank correlation coefficients of 0.84, 0.73 and 0.90 at 190 nm, 208 nm and 220 nm, respectively, are achieved. The slopes of the regression lines were 0.6, 1.22, 1.02, respectively. In particular, the agreement at 220 nm is encouraging, since the intensity at 220 nm is associated with the helical content of the proteins. Detailed analysis of the different class of proteins shows that much of the error arises from the class II of β-sheet proteins. For these proteins the computed CD are not even qualitatively correct, predicting bands of the wrong sign. One possible reason is that fluctuations in backbone conformation may result in large changes in the CD.

Other authors have also reported improved calculations of protein CD. Using an improved semi-empirical based parametrisation Woody and Sreerama[44] presented matrix method calculations of protein CD that showed good correlation with experiment at the three wavelengths. This was achieved by fixing the transition dipole moment of the $\pi_{nb}\pi^*$ transition to that measured by Clark[24] and improving the location of the point charges. Following an alternative procedure, Bode and Applequist[45] used the dipole interaction method to study the CD of a number of proteins. However, since this model does not include the $n\pi^*$ transition, modelling of the band at 220 nm is necessarily incomplete.

A logical progression for improving these parameters would be to use condensed phase MRCI calculations with the electrostatic potential fitting of charges. This may result in improved correlations but is unlikely to remedy the problem of the class II β-sheet proteins. However, there are more significant sources of error. In our calculations we have only considered the protein backbone. Electronic excitations on side-chain chromophores can contribute to protein CD in the far-UV.[46-47] Including these excitations within the matrix method framework is straightforward. Presently, high quality parameter sets for side-chains are not available. In principle these can be generated in an analogous way to NMA. However, the large number of side-chain chromophores makes this a considerable undertaking. A principal assumption in the matrix method is that electronic excitations cannot occur between backbone chromophores. The inclusion of charge transfer excitations within the matrix method formalism is possible. These avenues are currently being explored in our laboratory.

5 APPLICATIONS

The correlation of 0.9 at 220 nm suggests that the intensity at this wavelength can be studied with some confidence. The dependence of intensity with helix length has been expressed as[48]

$$[\theta]_{220} = [\theta]_{220}^{\infty} \frac{r-k}{r} \tag{10}$$

where $[\theta]_{220}$ is the ellipticity at 220 nm, $[\theta]_{220}^{\infty}$ the ellipticity at 220 nm for an infinite helix, r the helix length and k is a constant. From computed values of $[\theta]_{220}$ for Pauling-Corey helices both $[\theta]_{220}^{\infty}$ and k can be determined through a least squares fit. Values of $[\theta]_{220}^{\infty}$= -37000 degcm^{2}dmol^{-1} and k~2.8 were reported.[43] In addition, the dependence of the intensity at 220 nm with helix conformation was examined for 124 helical fragments extracted from the X-ray structures of the proteins in the data set.

In a recent study,[49] the influence of hydrogen bond length on the 220 nm intensity was investigated. This was motivated by a report[50] of unprecedented helicities for peptides of the form AcHe-(Ala$_4$Lys)$_n$Ala$_2$-NH$_2$ Calculation of $[\theta]_{220}$ for model helical fragments generated by constrained minimisation in a molecular mechanics force field showed that short hydrogen bond lengths resulted in unusually high intensities. Since the experimental measurements used a water/ethylene glycol solvent which is known to promote hydrogen bonding, this may rationalise the experimental observations.

6 CONCLUSIONS

The development of time-resolved CD spectroscopy provides an exciting new probe of the evolution of protein secondary structure. To realise the full potential of this technique requires theoretical simulations. In this paper, we have described recent progress towards quantitative calculations of protein CD form first principles. Through reparametrisation of the electronic excited states of NMA, near quantitative calculations of protein CD have been achieved within the matrix method formalism. In the near future, we anticipate that further refinement of these calculations coupled with molecular dynamic simulations will provide new insights into the structure and dynamics of proteins.

7 ACKNOWLEDGMENTS

Some of the work described here was carried out at The Scripps Research Institute under support from NSF, and ACS Petroleum Research Fund. Our current work is supported by EPSRC and BBSRC.

References

1 K. Nakanashi, N. Berova and R. W. Woody, *Circular Dichroism Principles and Applications*, VCH, New York, 1994.

2 *Circular Dichroism and the Conformational Analysis of Biomolecules*, ed. G. D. Fasman, Plenum Press, New York, 1996.

3 K. Rosenheck and P. Doty, *Proc. Natl. Acad. Sci. USA*, 1961, **47**, 1775.

4 G. Holzwarth and P. Doty, *Proc. Natl. Acad. Sci. USA*, 1965, **87**, 218.

5 P. Manavalan and W. C. Johnson Jr., *Nature*, 1983, **305**, 831.

6 B. A. Wallace, *Nature Str. Biol.*, 2000, **7**, 708.

7 E. Chen, P. Wittung-Stafshede and D. S. Kliger, *J. Am. Chem. Soc.*, 1999, **121**, 3811.

8 C.-F. Zang, J. W. Lewis, R. Cerpa, I. D. Kuntz and D. S. Kliger, *J. Phys. Chem.*, 1993, **97**, 5499.

9 Y. Duan and P. A. Kollman, *Science*, 1998, **282**, 740.

10 L. Rosenfeld, *Z. Phys.*, 1928, **52**, 161.

11 W. Moffitt and J. T. Yang, *Proc. Natl. Acad. Sci. USA*, 1956, **42**, 596.

12 W. Moffitt, *J. Chem. Phys.*, 1956, **25**, 467.

13 W. Moffitt, D. D. Fitts and J. G. Kirkwood, *Proc. Natl. Acad. Sci. USA*, 1957, **43**, 723.

14 R. W. Woody, *J. Chem. Phys.*, 1968, **49**, 4797.

15 P. M. Bayley, E. B. Nielsen and J. A. Schellman, *J. Phys. Chem.*, 1969, **73**, 228.

16 W. J. Goux and T. M. Hooker Jr, *J. Am. Chem. Soc.*, 1980, **102**, 7080.

17 I. Tinoco, *Adv. Chem. Phys.*, 1962, **4**, 113.

18. H. D. Hunt and W. T. Simpson, *J. Am. Chem. Soc.*, 1953, **75**, 4540.

19. D. L. Peterson and W. T. Simpson, *J. Am. Chem. Soc.* 1957, **79**, 2375.

20. K. Kaya and S. Nagakura, *Theor. Chim. Acta*, 1967, **7**, 124.

21. E. B. Nielsen and J. A. Schellman, *J. Phys. Chem.*, 1967, **71**, 2297.

22. H. Basch, H. B. Robin and N. A. Kuebler, *J. Chem. Phys.*, 1968, **49**, 5007.

23. J. M. Gingell, N. J. Mason, H. Zhao, I. C. Walker and M. R. F. Siggel, *Chem. Phys.*, 1997, **220**, 191.

24. L. B. Clark, *J. Am. Chem. Soc.*, 1995, **117**, 7974.

25. V. Pajcini and S. A. Asher, *J. Am. Chem. Soc.*, 1999, **121**, 10942.

26. J. D. Hirst, D. M. Hirst and C. L. Brooks III, *J. Phys. Chem.*, 1996, **100**, 13487.

27. J. D. Hirst, D. M. Hirst and C. L. Brooks III, *J. Phys. Chem.*, 1997, **101**, 4821.

28. L. Serrano-Andrés and M. P. Fülscher, *J. Am. Chem. Soc.*, 1996, **118**, 12190.

29. P. Szalay and P. Fogarasi, *Chem. Phys. Lett.*, 1997, **270**, 406.

30. N. L. Doltsinis and M. Sprik, *Chem. Phys. Lett.*, 2000, **330**, 563.

31. A. L. Sobolewski, *Photochem. Photobiol.*, 1995, **89**, 89.

32. M. Krauss and S. P. Webb, *J. Chem. Phys.*, 1997, **107**, 5771.

33. N. A. Besley and J. D. Hirst, *J. Phys. Chem. A*, 1998, **52**, 10797.

34. A. Bernhardsson, R. Lindh, G. Karlström and B. O. Roos, *Chem. Phys. Lett.*, 1996, **151**, 141.

35. N. A. Besley and J. D. Hirst, *J. Am. Chem. Soc.*, 1999, **121**, 8559.

36. N. A. Besley and J. D. Hirst, *J. Mol. Structure (THEOCHEM)*, 2000, **506**, 161.

37. J. D. Hirst and B. Joakim-Persson, *J. Phys. Chem. A*, 1998, **102**, 7519.

38. L. Serrano-Andrés and M. P. Fülscher, *J. Am. Chem. Soc.*, 1998, **120**, 10912.

39. N. A. Besley, M.-J. Brienne and J. D. Hirst, *J. Phys. Chem. B*, 2000, **104**, 12371.

40. M.-J. Brienne, J. Gabard, M. Leclercq, J.-M. Lehn, M. Cesario, C. Pascard, M. Chevé and G. Dutruc-Rosset, *Tetrahedron Letts.*, 1994, **35**, 8157.

41. J. D. Hirst, *J. Chem. Phys.*, 1998, **109**, 782.

42. J. D. Hirst and N. A. Besley, *J. Chem. Phys.*, 1999, **111**, 2846.

43. N. A. Besley and J. D. Hirst, *J. Am. Chem. Soc.*, 1999, **121**, 9636.

44. R. W. Woody and N. Sreerama, *J. Chem. Phys.*, 1999, **111**, 2844.

45. K. A. Bode and J. Applequist, *J. Am. Chem. Soc.*, 1998, **120**, 10938.

46. R. W. Woody, *Biopolymers*, 1978, **17**, 1451.

47. A. Chakrabarthy, T. Kortemme, S. Padmanabhan and R. L. Baldwin, *Biochemistry*, 1993, **32**, 5560.

48. Y.-H. Chen, J. T. Yang and H. M. Martinez, *Biochemistry*, 1972, **11**, 4120.

49. Z. Dang and J. D. Hirst, *Angew Chem. Intl. Ed.,* 2001, **40**, 3619.

50. P. Wallimann, R. J. Kennedy and D. S. Kemp, *Angew. Chem. Intl. Ed.*, 1999, **38**, 1290.

PROBING CELLULAR STRUCTURE AND FUNCTION BY ATOMIC FORCE MICROSCOPY

Michael A. Horton, Petri P. Lehenkari and Guillaume T. Charras

Department of Medicine, University College London, London WC1A 6JJ, U.K.
E-mail: m.horton@ucl.ac.uk

1 SUMMARY

Scanning probe microscopy techniques have been applied increasingly to cell biological problems that make use of their precision in spatial and force control rather than their ability to image at high resolution. We have developed a 'life sciences platform' that integrates a commercial atomic force microscope (AFM) with an inverted optical microscope equipped with bright field video recording, fluorescent imaging, and laser scanning confocal microscopy. This has enabled us to image and analyse live cells in real time by AFM including measurement of cell topography and material properties, force-distance analysis and binding site mapping, together with simultaneous monitoring of cell morphology and fluorescence by optical means. Further, we have used the AFM as a precise micro-manipulator to apply forces and deliver chemical signals to cells and have followed downstream phenotypic and signal transduction events with conventional microscopy. We suggest that this latter feature, for example, will have wide application to the analysis in real time of receptor-ligand interactions or enable the investigation of mechanically operated receptors and channels in live cells. Details of the experimental set-up are given with examples of the different analytical techniques.

2 INTRODUCTION

A large number of techniques have been developed to measure the chemical interactions upon which physiological phenomena in living organisms are based. Many of the most important molecular events occur at cell surfaces and several methods, of which atomic force microscopy (AFM) is an example (Table 1), have been developed specifically to interrogate surfaces. Interactions at cell membranes are inexorably linked to the intracellular processes, such as activation of signal transduction cascades or modification to the structure of the cellular cytoskeleton, that regulate, for example, tissue differentiation and function or cell survival. These latter phenomena have been already examined by traditional optical or electron-based microscopies and the development of methods that integrate approaches, such as scanning near-field optical microscopy (SNOM), scanning ion conductance microscopy (SICM), and multi-photon and fluorescence resonant energy transfer (FRET) methods, are now being applied to cell physiological processes. Cell surfaces also have been investigated by various micro-manipulative techniques, such as the surface force apparatus[1], pipette suction[2,3], laser tweezers or with magnetically trapped beads[4-6]. All of these are based either on the use of a physical, and often functionalised, probe. A classical example of such an approach is the use of laser tweezers in the study of the molecular basis for muscle contraction where myosin molecules move on actin filaments[7]. Many of these complementary methods are summarised in Table 1. In this review we will focus on AFM as a specific type of non-optical scanned probe microscopy that is developing into a method suitable for cell biology.

Table 1   *Techniques for cellular imaging at high resolution*

Technique	
Light microscopy	Standard wide-field methods, confocal (laser scanning) microscopy, evanascent wave microscopy, proximity techniques such as FRET
Electron and X-ray microscopy	Transmission, scanning, environmental (unfixed, FESEM); immunological contrast and specificity (gold, enzyme methods)
Scanning probe microscopy	**Atomic Force Microscopy (AFM)**: including topography, chemical imaging, molecular recognition (antibody on tip) microscopy, material property analysis, use as micro-manipulator
	High resolution AFM: topography down to molecular scale, measurement of intermolecular interaction forces and molecular mechanics
	Scanning near-field optical microscopy (SNOM): optical and fluorescence imaging below the visible light wave limit
	Scanning ion conductance microscopy (SICM): ion conductance feed-back for functional imaging and mapping of channels in cells

The AFM was devised by Binnig and colleagues in 1986[8] and first applied to the life sciences by Marti and co-workers[9]. AFM, in its present form, can be used to acquire images of intact cells at high, sub-micrometre scale, resolution and for topographic maps to be generated. Measurement of an increasing variety of cellular events under physiological conditions have become possible and Table 2 presents a summary of some of these applications to cell biology that have been reported in the literature. Intriguingly, the ability to apply forces via the AFM probe has guided its application increasingly towards use as a precise micro-manipulator rather than an imaging tool. Indeed, pure imaging applications have been limited as image capture rate is disappointingly slow and limited in comparison to optical and scanning electron

Table 2   *The use of AFM in cell biology*

Cellular property studied	References
Cell morphology, membrane structure, functional changes, etc.	10-32
Cytoskeleton and cytoskeletal elements	12,14,19,33-36
Cell division	37-39
Cell migration and locomotion	40-44
Organelles and organelle movement	35,41,45,46
Secretory structures	35,47,48
Viral budding, infected cells	49-52
Cell-matrix and cell-cell binding forces	53-55
Receptor-ligand binding forces on cells	54
Micromanipulation, microdissection	12,56,57
Elastic properties of cells	14,58-63
plasma membrane, nucleus, cytoskeletal fibres etc	34,36,43,47,60
and after cytoskeletal disruption	36,63,64
vinculin-deficient cells	65

microscopy. However, and importantly, AFM can be performed on living samples with the application of appropriate techniques and equipment design.

2.1   Principles of AFM

In AFM, a micro-fabricated cantilever (usually 100-200 nm long) with a very small tip (with a few nm^2 contact area) is raster scanned on the surface (for example, of a cell) whose topography is to be measured (Figure 1). A piezo-electric ceramic driver is used to raise or lower the cantilever in order to maintain a constant bending of the cantilever. A laser beam is reflected via the top of the cantilever towards a photo-detector that detects any movement or bending of the cantilever, thus enabling the position of the cantilever to be calculated and adjusted. As a result, the AFM records images of surface topography under a constant applied force in the low nN range (Figure 2).

Many different imaging modes have been developed (reviewed by Hansma et al.[66]). Contact mode imaging consists of scanning the whole surface while maintaining the tip in constant contact. In contact mode, it is also possible to measure the torque applied on the cantilever as it is scanned sideways over the surface. Differences in torque equate to differences in the friction of the surface and variation in material properties[67,68]. Non-contact, or 'tapping', mode consists of rapidly oscillating (mechanically or magnetically) the cantilever in the z direction whilst slowly lowering it towards the surface and scanning. When the cantilever comes close to the surface, the amplitude of the oscillations is dampened and the surface can be detected. This mode is much less damaging to samples[70]. Finally, in force-distance measurements, the cantilever is slowly lowered towards the surface and its deflection is constantly recorded. This yields a curve showing the

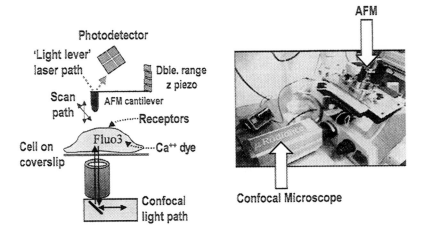

Figure 1   *Equipment setup of the bio-AFM*
(LHS), A schematic illustration of the AFM operation. The sample being analysed is a cell adhering to a glass coverslip and loaded with the calcium sensitive dye Fluo-3 (see Figure 5). The tip of the AFM cantilever approaches and scans the surface of the cell (Figure 2), its position being controlled and measured using a light lever and photodetector. The tip can also be used to interact with the cell surface, acting as a micromanipulator to interrogate cell surface receptors (see Figure 4) or to deliver a mechanical force (Figure 5). Fluorescence is measured from below via an integrated confocal microscope. (RHS), The hybrid AFM-confocal microscope is illustrated. The AFM is fitted onto a specially designed inverted microscope interface allowing the tip to be aligned into a desired position independent of the movement of the microscope stage and the sample holder. This configuration allows simultaneous light, phase contrast and video capture, and epifluoresecence confocal imaging. The equipment used was a Thermomicroscopes Explorer[TM] AFM with modifications[54] and a BioRad Radiance confocal microscope. (Adapted from (Lehenkari et al., 2000).

bending of the cantilever as a function of the distance traveled at a single point rather than under scanning conditions (see Figure 3). Using the spring constant of the cantilever, it is then possible to calculate the force needed to create that deflection, forming the basis for measurement of material properties of cells and of ligand-receptor binding forces. If such measurements are made across a surface, then a 3D map of binding events can be developed (Figure 4).

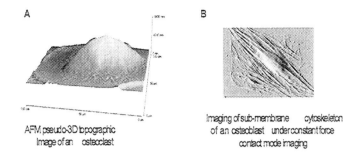

A

B

AFM pseudo-3D topographic
Image of an osteoclast

Imaging of sub-membrane   cytoskeleton
of an osteoblast   under constant force
contact mode imaging

Figure 2   *Topography on live cells*
(A), An exemplary three-dimensional surface image of the osteoclast cultured on a glass coverslip; this is a large cell, with dimensions exceeding 100x100x8μm. (B), A live osteoblast scanned in contact mode using constant force. As the tip of the cantilever is much stiffer than the cell, it deforms the cell membrane revealing intracellular cytoskeletal structures.

3 APPLICATIONS OF AFM

3.1 Topography measurements

AFM was initially designed for the study of surface topography, and its lateral resolution is on the atomic level when it is used in contact mode on surfaces that are harder than the probe[8]. By the same principle, AFM can also be used to create a height map at lower resolution on soft samples, such as cells, this being derived by measuring the cantilever position during scanning across the surface. From this map, it is possible to construct a three-dimensional profile of a cell surface (Fig. 2A). As the cell surface is softer than the tip, resolution is reduced due to tip indentation along the probed surface[61]. Indeed, on materials as soft as living cells, the resolution can be as low as on the order of 200 nm. In comparison, the resolution of confocal laser scanning microscopy is limited by the wavelength of the light used and is significantly lower than that of scanning electron microscopy. However, as resolution is proportional to the deformation caused by the force used for imaging, contact imaging with higher forces can reveal relatively 'hard' intracellular structures in soft, unfixed cells (Fig. 2B). Operation of the AFM in non-contact mode can enhance the resolution where it can reach the order of a few tens of nanometers[70]. Non-contact imaging is a very demanding mode to operate in liquids, particularly on living cells. Some topographic images of cellular plasma and nuclear membranes have, though, been published that reveal interesting sub-cellular details that would not be resolved by standard optical imaging (for example, [32]); it has though not reached molecular resolution on native plasma membranes of eukaryotic cells. With advances in equipment and software design, the ability to routinely image, at high resolution, live, underivatised cellular material by AFM may yet be realised.

Pure imaging by AFM can be used to reveal details of membrane and sub-surface biological processes in real time and under ambient conditions. Early examples included vesicle trafficking, and cytoskeletal rearrangement[12,33]. More recently the range of physiological responses that have been analysed by AFM have begun to rival those achieved by optical microscopy, and a number of these are given in Table 2. Lal et al[19] examined neurite outgrowth from cultured PC12 cells, Schneider and colleagues[26,30] used the atomic force microscope to measure aldosterone-induced volume increase in live endothelial cells, Quist et al[46] examined the role of cellular gap junctions by AFM, and Spudich and Braunstain[35] monitored basophil degranulation. Similarly, the group of Radmacher have examined physiological processes such as cell division[39], cardiomyocyte beating[42] and the control of cellular material properties[64].

3.2 Binding force measurements

Since its invention, AFM has proven its worth in the quantification of binding forces for interactions, for example, between proteins and other macromolecules, receptors and ligands, or cells and substrates (Table 3). Intermolecular receptor-ligand binding forces in the range 15-250 pN have been reported for protein interactions, and up to 220 nN for total cellular binding forces. Using the AFM, it is possible to evaluate the binding force between a single receptor molecule and its ligand directly[71] AFM has also been used extensively in 'molecular stretching' experiments (see [72-74]) with a wide range of linear biomolecules, such as the giant muscle protein titin, DNA and polysaccharides, and high resolution molecular imaging in purple membranes and artificial lipid bilayers (see [75-78]).

In order to achieve such a measurement, the tip of the AFM is 'functionalised' with a specific molecule. Different techniques have been developed to achieve this. One relies on molecules being randomly picked up from a surface prior to molecular pulling. Linkage of molecules to the AFM tip utilise either passive chemisorption of the molecule in solution onto the tip, often with polyethylene glycol (PEG) as a linker/adsorbent[54], or covalent coupling using various chemistries to directly link the molecule to the tip[79,80]. Force-distance curves are then recorded in different locations until an adhesion event occurs, being recognised by the shape of the force distance curve. The aspect and interpretation of a typical force-distance measurement are presented in Figure 3.

Table 3   *Inter-molecular interaction forces measured by AFM*

Ligand-Receptor	Force (pN)	Reference
Isolated molecules: disruption forces		
Meromyosin-actin	15-25	7
Avidin-biotin	160-200	3,81,82
Streptavidin-biotin	200-257	81,83
Cell adhesion proteoglycans	40-125	84,85
Antibody-antigen	49-244	86-88
P-selectin-PGSL-1	165	89
VE cadherin	35-55	90
Cholera toxin-ganglioside	90	91
Lectin-carbohydrate	35-65	92
Targeting proteins (αSNAP-NSF; VAMP1-syntaxin)		93,94
Isolated molecules: adhesion force mapping		
Antibody-lysozyme		95
Antibody-ICAM1		96
Enzymes on tips: substrate detection (ATP, glucose, acetyl cholinesterase)		97-100
Lectin-cell carbohydrate interaction		
Con A-yeast	75-200	101
Lectin-epithelium, fibroblast	50	102,103
Receptors in cells: forces		
RGD-integrin	35-120	54,104
Antibody-cell	40-127	54,105
Receptors in cells: mapping		
RGD-integrin		Lehenkari et al, in prep.
VIP receptor		106,107
Blood group A on red cells		108
Cell-to-cell/surface forces		
Dictyostelium on tip/csA adhesin single molecule interaction	23	109
Yeast on tip		110
Bacteria on tip		111,112
Trophoblast-uterine epithelium	1000-16000	53
Cell-uncoated surface	19000-100000	55
Cell-coated surface	100000-220000	55

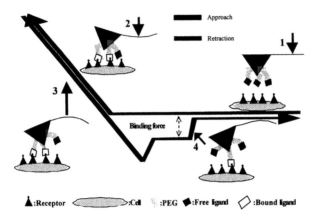

Figure 3   *The 'force-distance' curve*
A representation of a force-distance curve between a tip-bound ligand and a surface-bound receptor molecule is illustrated. The AFM tip approaches the surface (1) and contact is made. The cantilever bends (2) until it reaches the specified force limit which is to be applied and is then withdrawn (3). The tip then leaves contact with the surface and, as the ligand on the tip remains bound to surface receptor molecule, both are extended. With further application of retraction force, the molecule-ligand complex dissociates (4). The cantilever then returns into its resting position, (1), and is ready for another measurement. The maximum difference between the approach and retraction curves, and the curve shape in the region of (4), gives information on the interaction 'binding' force between the tip and surface and their physical properties.

    To enable these binding force measurements to be performed, the ligand can either be adsorbed to atomically smooth, cleaved mica or be present as a natural component of the cell membrane. Measurements on living cells are inherently difficult: the target molecule is easily lost from the cell membrane, can be at too low a density to make a successful interaction likely, or the tip may become contaminated by cellular debris during analysis. Recently, Lehenkari and Horton[54] have reported the first measurements of binding forces of an Arg-Gly-Asp (RGD) peptide to its cognate receptor, integrin, on living cells (Figure 4).

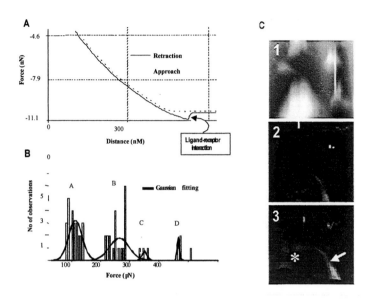

Figure 4   *Integrin receptors on osteoclasts analysed by AFM*
(A, B) Measuring interaction forces between ligands and cell surface receptors by AFM. Interaction forces were evaluated between F11 antibody molecules on the AFM cantilever tip and surface-expressed $\alpha_v\beta_3$ integrin dimers on freshly isolated rat osteoclasts. (A), Force-distance analysis revealed a typical single interaction (the 'jump' in the AFM retraction curve seen is marked). (B), Individual release forces were plotted as a histogram and accumulate around certain values; these can be further analysed by multi peak Gaussian curve fitting, as described in [54]. The results (B) showed multiples of a particular binding force between F11 antibody and osteoclast $\alpha_v\beta_3$ of 127 ± 16 pN (mean ± SD). (A, B adapted from Lehenkari and Horton, 1999[54], with permission). (C) Mapping of integrin receptor distribution with RGD peptide in living cells. A height map (C1, z-range 0-8μm) was created using a 20μm 'glass ball' cantilever tip producing a low resolution topographic image of a freshly isolated rat osteoclast (as in Lehenkari et al., 2000[69]). The interaction forces between a GRGDSP peptide, coated onto the AFM tip (Lehenkari and Horton, 1999[54]), and αvβ3 integrin receptors were analysed by taking repeated force-distance measurements across the entire surface of the cell. The distribution of the RGD-integrin receptor binding forces on the cell was thus evaluated (C2, force range 0-5nN). A negative control GRGESP peptide failed to show binding (not shown). The topography and force map were then merged into an image of the binding forces displayed on a pseudo-3D height image of a cell (C3). The RGD binding sites were located at the edge of the cell (arrow) and on the top of the osteoclast (asterisk).

3.2.1   Integrin-RGD recognition - an example of intermolecular force measurement in cells by AFM.

Members of the integrin family of cell adhesion receptors play a key role in cell-matrix recognition[113]. They are found in virtually all cell types as transmembrane receptors that consist of two non-covalently linked subunits that require divalent cations for optimal function. In cells adhering to extracellular matrix substrates, integrins are found concentrated at the underside of cells in focal adhesion plaques associated with intracellular linker proteins such as vinculin, the F-actin cytoskeleton and signalling molecules. Many integrins recognise the Arg-Gly-Asp (RGD) peptide consensus sequence. Ligands containing this sequence include several cell surface and extracellular matrix proteins. The affinity of the receptor for a given ligand depends not only on the integrin type but also on conformational changes in the receptor and hence receptor activation status. These changes are triggered by alterations in the extracellular microenvironment of the receptor and by interactions between the cytoplasmic part of the receptor and intracellular signalling molecules and the cytoskeleton. The properties of integrins suggest that, to evaluate their affinity for ligands reliably, this must be performed on the surface of intact cells[54]. It is likely that this is true for many other systems of cell surface membrane receptors.

RGD binding forces, measured by AFM in live bone cells, osteoclasts and osteoblasts, is dependent not only on the type of integrin expressed by a cell and its activation status but also on the spatial conformation of the sequence in which the RGD sequence is present[54] This may be because a more favorable orientation between the ligand and the receptor results in shorter interaction ranges for the molecules and thus higher interaction forces between RGD and the binding site. Surprisingly the dynamic range of RGD binding forces is small compared to affinity measurements, based upon inhibition of receptor-ligand interaction, such as IC_{50} estimates. Here, the efficacy of a linear RGD peptide and a potent snake venom inhibitor differs by approximately a factor of 10^5, whereas the binding forces measured by AFM vary by about 2-fold[54]. The relationship between these two values remains to be clarified and it is probable that fundamental differences exist between the parameters measured.

AFM measures binding force and binding probability separately, whereas 'affinity' measures a combination of these[86] and, thus, AFM can provide further information on the nature of molecular interactions. The recognition events observed in a force-distance curve do not represent unitary constants, as unbinding forces depend upon the rate of dissociation of the two species[3,89,114]. Ligand-receptor dissociation involves the rupture of multiple bonds and transitions between intermediates of varying stability and energies[115]. By measuring unbinding forces across a range of loading rates, information on the non-covalent bond strengths that mediate intermolecular interactions, the 'energy landscape', can be obtained[116]. Meanwhile at a practical level, the pull-out speeds and cantilever properties must be carefully taken into account when 'absolute' force values are cited.

3.3   Measuring the material properties of a cell

The intracellular cytoskeleton gives the cell its physical integrity and adapts to the environment and the activity of the cell. However, the exact role of each element of the cytoskeleton is still unclear. By measuring the material properties of the cell - such as stiffness, plasticity, visco-elasticity - and determining the effect of induced cytoskeletal changes on these, one can gain insight into the particular role of each cytoskeletal element.

If the slopes of force-distance curves recorded during indention with an AFM probe are analysed, the elastic modulus, that is stiffness, of the cell can be determined at a particular location by use of the theory of indentation[117] (see Table 2). Thus, AFM can be used to determine the cellular elasticity of tissues (reviewed by Radmacher[61]) and this information can be accumulated during raster scanning across a cell to generate 3D maps of elastic properties and height information.

Common materials such as steel or bone have stiffnesses of 200 GPa and 10 Gpa, respectively, whereas living cells are more compliant with stiffnesses between 1-150 kPa (see Table 2 for references). It appears that the different elements and regions of a cell have very different stiffnesses. The effect of cytoskeleton-disrupting drugs on the cell stiffness has been studied extensively[63,64] (Table 2). In particular, it was found that disrupting the F-actin network with cytochalasin reduced the cell stiffness, whereas chemical cross-linking increased cell stiffness. Together, these studies show that F-actin filaments participate in the maintenance of cellular elasticity. The F-actin cytoskeleton is linked to cell surface integrin receptors by vinculin in focal adhesion plaques. Recently, Goldmann et al[65] reported that the elastic modulus of vinculin-deficient cells was decreased compared to wild type; when vinculin expression was returned by gene transfer, a near wild-type elastic modulus value was attained. This points to the role of vinculin in stabilising focal adhesions and transferring mechanical stresses to the cytoskeletal network. Cellular cytoskeletal mechanics has been extensively reviewed[118]. Other material properties such as the visco-elasticity and the plasticity of cells have also been measured. Visco-elasticity can be measured by indenting the cell and following its recovery over a period of time[63]. Plasticity measurements consist of measuring the permanent deformations inflicted on the cell through indention. These fundamental measurements are essential if the response of cells to mechanical perturbation is to be understood at any level other than the purely phenomenological. Many types of cells and tissues, for example those of the skeleton or cardiovascular systems, are responsive to states of mechanical strain. Probing the molecular and physiological details of their control with be important for a full understanding of the pathologies that affect these tissues (see section 3.4.1).

3.4 AFM as a micro-manipulator

Several papers have reported experiments that fall under neither of the previous categories. These use the AFM as a micro-manipulator or a micro-detector. Domke et al[42] used the AFM to map the mechanical pulse of cultured cardiomyocytes. Thie et al[53] examined the adhesive forces between a trophoblast and uterine epithelium using whole cells to functionalise the tips instead of isolated molecules. The adhesion forces recorded between cells were around 3nN, which is an order of magnitude higher than the molecule-ligand adhesion forces. Recently, Sagvolden et al[55] developed a new use of AFM to quantify the adhesion forces of cells to a substrate. This is of particular interest to the fields of tissue engineering and implant biomaterial design, where knowledge of cellular adhesive interactions will be fundamental to the development of biocompatible orthopaedic and vascular grafts.

3.4.1 Mechanical stimulation of live osteoblasts.

Virtually all cell types have been reported to sense mechanical stimuli. Amongst these, bone cells are particularly interesting as they govern the adaptation of bone structure in response to mechanical usage. We have used AFM as an indentor to mechanically stimulate bone cells[119] (Figure 5). Although there are many other techniques of applying mechanical strain to single cells, AFM provides an unique combination of positional precision, control of the force applied, innocuity to cells, and the ability to estimate cellular elasticity and strain. The latter, combined with monitoring of intracellular calcium reactions by linked optical methods (Figure 1), enables the threshold strain needed to elicit intracellular calcium responses in cells to be determined. Examination of the strain distribution in cells allows one to speculate on possible mechanisms of strain detection.

Cells can sense mechanical stimuli in a number of ways: strain can be detected through stretch-activated ion channels, integrin trans-membrane receptors that link the extracellular matrix to the cell cytoskeleton or receptor type tyrosine kinases[120-122]. The exact mechanism involved and the downstream events that may be influenced by the type of mechanical stimulation are yet to be fully evaluated. Cells respond to mechanical stimuli by a rise in intracellular calcium concentration. Both the amplitude and the frequency of calcium transients modulate a number of downstream events such as gene expression[123].

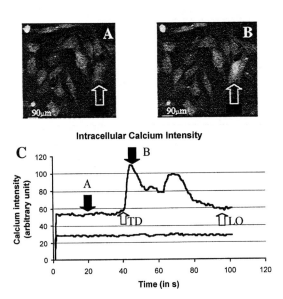

Figure 5   *Intracellular calcium responses in osteoblasts to strain applied via an AFM equipped with a 'glass ball' tip cantilever*
Rat osteoblasts were loaded with Fluo-3 prior to indentation as shown in Figure 1. (A), The cell about to be indented is indicated with an arrow. (B), The osteoblast after indentation has an increased intracellular calcium concentration. (C), Time course of the calcium intensity within the indented cell is shown. The times that correspond to the fluorescence images (A and B) are identified. TD indicates the time when the AFM cantilever enters in contact with the cell and LO indicates the time when the AFM cantilever is lifted out of contact of the cell surface. (Adapted from Charras et al, 2001[119] with permission).

Calcium can either originate from the extra-cellular environment or from intracellular stores and is intricately controlled (for a review, see [124]). Furthermore, having sensed a mechanical stimulus, the cell can transmit this information to neighbouring cells[125,126] to sensitise a larger area of tissue and give rise to a coherent tissue response.

The precision of prediction of cellular strain using AFM has been estimated and a good correlation between the predicted values for the radius of indentation and the experimentally measured ones was found[119]. Although part of the overall error is imputable to errors in optical measurement, it still remains necessary to find a more realistic model than the current linear elastic isotropic models in order to describe living cells (for a review, see Radmacher[61]). Several limitations can be pointed out. Firstly, the cell is a complex structure, and it is best described as a fluid bilayer that is non-uniformly tethered to an intricate network of interconnected struts (the intracellular microtubular network) and cables (the F-actin cytoskeleton and intermediate filaments)[127]. Secondly, living cells exhibit viscoelastic and plastic properties[63] and can actively adapt their cytoskeleton after long term exposure to mechanical strain[128].

The strain distributions determined throughout the cell thickness using finite element modelling have enabled a number of hypotheses regarding the way in which cells may sense applied strains to be formulated[119]. Maximal and minimal radial strains were situated on the cell surface, suggesting a detection mediated by components in the cell membrane, such as mechano-sensitive cation channels that open in response to membrane strain and let extracellular cations into the cell ([121] for a review). As there is a high strain gradient at the interface between the indented and un-indented regions, the strain would vary greatly in amplitude over a short distance, possibly stretching the membrane sufficiently to open mechano-sensitive cation channels. The vertical strain distribution shows a large compressive strain just under the area of indentation. This sub-membranous area is a prime area for connection between the cell membrane and the cellular cytoskeleton. Hence, the signal may be detected by a direct deformation of cellular cytoskeleton, which may in turn activate mechanisms that increase intracellular calcium levels, for example phosphoinositides, ryanodine receptor and chloride channel mediated responses[124].

AFM, via its use both as an engineering tool and a micro-manipulator, shows great promise in helping our understanding of mechano-transduction pathways. A nagging question, though, comes to mind - do all cells react to similar applied mechanical strains? Through its capacity to quantify biological phenomena in engineering terms, AFM may bring biology into the era of solution engineering and enable such a question to be answered.

4 CONCLUSIONS: OUTSTANDING RESEARCH QUESTIONS

The use of AFM in biology in the future might, intriguingly, be based not so much on its atomic scale imaging capabilities, but rather on the opportunity it offers to apply and measure forces between biological entities, such as molecules, cells or biomaterials. As a consequence, AFM is being used increasingly as a precise micro-manipulator rather than an imaging tool. Interesting work using this capability has focused not only on quantifying inter-molecular, but also intra-molecular, bond strengths. For example, unfolding forces for large linear molecules have been measured and modelled[72,74]. As equipment improves and the range of different commercially available systems widens, it is predicted that measurements of molecular interaction forces by AFM will become a routine procedure for many cell biologists.

Of the many applications that we have described, chemical force or molecular recognition microscopy has the potential to become used industrially in the evaluation process of candidate pharmaceuticals as an adjunct to standard pharmacological tools. Likewise, affinity mapping of receptor distribution is particularly useful for examining the location of receptors to which antibodies may not exist or form the large group of 'orphan receptors' in the human genome, as well as for evaluating the functionality of receptors present within the cell membrane; with development, an AFM-based method could replace the current gold standard technique of receptor autoradiography.

Adhesion measurement of whole cells, as described by Sagvolden et al[55] or Thie et al[53], may enable new materials to be characterised for implantation in patients. Thus, new orthopaedic implants could be evaluated to select materials for their specific purpose so that, for example, they would promote adhesion of osteoblasts over other cell types and promote ideal integration into the body.

Cellular bio-mechanics has been hindered by the lack of a precise tool to enable hypotheses to be experimentally verified. AFM may help us understand how cells react to strain, by which mechanisms they adapt to life in strained environments, or how mechanical and signalling pathways interact at the cellular level, as has been hypothesised (for example, percolation[129] and tensegrity[127]). Furthermore, in conjunction with finite element modelling, AFM may help to answer some of the more intriguing questions posed by biology. For example, how do erythrocytes manage to pass through capillaries whose diameter is smaller than their own?

Currently, there is also a lack of suitable methods to analyse the 3D structure of membrane glycoproteins at high resolution in their native context and configuration. The pioneering work of Muller and colleagues[77,78] used proteins of bacterial purple membranes, which are naturally present as 2D crystals, such as bacteriorhodopsin and Ompf, or eukaryotic membranes naturally enriched in specific proteins[75,76]. This makes equivalent methods for molecules present in the membranes of eukaryotic cells particularly attractive, especially if high resolution, 'soft' imaging techniques can be developed. Here, membrane glycoproteins are typically present at much lower densities and below levels that would be expected to form crystalloid features. By refining such experiments on eukaryotic cells, one may be able to gain a definitive insight into the structure and function of, for example, ion channels or receptors in their native membrane macro-molecular complexes.

Finally, in order to realise its full potential in cell biology, a number of design and technical problems still need to be optimised or solved and made available in a commercial setting. The AFM has to be interfaced to an optical microscope to be able to choose the cell to be examined, maintained in a near physiological condition which must be easily changeable during imaging. As certain cell types are very tall and certain substrates very uneven, the z-range of the AFM has to be extended beyond the range that is generally available commercially. Phase-contrast and fluorescence imaging of cells examined using AFM during or post-experimentation must be possible and of high optical quality and resolution. For chemical force AFM, a robust, easy and reliable way of functionalising tips needs to be devised and these need to be optimised for cellular imaging. In order to reliably examine biological phenomena in real time, an increased scanning speed applying less force would be desirable[130]. In order to realise the full potential of integration with the whole range of biological examination techniques, the AFM apparatus must leave easy access to the sample to enable simultaneous AFM and micro-manipulation, micro-injection or electrophysiology. Finally, serious attention must be paid to ease of use, both in terms of the hard- and soft-ware, for AFM to gain as wide acceptance in the biological community as the confocal microscope.

Acknowledgements

This work was supported by a Programme Grant to M.A.H. from The Wellcome Trust. P.P.L. was in receipt of an EMBO fellowship and G.T.C. was funded by a Johnson and Johnson COSAT award.

References

1   D.E. Leckband, J.N. Israelachvili, F.J. Schmitt and W. Knoll, *Science*, 1992, **255**, 1419.
2   E. Evans, D. Berk, A. Leung and N. Mohandas, *Biophys. J.,* 1991 **59**, 849.
3   R. Merkel, P. Nassoy, A. Leung, K. Ritchie and E. Evans, *Nature*, 1999, **397**, 50.
4   K. Svoboda , C.F. Schmidt, B.J. Schnapp and S.M. Block, *Nature*, 1993, **365**, 721.
5   E.L. Florin, A. Pralle, J.K. Horber and E.H. Stelzer, *J. Struct Biol.*, 1997, **119**, 202.
6   M. Glogauer, P. Arora, G. Yao, I. Sokholov, J. Ferrier and C.A. McCulloch, *J. Cell Sci.*, 1997, **110**, 11.
7   H. Nakajima, Y. Kunioka, K. Nakano, K. Shimizu, M. Seto and T. Ando, *Biochem. Biophys. Res. Commun.*, 1997, **234**, 178.
8   G. Binnig, C.F. Quate and C. Gerber, *Phys. Rev. Lett.,* 1986, **56**, 930.
9   O. Marti, V. Elings, M. Haugan, C.E. Bracker, J. Schneir, B. Drake, S.A. Gould, J. Gurley, L. Hellemans, K. Shaw, et. al., *J. Microsc.*, 1988, **152**, 803.
10   J.K. Horber, W. Haberle, F. Ohnesorge, G. Binnig, H.G. Liebich, C.P. Czerny, H. Mahnel and A. Mayr, *Scanning Microsc.*, 1992, **6**, 919.
11   J.H. Hoh, G.E. Sosinsky, J.P. Revel and P.K. Hansma, *Biophys J.*, 1993, **65**, 149.
12   V. Parpura, P.G. Haydon and E. Henderson, *J. Cell Sci.*, 1993, **104**, 427.
13   M. Beckmann, H.A. Kolb and F. Lang, *J. Membr. Biol.*, 1994, **140**, 197.
14   J.H. Hoh and C.A. Schoenenberger, *J. Cell Sci.*, 1994, **107**, 1105.
15   C. Le Grimellec, E. Lesniewska, C. Cachia, J.P. Schreiber, F. de Fornel and J.P. Goudonnet, *Biophys. J.*, 1994, **67**, 36.
16   H. Oberleithner, E. Brinckmann, A. Schwab and G. Krohne, *Proc. Natl. Acad. Sci., USA,* 1994, **91**, 9784.
17   K.A. Barbee, *Biochem. Cell Biol.*, 1995, **73**, 501.
18   R. Lal, S.A. John, D.W. Laird and M.F. Arnsdorf, *Am. J. Physiol.*, 1995, **268**, C968.
19   R. Lal, B. Drake, D. Blumberg, D.R. Saner, P.K. Hansma and S.C. Feinstein, *Am. J. Physiol.*, 1995, **269**, C275.
20   C. Le Grimellec, E. Lesniewska, M.C. Giocondi, C. Cachia, J.P. Schreiber and J.P. Goudonnet, *Scanning Microsc.*, 1995, **9**, 401.
21   D.J. Muller, F.A. Schabert, G. Buldt and A. Engel, *Biophys. J.*, 1995, **68**, 1681.
22   A. Schwab, K. Gabriel, F. Finsterwalder, G. Folprecht, R. Greger, A. Kramer and H. Oberleithner, *Pflugers Arch.*, 1995, **430**, 802.
23   F. Braet, R. De Zanger, W. Kallem, A. Raap, H. Tanke and E. Wisse, *Scanning Microsc. Suppl.*, 1996, **10**, 225.
24   T. Danker, M. Mazzanti, R. Tonini, A. Rakowska and H. Oberleithner, *Cell Biol. Int.*, 1997, **21**, 747.
25   U. Ehrenhofer, A. Rakowska, S.W. Schneider, A. Schwab and H. Oberleithner, *Cell Biol. Int.*, 1997, **21**, 737.
26   S.W. Schneider, Y. Yano, B.E. Sumpio, B.P. Jena, J.P. Geibel, M. Gekle and H. Oberleithner, *Cell Biol. Int.*, 1997, **21**, 759.

27 R.A. Garcia, D.E. Laney, S.M. Parsons and H.G. Hansma, *J. Neurosci. Res.*, 1998, **52**, 350.

28 C. Le Grimellec, E. Lesniewska, M.C. Giocondi, E. Finot, V. Vie and J.P. Goudonnet, *Biophys. J.*, 1998, **75**, 695.

29 A. Rakowska, T. Danker, S.W. Schneider and H. Oberleithner, *J. Membr. Biol.*, 1998, **163**, 129.

30 S.W. Schneider, P. Pagel, J. Storck, Y. Yano, B.E. Sumpio, J.P. Geibel and H. Oberleithner, *Kidney Blood Press Res.*, 1998, **21**, 256.

31 H. Schillers, T. Danker, H.J. Schnittler, F. Lang and H. Oberleithner, *Cell Physiol. Biochem.*, 2000, **10**, 99.

32 V. Vie, M.C. Giocondi, E. Lesniewska, E. Finot, J.P. Goudonnet and C. Le Grimellec, *Ultramicroscopy*, 2000, **82**, 279.

33 E. Henderson, P.G. Haydon and D.S. Sakaguchi, *Science*, 1992, **257**, 1944.

34 S.G. Shroff, D.R. Saner and R. Lal, *Am. J. Physiol.*, 1995, **269**, C286.

35 A. Spudich and D. Braunstein, *Proc, Natl. Acad. Sci., USA*, 1995, **92**, 6976.

36 U.G. Hofmann, C. Rotsch, W.J. Parak and M.J. Radmacher, *Struct. Biol.*, 1997, **119**, 84.

37 Y.G. Kuznetsov, A.J. Malkin and A. McPherson, *J. Struct. Biol.*, 1997, **120**, 180.

38 J.A. Dvorak and E. Nagao, *Exp. Cell Res.*, 1998, **242**, 69.

39 R. Matzke, K. Jacobson and M. Radmacher, *Nat. Cell Biol.*, 2001, **3**, 607.

40 H. Oberleithner, G. Giebisch and J. Geibel, *Pflugers Arch.*, 1993, **425**, 506.

41 M. Fritz, M. Radmacher and H.E. Gaub, *Biophys. J.*, 1994, **66**, 1328.

42 J. Domke, W.J. Parak, M. George, H.E. Gaub and M. Radmacher, *Eur. Biophys. J.*, 1999, **28**, 179.

43 C. Rotsch, K. Jacobson and M. Radmacher, *Proc. Natl. Acad. Sci., USA*, 1999, **96**, 921.

44 S.W. Schneider, P. Pagel, C. Rotsch, T. Danker, H. Oberleithner, M. Radmacher and A. Schwab, *Pflugers Arch.*, 2000, **439**, 297.

45 B.P. Jena, S.W. Schneider, J.P. Geibel, P. Webster, H. Oberleithner and K.C. Sritharan, *Proc. Natl. Acad. Sci., USA*, 1997, **94**, 13317.

46 A.P. Quist, S.K. Rhee, H. Lin and R. Lal, *J. Cell Biol.*, 2000, **148**, 1063.

47 V. Parpura and J.M. Fernandez, *Biophys. J.*, 1996, **71**, 2356.

48 V. Parpura, R.T. Doyle, T.A. Basarsky, E. Henderson and P.G. Haydon, *Neuroimage*, 1995, **2**, 3.

49 W. Haberle, J.K. Horber, F. Ohnesorge, D.P. Smith and G. Binnig, *Ultramicroscopy*, 1992, **42-44**, 1161.

50 F.M. Ohnesorge, J.K. Horber, W. Haberle, C.P. Czerny, D.P. Smith and G. Binnig *Biophys. J.*, 1997, **73**, 2183.

51 J.A. Dvorak, S. Kobayashi, K. Abe, T. Fujiwara, T. Takeuchi and E.J. Nagao, *Electron. Microsc.*, (Tokyo), 2000, **49**, 429.

52 E. Nagao, O. Kaneko and J.A. Dvorak, *J. Struct. Biol.*, 2000, **130**, 34.

53 M. Thie, R. Rospel, W. Dettmann, M. Benoit, M. Ludwig, H.E. Gaub and H.W. Denker, *Hum. Reprod.* 1998, **13**, 3211.

54 P.P. Lehenkari and M.A. Horton, *Biochem. Biophys. Res. Commun.*, 1999, **259**, 645.

55 G. Sagvolden, I. Giaever, E.O. Pettersen and J. Feder, *Proc. Natl. Acad. Sci., USA*, 1999, **96**, 471.

56 J.H. Hoh, R. Lal, S.A. John, J.P. Revel and M.F. Arnsdorf, *Science*, 1991, **253**, 1405.

57 M.G. Langer, A. Koitschev, H. Haase, U. Rexhausen, J.K. Horber and J.P. Ruppersberg, *Ultramicroscopy*, 2000, **82**, 269.

58 M. Sato, K. Nagayama, N. Kataoka, M. Sasaki and K. Hane, *J. Biomech.,* 2000, **33,** 127.

59 A.L. Weisenhorn, M. Khorsandi, S. Kasas, V. Gotzos and H.J. Butt, *Nanotechnology,* 1993, **4,** 106.

60 M. Radmacher, M. Fritz, C..M. Kacher, J.P. Cleveland and P.K. Hansma, *Biophys. J.,* 1996, **70,** 556.

61 M. Radmacher, *IEEE Eng. Med. Biol. Mag.,* 1997, **16,** 47.

62 E. A-Hassan, W.F. Heinz, M.D. Antonik, N.P. D'Costa, S. Nageswaran, C.A. Schoenenberger and J.H. Hoh, *Biophys. J.,* 1998, **74,** 1564.

63 H.W. Wu, T. Kuhn and V.T. Moy, *Scanning,* 1998, **20,** 389.

64 C. Rotsch and M. Radmacher, *Biophys. J.,* 2000, **78,** 520.

65 W.H. Goldmann, R. Galneder, M. Ludwig, W. Xu, E.D. Adamson, N. Wang and R.M. Ezzell, *Exp. Cell Res.,* 1998, **239,** 235.

66 H.G. Hansma, K.J. Kim, D.E. Laney, R.A. Garcia, M. Argaman, M.J. Allen and S.M. Parsons, *J. Struct. Biol.,* 1997, **119,** 99.

67 G. Shang, X. Qiu, C. Wang and C. Bai, *Appl. Phys. A.,* 1998, **66,** S333.

68 M. Benoit, T. Holstein and H.E. Gaub, *Eur, Biophys. J.,* 1999, **26,** 283.

69 P.P. Lehenkari, G.T. Charras, A. Nykänen and M.A. Horton, *Ultramicroscopy,* 2000, **82,** 289.

70 C.A. Putman, K.O. van der Werf, B.G. de Grooth, N.F. van Hulst and J. Greve, *Biophys. J.,* 1994, **67,** 1749.

71 A. Chilkoti, T. Boland, B.D. Ratner and P.S. Stayton, *Biophys. J.,* 1995, **69,** 2125.

72 T.E. Fisher, P.E. Marszalek and J.M. Fernandez, *Nat. Struct. Biol.,* 2000, **7,** 719.

73 J. Zlatanova, S.M. Lindsay and S.H. Leuba, *Prog. Biophys. Mol. Biol.,* 2000, **74,** 37.

74 M. Carrion-Vazquez, A.F. Oberhauser, T.E. Fisher, P.E. Marszalek, H. Li and J.M. Fernandez, *Prog. Biophys. Mol. Biol.,* 2000, **74,** 63.

75 T. Danker and H. Oberleithner, *Pflugers Arch.* 2000, **439,** 671.

76 H. Schillers, T. Danker, M. Madeja and H.J. Oberleithner, *Membr. Biol.,* 2001, **180,** 205.

77 A. Engel and D.J. Muller, *Nat. Struct. Biol.,* 2000, **7,** 715.

78 D.J. Muller, J.B. Heymann, F. Oesterhelt, C. Moller, H. Gaub, G. Buldt, A. Engel, *Biochim. Biophys. Acta.,* 2000, **1460,** 27.

79 O.H. Willemsen, M.M. Snel, A. Cambi, J. Greve, B.G. De Grooth and C.G. Figdor, *Biophys. J.,* 2000, **79,** 3267.

80 G.J. Schutz, M. Sonnleitner, P. Hinterdorfer and H. Schindler, *Molec. Memb. Biol.,* 2000, **17,** 17.

81 V.T. Moy, E.L. Florin and H.E. Gaub, *Science,* 1994, **266,** 257.

82 E.L. Florin, V.T. Moy and H.E. Gaub, *Science,* 1994, **264,** 415.

83 S.S. Wong, E. Joselevich, A.T. Woolley, C.L. Cheung and C.M. Lieber, *Nature,* 1998, **394,** 52.

84 U. Dammer, O. Popescu, P. Wagner, D. Anselmetti, H.J. Guntherodt and G.N. Misevic, *Science,* 1995, **267,** 1173.

85 G.N. Misevic, *Microsc. Res. Tech.,* 1999, **44,** 304.

86 P. Hinterdorfer, W. Baumgartner, H.J. Gruber, K. Schilcher and H. Schindler, *Proc. Natl. Acad. Sci., USA,* 1996, **93,** 3477.

87 S. Allen, X. Chen, J. Davies, M.C. Davies, A.C. Dawkes, J.C. Edwards, C.J. Roberts, J. Sefton, S.J. Tendler and P.M. Williams, *Biochemistry,* 1997, **36,** 7457.

88 U. Dammer, M. Hegner, D. Anselmetti, P. Wagner, M. Dreier, W. Huber and H.J. Guntherodt, *Biophys. J.,* 1996, **70,** 2437.

89   J. Fritz, A.G. Katopodis, F. Kolbinger and D. Anselmetti, *Proc. Natl. Acad. Sci. USA*, 1998, **95**, 12283.

90   W. Baumgartner, P. Hinterdorfer, W. Ness, A. Raab, D. Vestweber, H. Schindler and D. Drenckhahn, *Proc. Natl. Acad. Sci., USA*, 2000, **97**, 4005.

91   P.F. Luckham and K. Smith, *Faraday Discuss,* 1998, (111), 307.

92   W. Dettmann, M. Grandbois, S. Andre, M. Benoit, A.K. Wehle, H. Kaltner, H.J. Gabius and H.E. Gaub, *Arch. Biochem. Biophys.*, 2000 **383**, 157.

93   K.C. Sritharan, A.S. Quinn, D.J. Taatjes and B.P. Jena, *Cell Biol. Int.*, 1998, **22**, 649.

94   D.J. Ellis, T. Berge, J.M. Edwardson and R.M. Henderson, *Microsc. Res. Tech.*, 1999, **44**, 368.

95   A. Raab, W. Han, D. Badt, S.J. Smith-Gill, S.M. Lindsay, H. Schindler and P. Hinterdorfer, *Nat. Biotechnol.*, 1999, **17**, 901.

96   O.H. Willemsen, M.M. Snel, K.O. van der Werf, B.G. de Grooth, J. Greve, P. Hinterdorfer, H.J. Gruber, H. Schindler, Y. van Kooyk and C.G. Figdor, *Biophys. J.*, 1998, **75**, 2220.

97   S.W. Schneider, M.E. Egan, B.P. Jena, W.B. Guggino, H. Oberleithner and J.P. Geibel, *Proc. Natl. Acad. Sci., USA*, 1999, **96**, 12180.

98   Z. Yingge, Z. Delu, B. Chunli and W. Chen, *Life Sci.*, 1999, **65**, PL253.

99   R. de Souza Pereira, *FEBS Lett.*, 2000, **475**, 43.

100 M. Fiorini, R. McKendry, M.A. Cooper, T. Rayment and C. Abell, *Biophys J.*, 2001, **80**, 2471.

101 M. Gad, A. Itoh and A. Ikai, *Cell Biol. Int.*, 1997, **21**, 697.

102 T. Osada, S. Takezawa, A. Itoh, H. Arakawa, M. Ichikawa and A. Ikai, *Chem. Senses,* 1999, **24**, 1.

103 A. Chen and V.T. Moy, *Biophys. J.*, 2000, **78**, 2814.

104 N.B. Holland, C.A. Siedlecki and R.E. Marchant, *J. Biomed. Mater. Res.*, 1999, **45**, 167.

105 G. Kada, L. Blayney, L.H. Jeyakumar, F. Kienberger, V.P. Pastushenko, S. Fleischer, H. Schindler, F.A. Lai and P. Hinterdorfer, *Ultramicroscopy,* 2001, **86**, 129.

106 P. Lundberg, A. Lie, A. Bjurholm, P.P. Lehenkari, M.A. Horton, U.H. Lerner and M. Ransjo, *Bone,* 2000, **27**, 803.

107 P. Lundberg, I. Lundgren, H. Mukohyama, P.P. Lehenkari, M.A. Horton and U.H. Lerner, *Endocrinology,* 2001, **142**, 339.

108 M. Grandbois, W. Dettmann, M. Benoit and H.E. Gaub, *J. Histochem. Cytochem.*, 2000, **48**, 719.

109 M. Benoit, D. Gabriel, G. Gerisch and H.E. Gaub, *Nat. Cell Biol.*, 2000, **2**, 313.

110 W.R. Bowen, R.W. Lovitt, C.J. Wright, *J. Colloid. Interface Sci.*, 2001, **237**, 54.

111 A. Razatos, Y.L. Ong, M.M. Sharma and G. Georgiou, *Proc. Natl. Acad. Sci., USA*, 1998, **95**, 11059.

112 S.K. Lower, M.F. Jr Hochella, T.J. Beveridge, *Science,* 2001, **292**, 1360.

113 C.M. Isacke and M.A. Horton, The Adhesion Molecule FactsBook, 2nd. Ed. *Academic Press, London, UK,* 2000.

114 T. Strunz, K. Oroszlan, R. Schafer and H.J. Guntherodt, *Proc. Natl. Acad. Sci., USA*, 1999, **96**, 11277.

115 B. Isralewitz, M. Gao and K. Schulten, *Curr. Opin. Struct. Biol.*, 2001, **11**, 224.

116 E. Evans, *Annu. Rev. Biophys. Biomol. Struct.*, 2001, **30**, 105.

117 K.L. Johnson, Contact Mechanics, *Cambridge University Press*, Cambridge, UK, 1985.

118 D.E. Ingber, *Sci. Am.*, 1998, **278**, 48.

119 G.T. Charras, P.P. Lehenkari and M.A. Horton, *Ultramicroscopy,* 2001, **86**, 85.

120 A.J. Banes, M.Tsuzaki, J. Yamamoto, T. Fischer, B. Brigman, T. Brown and L. Miller, *Biochem. Cell Biol.*, 1995, **73,** 349.

121 F. Sachs and C.E. Morris, *Rev. Physiol. Biochem. Pharmacol.*, 1998, **132**, 1.

122 J. Sadoshima, S. Izumo, *Heart Vessels*, 1997, Suppl **12**, 194.

123 M.J. Berridge, *Neuron.*, 1998, **21**, 13.

124 M.J. Berridge, P. Lipp and M.D. Bootman,. *Nat. Rev. Mol. Cell Biol.*, 2000, **1**, 11.

125 S.L. Xia and J. Ferrier, *Biochem. Biophys. Res. Commun.*, 1992, **186**, 1212.

126 N.R. Jorgensen, Z. Henriksen, C. Brot, E.F. Eriksen, O.H. Sorensen, R. Civitelli and T.H. Steinberg, *J. Bone Miner. Res.*, 2000, **15**, 1024.

127 D.E. Ingber,. *J. Cell Sci.*, 1993, **104**, 613.

128 P.R. Girard and R.M. Nerem, *Front Med. Biol. Eng.*, 1993, **5**, 31.

129 G. Forgacs, *J. Cell Sci.*, 1995, **108**, 2131.

130 J. Tamayo, A.D. Humphris, R.J. Owen and M.J. Miles, *Biophys. J.*, 2001, **81**, 526.

PHYSICAL CHARACTERIZATION OF WILD TYPE AND *mnn9* MUTANT CELLS OF *Saccharomyces cerevisiae* BY ATOMIC FORCE MICROSCOPY (AFM)

A.Méndez-Vilas[1], I.Corbacho[2], M.L.González-Martín[1] and M.J.Nuevo[1]

[1]Department of Physics. University of Extremadura. Avda. Elvas s/n. 06071 Badajoz, SPAIN. E-Mail: maria@unex.es
[2]Department of Microbiology. University of Extremadura. Avda. Elvas s/n. 06071 Badajoz, SPAIN.

1 INTRODUCTION

The study of glycosylation processes in eukaryotic cells is a very relevant field of Cell and Molecular Biology. Glycosylation is the main covalent modification which proteins secreted by eukaryotic membranes undergo. Glycosylation mechanisms are well preserved in eukaryotic cells and they all follow the same fundamental scheme. The percentage of the glucidic part in a glycoprotein can range from 2 to 90%. The carbohydrate contribution to the proteic function is variable. *Saccharomyces cerevisiae* was used as a model, for different reasons: (a) it is unicellular and can be easily grown in cheap culture media; (b) it is a well known species; (c) it is easy to obtain glycosylation mutants; (d) it is surrounded by a thick cell wall of about 200 nm, which represents 15-30% of the overall weight. This cell wall is composed of mannoproteins (40%), glucans and chitin. Mannoproteins are glycoproteins whose glucidic portion is almost exclusively mannose. Glucan and chitin are structural, and the mannoproteins are the filling. In this work, among the three types of links in glycoproteins, the N-glycosidic links were those analyzed. A review of the significance of N-glycosylation can be found on Ref 1.

In order to study the N-glycosidic remains structure, several kinds of mutants in glycosylation processes were isolated, including the mnn mutants.[2-4]. The difference between the distinct kinds of mutants is the glycosylation process phase related to the muted gen. For example, the *mnn1* mutant lacks á (1,3) linked terminal mannoses. The *mnn6* mutant presents a minor inclusion of mannosylphosphate groups, and the *mnn9*, which is the most dramatically affected, lacks most of the external chain, and presents a lower growth rate.

The morphology and physico-chemical properties (adhesion, mechanical properties, etc.) of these cells, is a near-total-darkness field. The influence of the defects of the N-glycosylation processes on the physical properties of the cell wall is of great interest, and Atomic Force Microscopy offers new possibilities. AFM permits an extremely high lateral resolution, without previous manipulation of the sample, unlike other imaging techniques such as Scanning Electron Microscopy (SEM). Moreover, different physical properties can be also measured by AFM, such as surface roughness, friction, adhesion force and mechanical properties, with the same lateral resolution as topography.

AFM scans the surface of a sample with a sharp silicon or Si_3N_4 tip (a few μm long and less than 0.01 μm in diameter) located at the free end of a cantilever (about 100 μm long). Forces between the tip and the sample surface cause the cantilever to bend or deflect. A detector measures the deflection of the cantilever and produces a map of the surface topography of the sample scanned under the tip. AFM can be used to study insulators or conductors in two general modes (contact or non-contact regimes), depending on the way the tip interacts with the surface of the sample. In the AFM contact mode, additional studies can be performed by lateral force microscopy (LFM). This technique measures lateral deflections (twisting) of the cantilever that arise from forces on the cantilever parallel to the plane of the sample surface[5]. It is useful for imaging variations in surface friction that can arise from inhomogeneities in the surface material[6,7] and also for obtaining edge-enhanced images of any surface[8]. The LFM and AFM images are, in general, collected simultaneously, so that changes in the composition of the sample surface can be observed.

The effect of different kinds of mutations on the morphology and physical surface properties of cells have been previously studied by some authors by Atomic Force Microscopy. Ishijima *et al*[9] studied the ultrastructure of the cell wall of the *cps8* actin mutant cell in *Schizosaccharomyces pombe*. This mutant was used to determine the role of the actin cytoskeleton in cell wall formation. In addition to the studies about the ultrastructure of cells, AFM can also be used to measure surface forces such as adhesion, as well as mechanical surface properties. Differences in adhesion to biomaterials by *E.coli* K12 mutant (cells with truncated LPS molecules) and its parental strain D21 (which synthesize the complete carbohydrate chain) were studied by Y.L.Ong *et al*[10].

In order to analyze the influence of this mutation on the morphology of the cells and their surface structure, contact and non-contact images were obtained on both wild and *mnn9* mutant cells of *S.cerevisiae*. Moreover, the surface ultrastructure of both surfaces was characterized by measuring their so-called root mean square (rms) surface roughness.

2 METHOD AND RESULTS

2.1 Sample Preparation

Both wild type and *mnn9* strains were grown in YEPD (Yeast Extract – Peptone – Dextrose) solid medium, at 30 °C for 48 hours. A small portion of cells were then collected using a toothpick and extended over a microscope slide with a drop of water and allowed to dry at room temperature. The strains were kindly supplied by LM Hernández and I. Olivero from the Department of Microbiology of the University of Extremadura.

2.2 Atomic Force Microscopy

All the experiments were carried out with an Autoprobe CP atomic force microscope (Park Scientific Instruments, Geneva, Switzerland), equipped with a scanner of maximum ranges of 100 μm and 7 μm in *xy* and *z* directions, respectively. This instrument has an optical microscope for easy location the region of interest, by monitoring the sample on a TV screen. The images were acquired using silicon nitride cantilevers with a nominal force constant of 0.4 N/m and a typical probe curvature radio of 10nm, supplied by the manufacturer. The scanner speed ranged between 1 and 5 μm/s and the images were acquired at 512×512 pixels. The maximum applied force was ≈ 10 nN. The non-contact images were taken at the cantilever resonance frequency of 112 kHz.

2.3 Wild Type Cells

In order to study the cells morphology, contact (repulsive imaging regimen) and non-contact (attractive regimen) images were taken. Fig.1(a) shows a contact mode image of a set of wild type cells. As can be derived from these images, we have piles of cells with an average diameter of around 4 μm. Fig.1(b) also shows a non-contact image, taken at a slightly lower frequency than the resonance frequency of the cantilever (referred to as Tapping Mode). As can be seen equivalent information is obtained from both imaging modes, which indicates that in contact mode (the most aggressive imaging mode) the tip is not damaging the cell surface. Both types of image show a very smooth cell surface, without any noticeable characteristics.

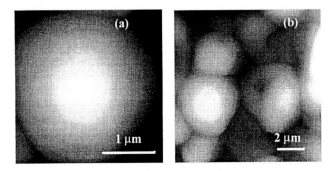

Figure 1   *Contact and non-contact images of wild type cells.*

Friction images were also made in order to gain more insight into the physical properties of these cell surfaces, in particular regarding surface composition homogeneity. Friction can also be used in many cases for enhancing topographical features. Fig.2(a)-(b) show a forward and backward friction image, respectively, of a wild type cell. No contrast between the two images is appreciated, which seems to indicate the high composition homogeneity of the cell surface, even at sub-micrometer scales. This coincides with what was expected from the bibliography.

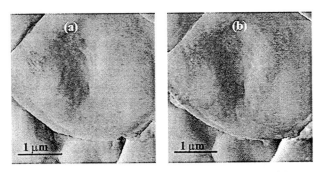

Figure 2   *(a)Forward and (b)backward friction images of wild type cells.*

An example of the utility of friction image for enhancing topographical features can be seen in Fig. 3 and 4, where (a) is a contact topographical image and (b)-(c) are the corresponding forward and backward friction images, for two different samples,

3 CONCLUSION

Atomic Force Microscopy is revealed as a powerful technique for studying the influence of N-glycosylation process defects on the physical cell wall properties of *S. cerevisiae* cells. Non-contact images show highly rough and irregular structures on the mutant cells, this observation being confirmed by the surface roughness measurements over wild type and *mnn9* cells. The fact that AFM is capable of probing biological surfaces without previous treatment and its unprecedented lateral and vertical resolution, make this instrument a unique tool for nanometric studies of membranes and cell surfaces.

Acknowledgements

This work was supported by Project IPR00A083 from Junta de Extremadura (Consejería de Educación, Ciencia y Tecnología).

References

1   S.Munro. *FEBS Letters*, 2001, **498**, 223.
2   C.E. Ballou. *Methods Enzymol.*, 1990, **185**, 440.
3   L. Ballou, R.E. Cohen, C.E. Ballou. *J. Biol. Chem.*, 1980, **255**, 5986.
4   P. Mañas, I. Oliveros, M. Ávalos y L.M. Hernández. *Glycobiology*, 1997, **4**, 487
5   S.N. Magonov, M.-H. Whangbo, in *Surface analysis with STM and AFM*, VCH, Weinheim, 1996, 323.
6   R.M. Overney, E. Meyer, J. Frommer, D. Brodbeck, R. Lüthi, L. Howald, H.-J. Güntherodt, M. Fujihira, H. Takano, Y. Gotoh, *Nature*, 1992, **359**, 133.
7   J. Colchero, H. Bielefeldt, A. Ruf, M. Hipp, O. Marti, and J. Mlynek, *Phys. Stat. Sol. (a)*, 1992, **131**, 73.
8   H.-Y. Nie, M.J. Walzak, N.S. McIntyre, A.M. EL-Sherik, *Appl. Surf. Sci.,*1999, **144-145**, 633.
9   S.A. Ishijima, M. Konomi, T. Takagi, M. Sato, J. Ishiguro, M. Osumi, *FEMS Microbiol. Lett.*, 1999, **180(1)**, 31.
10 Y-L. Ong, A. Razatos, G. Georgiou, M.M. Sharma, *Langmuir, 1999,* **15**, 2719.
11 J.D. Kiely, D.A. Bonnell, *J. Vac. Sci. Technol B*, 1997, **15(4)**, 1483.
12 P.K. Tsai, J. Frevert, C.E. Ballou. *J. Biol. Chem.*, 1984, **259(6)**, 3805.

PROBING SUPRAMOLECULAR ORGANISATION AT IMMUNE SYNAPSES.

Fiona E. McCann[1], Klaus Suhling[1], Leo M. Carlin[1], Konstantina Eleme[1], Kumiko Yanagi[1], Paul M.W. French[2], David Phillips[3], Daniel M. Davis[1, 4].

[1] Department of Biological Sciences, Sir Alexander Fleming Building, [2] Department of Physics, [3] Department of Chemistry, Imperial College of Science, Technology and Medicine, London, SW7 2AZ, UK. [4] E-Mail: d.davis@ic.ac.uk

1 INTRODUCTION: MOLECULAR RECOGNITION OF DISEASE BY NATURAL KILLER CELLS

Natural Killer (NK) cells are large granular lymphocytes that have the ability to lyse certain target cells that have undergone viral infection or tumour transformation.[1] One way that NK cells can discriminate infected cells from healthy cells is by detecting a loss or downregulation of self proteins, in particular major histocompatibility complex (MHC) class I molecules on the target cell surface.[2] Viral infection and tumourigenesis are frequently associated with alterations in expression of MHC class I molecules as a means to avoid detection by cytotoxic T cells.[3] NK cell recognition of self protein is mediated by groups of receptors that bind specific MHC class I molecules that transduce an inhibitory signal, thus preventing the natural cytotoxicity incurred with NK cell docking. In particular, human NK cells detect MHC class I protein via a variety of immunoglobulin-like and C-type lectin receptors.[4-6] Killer Immunoglobulin-like Receptors (KIR) with two immunoglobulin (Ig) domains, KIR2DL1 and KIR2DL2 recognise the MHC class I proteins, HLA-Cw4 or -Cw6, and HLA-Cw3 or -Cw7, respectively.[7-11] However, our understanding of how NK cells detect diseased cells is further complicated by the expression of stimulatory as well as inhibitory NK receptors for MHC class I protein.[12,13] In addition, several classes of NK receptors have recently been identified that recognise non-MHC proteins on target cell surfaces.[14-19] In short, regulation of NK cell cytotoxicity occurs through a complex and little understood integration of signals from stimulatory and inhibitory cell surface receptors. Here, we first review how fluorescence imaging has elucidated much about the organisation of proteins at immune synapses. We then discuss some molecular mechanisms that may influence this organisation of proteins at immune synapses. Next we discuss the recent unexpected observation that cell surface proteins transfer between cells at immune synapses. Finally, as an example of a new technique that could be employed in the study of immune synapses, we discuss the use of fluorescence lifetime imaging (FLIM) to probe the environment of cell surface proteins.

2 SUPRAMOLECULAR ORGANISATION AT IMMUNE SYNAPSES

It has recently emerged that immunological intercellular contacts, where information is exchanged between effector and target cells, consists of specialised membrane domains reminiscent of the neuronal synapse. This analogy was first suggested by Norcross,[20] and the term 'the immunological synapse' was proposed by Paul and Seder a decade later with reference to T lymphocytes and antigen presenting cells.[21] Although it is largely unknown how this organisation of cell surface proteins is orchestrated, several mechanisms have been proposed. These include lateral segregation of membrane proteins by lipid rafts, cytoskeletal involvement and distribution of proteins according to the size of their extracellular domains.

One of the most striking recent developments in molecular immunology is the discovery that T cell receptors and adhesion molecules cluster in unexpectedly large segregated membrane domains at immune synapses.[22-24] In a model system using T cells interacting with protein-rich two dimensional lipid bilayers, Grakoui et al [24] report on protein dynamics at the immune synapse following TCR ligation. They propose that after contact between the T cell and the protein-rich lipid bilayers, a bull's-eye pattern of protein domains forms with a central region containing integrins surrounded by a ring containing MHC protein. After a few minutes the pattern inverts so that the central region is predominantly occupied by T cell receptor (TCR) and MHC protein while the integrins, LFA-1 and ICAM-1 are segregated to the periphery.[24] These discrete domains have been termed supramolecular activating clusters (SMAC), subdivided into central, i.e. cSMAC, and peripheral, i.e. pSMAC, regions.[22] Subsequent stabilisation of this arrangement occurs which may serve to sustain the T cell activation.

The transmembrane phosphatase CD45 can dephosphorylate Lck and therefore block T cell activation during the early stages following TCR ligation. After about 3 min. of contact between a T cell and APC, CD45 is excluded from the developing T cell immune synapse, which is consistent with the notion that exclusion of tyrosine phosphatases from membrane compartments is required for triggering the kinase cascade. Intriguingly, after an initial exclusion of CD45 from the T cell immune synapse some CD45 relocates close to the cSMAC in the vicinity of the TCR/MHC complex within about 10 min.[25] These and other observations [23,26-30] suggest an important role for immune synapses in controlling T cell signalling (reviewed in references [31-33,34]).

Redistribution of receptors and ligands at NK cell immune synapses were first imaged by time-lapse laser scanning confocal microscopy (LSCM).[35] Interestingly, the inhibitory NK cell immune synapse has inverted arrangements of MHC, ICAM-1 and LFA-1 in comparison to the mature activating T cell synapse.[35] Strikingly, the NK inhibitory immune synapse prevails in the presence of drugs that disrupt cytoskeleton or deplete ATP, which suggests that synapse formation is independent of cytoskeleton and/or ATP mediated events.[35]

The different arrangements of proteins in a T cell activating and an NK cell inhibitory immune synapse are schematically represented in Figure 1. In line with the terminology used to first describe T cell immune synapses, discrete regions at inhibitory immune synapses have been termed p- and c-supramolecular inhibitory clusters (SMIC).[36] Recent high-resolution fluorescence imaging by confocal microscopy of target cells transfected with enhanced green fluorescent protein (GFP)-tagged MHC protein revealed that the patterning of p- and c-SMIC is fluid, i.e. the patterning of membrane domains can alter during contact with the NK cell.[36]

In the case of NK cells where inhibitory and activating receptors co-exist on the cell surface, segregation of proteins into discrete membrane domains may be critical in integrating stimulatory and inhibitory signals. The location of activating receptors and ligands at NK immune synapses is one major unknown in this area and more broadly, virtually nothing is known of the biophysical processes governing formation of immune synapses. Possible molecular mechanisms that may influence lateral segregation and redistribution of membrane proteins and ultimately synapse formation include lipid rafts (2.1), cytoskeleton (2.2), and distribution of proteins according to the size of their extracellular domains (2.3).

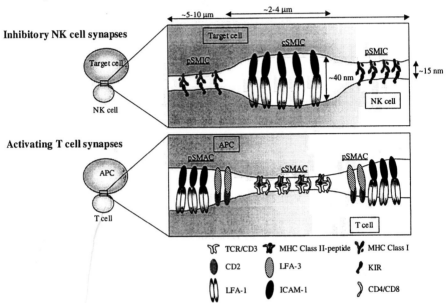

Figure 1: *Comparison of activating and inhibitory immune synapses.* To date activating T cell synapses and inhibitory NK cell synapses have been imaged and are schematically represented above. Both inhibitory T cell synapses, e.g. with anergised T cells, or activated NK cell synapses e.g. upon recognition of CMV- infected cells remain to be imaged.

2.1 Lipid Rafts

One important method by which proteins within plasma membranes can be separated is by association with different lipids (reviewed in references [37-39]). Early evidence for membrane domains came from fluorescence lifetime measurements that identified two distinct environments within the plasma membranes of living cells. These distinct populations corresponded to different lipid phases with different degrees of order of their hydrocarbon chains.[40] The saturated tails of sphingolipids associated with cholesterol are proposed to exist in a liquid-ordered phase separated from the liquid-disordered phase of the phospholipid containing plasma membrane. These sphingolipid, cholesterol-rich lipid rafts drift amongst the phospholipids, and association of proteins with the rafts segregates them from other non-associated proteins distributed within the phospholipid rich plasma membrane.[37,38] Rafts have been shown to exclude most membrane proteins and are thought to constitute approximately 10% of the cell surface.[33] Proteins known to associate with lipid rafts include glycosylphosphatidylinositol (GPI)-linked, myristylated and acylated proteins. Addition of a GPI anchor inserts the protein into the outer membrane leaflet while acylation normally promotes association with the cytoplasmic membrane leaflet,[41,42] one exception being the transmembrane transferrin receptor, which is excluded from lipid rafts despite being multiply acylated.[42] Examples of proteins constitutively associated with lipid rafts include the Src-family kinases, and G proteins. Others such as the T and B cell receptors, along with some downstream signalling molecules are recruited into rafts upon ligand binding, demonstrating the dynamic nature of events occurring within these membrane microdomains.[37-39,42]

Several studies of lipid rafts have relied on their insolubility in the detergent Triton X-100 at 4°C, in which they form glycolipid-enriched complexes, and are thus often referred to as detergent insoluble glycolipid enriched complexes (DIGs).[43] Due to their high lipid content, rafts 'float' to a low density during gradient centrifugation enabling the isolation and identification of associated proteins. This method cannot be used to quantitatively report lipid raft composition in the intact cell. Milder detergents such as octyglucoside will solubilise the rafts, and thus a larger than expected fraction of membrane proteins are insoluble in Triton-X-100.[44] Finally, the lipid composition of the rafts has not been proven to be distinct from the plasma membrane. Thus, the question can still be raised whether this methodology induces artefacts in identification of the contents of lipid rafts and this highlights the importance of employing detergent-independent methods to study lipid rafts in cell membranes at physiological temperatures.

Bioimaging can be used to reveal lipid rafts in living cells using fluorescent labels of proteins or lipids enriched in these domains. In resting cells, lipid rafts are not clearly identifiable and only when they are induced to coalesce into larger structures following crosslinking of raft components are they visible by microscopy.[45] Fluorophore-conjugated cholera toxin β subunit, which binds GM1, is commonly used as a stain for cholesterol-rich lipid domains.[46, 47] That clustering of cholera-toxin staining was induced by co-stimulation of T cells suggests an immunological importance of lipid raft motility.[47] Using fluorescence resonance energy transfer (FRET), Varma and Mayor[48] found cholesterol dependent clustering of GPI-anchored proteins into domains sized less than 70 nm, but using a similar methodology, another group found no evidence for such clustering.[49]

Current evidence supports a model in which lipid rafts play a role in immune cell activation.[50,51] In resting T cells, TCR are excluded from lipid rafts but ligand interaction may result in oligomerisation of the TCR and subsequent recruitment to lipid rafts where they are brought into contact with Src-family kinases.[39] The first biochemical event that can be detected following TCR ligation is Lck-mediated phosphorylation of tyrosine residues within the immunoregulatory tyrosine activating motifs (ITAMs).[52] T cell signalling was abolished in cells expressing only a mutant, non-palmitoylated isoform of Lck that is excluded from lipid rafts.[46] Strikingly, signalling was rescued when separate clusters of the mutant Lck and gangliosides within rafts were brought together using bridging antibodies, demonstrating that Lck was not inactivated by the mutation. This suggested that Lck must be physically close to its signalling partners for effective signalling and this association normally required clustering of the proteins within lipid rafts.[46]

Phosphorylation of ITAMs is immediately followed by recruitment of cytoplasmic ZAP-70 to lipid rafts where Lck constitutively resides (reviewed in reference 53). ZAP-70 is then in close proximity to linker for activation of T cells (LAT), which is subsequently phosphorylated triggering recruitment of further signalling proteins to lipid rafts. Despite localisation within the plasma membrane, a mutant LAT that is not palmitoylated and is excluded from cholesterol-rich lipid rafts was unable to be phosphorylated by ZAP-70 leading to the disruption of functional T cell signalling.[54] Clearly, segregation of membrane proteins by association with lipid rafts has the potential to selectively promote or exclude specific protein interactions. However, the relationship between lipid rafts and the supramolecular organisation at immune synapses remains to be directly investigated. Importantly, Lou et al [55] have demonstrated that lipid rafts, revealed by staining with cholera toxin, are not clustered at inhibitory NK cell immune synapses, inferring that lipid rafts are at least not necessary to create distinct protein domains at immune synapses.

2.2 The Role of the Cytoskeleton in Immune Synapse Formation.

In opposition to segregation of membrane proteins by association with lipids, several groups have suggested that protein organisation at immune synapses is controlled by association with intracellular cytoskeletal proteins.[32-56] Evidence for active transport of receptors to the immune synapse was first described using the large beads to observe the movement of the cortical actin cytoskeleton and linked receptors.[23] Upon TCR engagement, beads that are either specifically attached to ICAM-1, or non-specifically attached via streptavidin, translocate to the T cell immune synapse. The movement of non-specifically attached beads was abolished in the absence of ICAM-1, indicating the process to be activation dependent. Evidence that myosin motor proteins controlled this movement was that the drug butanedione monoxime (BDM) impaired bead movement across the surface of the T cell. It should be noted however that BDM is not a specific inhibitor of myosin motor proteins and the involvement of other processes cannot be ruled out.[57]

It is the T cell cytoskeleton that is thought to drive formation of the T cell immune synapse. When B cells are used as APC, no polarisation of their cytoskeleton has been observed during interaction with T cells, and inhibition of APC cytoskeleton rearrangement had no effect on T cell activation. However in contrast, in immune synapses between dendritic cells (DC) and T cells, Al-Alwan et al[58] observed polarization of filamentous (f-) actin and fascin, an actin bundling protein unique to dendritic cells. Pretreatment of DC with cytochalasin D, which disrupts assembly of filamentous actin, reduced the conjugation of DC with T cells by 76%.

2.3 The Role of Protein Size in Immune Synapse Formation

The current dogma[33] amongst immunologists is that supramolecular organisation at immune synapses is actively driven by ATP-dependent processes including myosin driven movement of the effector cell's cytoskeleton. However, our own experimental evidence suggests that the actin cytoskeleton is not necessary, at least for formation of the inhibitory NK immune synapse.[35] Another mechanism that could drive the formation of segregated protein domains is that they thermodynamically separate according to the sizes of their extracellular domains, as schematically represented in Figure 1. Initial adhesion of T cell/APC or NK /target cell occurs via binding of the large integrins LFA-1 and ICAM-1, which form a complex bridging approximately 40 nm across the synapse. This interaction is thought to initiate binding of smaller molecules e.g. TCR or KIR binding MHC protein, which would span approximately 15 nm. If it is energetically favourable for the membrane to bend rather than the proteins, segregated protein domains are created (Figure 1). Separation of proteins in this manner has been postulated to occur at a sub-micron scale, [56,59] but it is possible that such microdomains coalesce to create the supramolecular organisation of proteins seen by LSCM.

Molecular modelling suggests that such spontaneous organisation of immune synapses could indeed occur, based on thermodynamics governing the lateral mobility of proteins in the membrane, membrane mechanics, and the sizes of the extracellular portions of the cell surface proteins.[60] Thus, if membrane biophysics contributes, at least in part, to the organisation at immune synapses, membrane fluidity should be different in comparison to elsewhere at the cell surface. Furthermore, membrane viscosity should vary across the immune synapse such that domains in varying viscosity segregate in concordance with the domains rich in different proteins. Such aspects of the biophysical chemistry at immune synapses are wholly uncharted.

3 INTERCELLULAR TRANSFER OF CELL SURFACE PROTEINS AT IMMUNE SYNAPSES

Fluorescent probes along with use of LSCM have proved to be invaluable in visualising segregation of proteins at immune synapses with high temporal and spatial resolution. Unexpectedly, evidence is emerging that several cell surface proteins transfer between contacting cells at immune synapses.

Several groups have shown the receptor dependent transfer of antigen presenting MHC protein to T cells [61-65] by a number of experimental techniques. Other cell surface proteins also transfer from APC to T cells, namely CD80 (B7-1), OX40L and CD54 (ICAM-1).[66] T cells have even been observed to capture membrane fragments from APC, reliant upon TCR signalling.[64] The intercellular transfer of proteins may serve a number of immunoregulatory functions. For example, it has been demonstrated that T cells that have 'captured' target cell MHC protein onto their own cell surfaces can then stimulate neighbouring T cells to either proliferate [61] or be killed by cytotoxic T cells with the same specificity, termed fratricide.[63] T cell fratricide might support the notion of lymphocyte 'exhaustion' where T cell responses are reduced after prolonged exposure to antigen. Recently, B cells have also been shown to acquire antigen from professional APC after immune synapse formation.[67] In B cells intercellular transfer of antigen is suggested as a way that B cells can process and present membrane bound antigen.[67]

Our own study shows that NK cells expressing the KIR2DL1 receptor can acquire cognate HLA-C ligands from target cells at inhibitory NK cell immune synapses. A vesicle or group of vesicles of Green Fluorescent Protein (GFP) tagged HLA-C can clearly be seen moving from a target cell into an NK cell in Figure 2. Surprisingly large amounts, e.g. up to 30%, of the MHC protein can be 'captured' from target cells by NK cells.[36] An immunological function for this intercellular transfer is not known but we speculate that in NK cells intercellular transfer of MHC may serve to affect the threshold at which inhibitory signals are initiated, possibly by sequestering KIR or key signalling molecules away from the NK immune synapse.

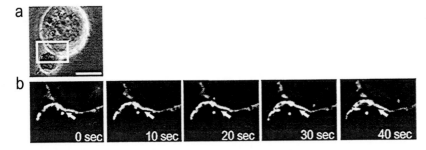

Figure 2. *Intercellular transfer of GFP tagged HLA-Cw6 from a 721.221 target cell to a human blood NK cell.* **(a)** Transmitted light image showing a target cell (721.221, EBV transformed B cell line) expressing GFP-tagged HLA-Cw6 in contact with a KIR2DL1 expressing human blood NK cell (smaller cell). **(b)** A series of confocal images showing the GFP fluorescence within the white box in **(a)** over 40 seconds. Each is a single 0.3 μm slice taken from the same focal plane. The white arrowhead in the first frame indicates a vesicle, or group of vesicles containing GFP-tagged HLA-Cw6 moving away from the HLA-Cw6 clustered at the NK inhibitory immune synapse and into the NK cell cytoplasm. The position of the white arrowhead remains constant throughout to indicate the movement of the vesicle, or group of vesicles. The zero time-point is arbitrarily defined. The scale bar in the lower right corner of **(a)** represents 10 μm. This figure is representative of at least three independent experiments (after reference 36).

The half-life of soluble KIR2DL1 binding to soluble HLA-C is around 0.3 s, more than an order of magnitude faster than the binding of TCR to MHC ligands and at least two orders of magnitude faster than mAb binding.[68] Interestingly however, the half-life of KIR2DL1/HLA-Cw6 has been measured to be similar to TCR/MHC binding in the presence of zinc.[69] The few minutes that NK cells remain in contact with target cells seems unlikely to be long enough for appreciable amounts of HLA-C to be captured without the induction of an active process by KIR. This is consistent with the requirement for NK cell ATP. Thus, the half-life of KIR/MHC and TCR/MHC complexes may be similar in order to facilitate a common molecular mechanism of capturing MHC protein from target cells. Possible molecular mechanisms that may influence intercellular cell surface protein transfer include (3.1) protein cleavage or ectodomain shedding, (3.2) exosomes, (3.3) protein and lipid mixing, (3.4) cytoskeleton-mediated transport.

3.1 Protein Cleavage or Ectodomain Shedding

Extracellular domains of cell surface receptors can be cleaved by proteases. This method of protein release has been documented in lymphocytes previously, for example in shedding TNF receptor and L-selectin.[70] However, this seems an unlikely mechanism for intercellular transfer of proteins at immune synapses for several reasons. Firstly, in our own studies, the MHC protein seen to transfer into NK cells was tagged with GFP on the cytoplasmic tail of the protein. Thus, for GFP fluorescence to be observed on the NK cell, the whole GFP-MHC construct is likely to have been transferred and certainly not just the extracellular portion.[36]

More directly, Western blotting the lysates of NK cells purified after incubation with target cells revealed that the NK cell acquired a protein of the size corresponding to the whole of the GFP-tagged HLA-C, indicating that the transferred protein was not first cleaved.[36] Finally, the transfer of OX40L in the presence of protease inhibitors [66] excludes the action of ectodomain shedding enzymes in facilitating intercellular protein transfer at immune synapses.

3.2 Exosomes

Exosomes, the product of fusion of late endosomal or lysosomal vesicles with the plasma membrane, are released by a number of lymphocytes, including cytotoxic T cells, B cells and dendritic cells. These exosomes can contain several cell surface proteins in the Ig superfamily, including CD54, CD86 (B7-2) and MHC class I and II proteins, reviewed by Denzer et al.[71] Antigen-presenting MHC complexes carried in the vesicles are accessible, as purified exosomes can stimulate T cells *in vitro*.[72,73] This could explain why the transfer of certain proteins is dependent on receptor recognition, as docking and internalisation of vesicles into the effector cell may depend on receptor ligation and/or signalling. However, constitutive release of MHC protein in APC derived exosomes is too slow to account for the large amounts of MHC protein transferred to NK cells, unless exosomal secretion is specifically triggered at immune synapses.

3.3 Protein and Lipid Mixing at Immune Synapses

The plasma membrane itself is highly dynamic in structure; lipids and proteins can relocate to various points at the cell surface. It could be that when two such membranes come into close proximity at the immune synapse, and receptor ligation occurs, lipids and proteins could transfer between the opposing lipid bilayers. In fact, there is evidence that lipids are transferred between cells along with proteins at the T cell immune synapse.[64] When the fluorescent lipid PKH26 was incorporated into target cell membranes it was shown to transfer to T cells only if TCR signalling has occurred. However CFSE, a stable internal protein dye used to label target cell cytosol did not transfer. Receptor specificity could play a part in a process like this, whereby receptors internalise and drag their ligands along with membrane fragments into the effector cell.

3.4 Cytoskeletal-mediated Transport at Immune Synapses

Tubulin cytoskeleton is often involved in intracellular vesicular transport, and as can be seen for example in Figure 2, the movement of HLA-C containing vesicles is directional, reminiscent of vesicles travelling along the cytoskeleton, rather than moving by Brownian motion. Thus, it is possible that the cytoskeleton is required to move HLA-C that has already transferred to the effector cell away from the immune synapse. However, treatment of the cells with cytochalasin D and BDM had no affect on the number cell conjugates in which intercellular protein transfer could be seen,[36] suggesting that polymerisation of the actin cytoskeleton, and myosin motors are not necessary for intercellular protein transfer. Studies using sodium azide to deplete ATP showed that ATP is required by the NK cell, and not the target cell, for transfer to occur. Thus, some active transport system must be invoked to facilitate intercellular protein transfer.

4 PROBING THE PROTEIN ENVIRONMENT BY FLUORESCENCE LIFETIME IMAGING

Immune synapse formation involves a complex rearrangement of proteins and lipids at intercellular contacts creating distinct local environments of each protein type. Clearly, new techniques are needed to probe the environment of protein domains within immune synapses. Fluorescence intensity imaging only reveals the location of proteins at immune synapses. But fluorescence emission is a multi-parameter signal that can report on, for example, the local viscosity,[74] refractive index,[75-77] pH,[78] solvent polarity[79,80] or the orientation of the fluorophore and its interaction with either the environment or other fluorophores.[81,82] Fluorescence lifetime imaging (FLIM)[83-86] is a technique that, in addition to position and intensity, also captures the fluorescence lifetime, i.e. the average time needed for the fluorescence to decay after an excitation pulse. This allows the local distribution and physical environment of the fluorescence probe or fluorophore-tagged proteins to be studied.[87] FLIM can probe the physical properties of each protein microenvironment and this information will help elucidate the molecular mechanisms of immune synapse formation. However, due to the number of parameters on which the fluorescence lifetime depends, the interpretation of data can be complex, particularly in a biological or biomedical context where many of these parameters may be unknown. Thus, in order to interpret FLIM of GFP in cells, we set out to determine the parameter that affects the fluorescence lifetime of GFP tagged receptors.

4.1 The Fluorescence Lifetime of GFP Reports the Local Refractive Index

In the absence of fluorescence quenching processes, the fluorescence lifetime τ can be calculated from the absorption and emission spectra using the Strickler-Berg formula,[88] taking into account the fluorescence quantum yield ϕ :

$$\frac{1}{\tau} = \frac{2.88 \times 10^{-9}}{\phi} n^2 \frac{\int I(\tilde{v})d\tilde{v}}{\int I(\tilde{v})\tilde{v}^{-3}d\tilde{v}} \int \frac{\varepsilon(\tilde{v})}{\tilde{v}} d\tilde{v}$$

Formula 1

where n is the refractive index of the medium, I is the intensity of the fluorescence emission, ε is the molar extinction coefficient and $\tilde{v}$ is the wave number (inverse of the wavelength λ, i.e. $\tilde{v} = \lambda^{-1}$). The relationship is calculated from the Einstein A and B coefficients for sharp atomic transitions and is applicable to molecules where the absorption band is broad and the emission Stokes shifted. The refractive index dependence is due to the medium in which the absorption and emission processes occur.[77]

Figure 3 shows three decays of GFP fluorescence upon excitation with a femtosecond laser. The fluorescence intensity (vertical axis, logarithmic) decays with time (horizontal axis) within nanoseconds after the excitation. The upper fluorescence decay is GFP in PBS buffer, the middle decay GFP in 50% glycerol/50% buffer and the lower decay 90% glycerol/10% buffer. A gradual decrease of the fluorescence lifetime as the glycerol content increases is clearly evident. These fluorescence decays were obtained by the time-correlated single photon counting method,[89] with excitation at 470 nm and the emission collected at 510 nm using a monochromator. An emission polarizer oriented at the magic angle 54.7° was used to eliminate rotational depolarization effects. A deconvolution of the decay curves with the instrumental response function yields the fluorescence lifetime τ. It ranges from 2.80 ns in PBS buffer (refractive index $n_{510nm}= 1.337$) to 2.20 ns in 90% glycerol/10% buffer ($n_{510nm}= 1.463$).

A plot of the inverse lifetime τ^{-1} versus the square of the refractive index of solvent solutions with an increasing amount of glycerol, n^2, is shown in Figure 4. It is a linear relationship as predicted by the Strickler-Berg formula assuming ε, I and ϕ are constant. A straight line fit yields a gradient of 0.239 ± 0.010 ns^{-1}, which is in good agreement with the value of 0.223 ± 0.020 ns^{-1} calculated directly from the absorption and emission spectra and the quantum yield ϕ using $\varepsilon =50000$ l mol^{-1} cm^{-1} and $\phi =0.60$ (Formula 1).[90] These data are not particular to using mixtures of glycerol and water as a solvent (Suhling et al., manuscript in preparation).

We have shown that in homogeneous and isotropic media the GFP fluorescence decay reports on the refractive index of the environment. Our aim now is to use FLIM and the GFP fluorescence decay directly, to report on the refractive index in different domains of immune synapses. This raises the question what biologically relevant parameter the refractive index reports on. We hypothesise that, for example, a cholesterol rich lipid microdomain [91] will have a distinct refractive index.

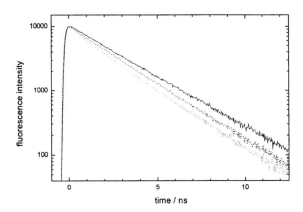

Figure 3: *Fluorescence decays of GFP in PBS buffer/glycerol mixtures.* The upper curve (continuous line) is the GFP fluorescence decay in PBS buffer, the middle curve (dashed line) in 50% glycerol/50% PBS buffer and the lower curve (dotted line) in 90% glycerol/10% PBS buffer. As the amount of glycerol is increased, the refractive index increases and the fluorescence decay is shortened. The excitation wavelength $\lambda_{ex}=470$ nm, and the fluorescence emission wavelength was $\lambda_{em}=510$ nm collected at the magic angle.

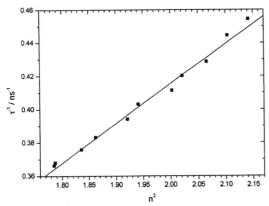

Figure 4: *The inverse fluorescence lifetime τ^{-1} versus the square of the refractive index, n^2. It is a linear relationship as predicted by the Strickler-Berg formula (Formula 1) assuming ε, I and ϕ are constant. The straight line is a fit to the data points and yields a gradient of 0.239 ± 0.010 ns^{-1}.*

5 SUMMARY

Understanding of the formation and function of immune synapses is a foundation from which new pharmaceutically active compounds can be designed to augment immune responses to viruses or suppress autoimmune responses. For example, candidate novel drugs could be identified by screening for small molecules that interfere with specific patterns of protein domains at immune synapses. To date, limited techniques have been used to probe supramolecular organisation at cell surfaces. Fluorescence Lifetime Imaging reporting on the physical environment of proteins as they redistribute into immune synapses is a highly promising technique to apply to studying immune synapse formation.

Acknowledgements

We are grateful to Jan Siegel for critical reading of this manuscript. Work in our laboratory is supported by grants from the Medical Research Council, the Biotechnology and Biological Sciences Research Council and The Royal Society.

References

1   G. Trinchieri, *Adv Immunol,* 1989, **47,** 187.
2   H. G. Ljunggren and K. Karre, *Immunol Today,* 1990, **11,** 237.
3   L. L. Lanier, *Annu Rev Immunol,* 1998, **16,** 359.
4   E. O. Long and S. Rajagopalan, *Semin Immunol,* 2000, **12,** 101.

5   E. O. Long, D. F. Barber, D. N. Burshtyn, M. Faure, M. Peterson, S. Rajagopalan, V. Renard, M. Sandusky, C. C. Stebbins, N. Wagtmann and C. Watzl, *Immunol Rev,* 2001, **181,** 223.

6   D. H. Raulet, R. E. Vance and C. W. McMahon, *Annu Rev Immunol,* 2001, **19,** 291.

7   M. Colonna, G. Borsellino, M. Falco, G. B. Ferrara and J. L. Strominger, *Proc Natl Acad Sci U S A,* 1993, **90,** 12000.

8   M. Colonna, E. G. Brooks, M. Falco, G. B. Ferrara and J. L. Strominger, *Science,* 1993, **260,** 1121.

9   J. C. Boyington, S. A. Motyka, P. Schuck, A. G. Brooks and P. D. Sun, *Nature,* 2000, **405,** 537.

10   Q. R. Fan, E. O. Long and D. C. Wiley, *Nat Immunol,* 2001, **2,** 452.

11   J. C. Boyington, A. G. Brooks and P. D. Sun, *Immunol Rev,* 2001, **181,** 66.

12   O. Mandelboim, H. T. Reyburn, M. Vales-Gomez, L. Pazmany, M. Colonna, G. Borsellino and J. L. Strominger, *J Exp Med,* 1996, **184,** 913.

13   A. Moretta, S. Sivori, M. Vitale, D. Pende, L. Morelli, R. Augugliaro, C. Bottino and L. Moretta, *J Exp Med,* 1995, **182,** 875.

14   R. Biassoni, C. Cantoni, D. Pende, S. Sivori, S. Parolini, M. Vitale, C. Bottino and A. Moretta, *Immunol Rev,* 2001, **181,** 203.

15   D. Pende, S. Parolini, A. Pessino, S. Sivori, R. Augugliaro, L. Morelli, E. Marcenaro, L. Accame, A. Malaspina, R. Biassoni, C. Bottino, L. Moretta and A. Moretta, *J Exp Med,* 1999, **190,** 1505.

16   M. Vitale, C. Bottino, S. Sivori, L. Sanseverino, R. Castriconi, E. Marcenaro, R. Augugliaro, L. Moretta and A. Moretta, *J Exp Med,* 1998, **187,** 2065.

17   S. Sivori, M. Vitale, L. Morelli, L. Sanseverino, R. Augugliaro, C. Bottino, L. Moretta and A. Moretta, *J Exp Med,* 1997, **186,** 1129.

18   A. Pessino, S. Sivori, C. Bottino, A. Malaspina, L. Morelli, L. Moretta, R. Biassoni and A. Moretta, *J Exp Med,* 1998, **188,** 953.

19   C. Cantoni, S. Verdiani, M. Falco, A. Pessino, M. Cilli, R. Conte, D. Pende, M. Ponte, M. S. Mikaelsson, L. Moretta and R. Biassoni, *Eur J Immunol,* 1998, **28,** 1980.

20   M. A. Norcross, *Ann Immunol (Paris),* 1984, **135D,** 113.

21   W. E. Paul and R. A. Seder, *Cell,* 1994, **76,** 241.

22   C. R. Monks, B. A. Freiberg, H. Kupfer, N. Sciaky and A. Kupfer, *Nature,* 1998, **395,** 82.

23   C. Wülfing and M. M. Davis, *Science,* 1998, **282,** 2266.

24   A. Grakoui, S. K. Bromley, C. Sumen, M. M. Davis, A. S. Shaw, P. M. Allen and M. L. Dustin, *Science,* 1999, **285,** 221.

25   K. G. Johnson, S. K. Bromley, M. L. Dustin and M. L. Thomas, *Proc Natl Acad Sci U S A,* 2000, **97,** 10138.

26   M. A. McCloskey and M. M. Poo, *J Cell Biol,* 1986, **102,** 2185.

27   B. C. Schaefer, M. F. Ware, P. Marrack, G. R. Fanger, J. W. Kappler, G. L. Johnson and C. R. Monks, *Immunity,* 1999, **11,** 411.

28   C. Wülfing, M. D. Sjaastad and M. M. Davis, *Proc Natl Acad Sci U S A,* 1998, **95,** 6302.

29   C. Wülfing, A. Bauch, G. R. Crabtree and M. M. Davis, *Proc Natl Acad Sci U S A,* 2000, **97,** 10150.

30   M. F. Krummel, M. D. Sjaastad, C. Wülfing and M. M. Davis, *Science,* 2000, **289,** 1349.

31   J. Delon and R. N. Germain, *Curr Biol,* 2000, **10,** R923.

32   M. L. Dustin and J. A. Cooper, *Nat Immunol,* 2000, **1,** 23.

33   S. K. Bromley, W. R. Burack, K. G. Johnson, K. Somersalo, T. N. Sims, C. Sumen, M. Davis, A. S. Shaw, P. M. Allen and M. L. Dustin, *Annu Rev Immunol,* 2001, **19,** 375

34   E. Donnadieu, P. Revy and A. Trautmann, *Immunology,* 2001, **103,** 417.

35   D. M. Davis, I. Chiu, M. Fassett, G. B. Cohen, O. Mandelboim and J. L. Strominger, *Proc Natl Acad Sci U S A,* 1999, **96,** 15062.

36   L. M. Carlin, K. Eleme, F. E. McCann and D. M. Davis, manuscript submitted.

37   D. A. Brown and E. London, *J Biol Chem,* 2000, **275,** 17221.

38   K. Simons and E. Ikonen, *Nature,* 1997, **387,** 569.

39   A. Cherukuri, M. Dykstra and S. K. Pierce, *Immunity,* 2001, **14,** 657.

40   R. D. Klausner, A. M. Kleinfeld, R. L. Hoover and M. J. Karnovsky, *J Biol Chem,* 1980, **255,** 1286.

41   D. A. Brown and E. London, *Annu Rev Cell Dev Biol,* 1998, **14,** 111.

42   P. W. Janes, S. C. Ley, A. I. Magee and P. S. Kabouridis, *Semin Immunol,* 2000, **12,** 23.

43   D. A. Brown and J. K. Rose, *Cell,* 1992, **68,** 533.

44   R. Schroeder, E. London and D. Brown, *Proc Natl Acad Sci U S A,* 1994, **91,** 12130.

45   E. D. Sheets, D. Holowka and B. Baird, *J Cell Biol,* 1999, **145,** 877.

46   P. W. Janes, S. C. Ley and A. I. Magee, *J Cell Biol,* 1999, **147,** 447.

47   A. Viola, S. Schroeder, Y. Sakakibara and A. Lanzavecchia, *Science,* 1999, **283,** 680.

48   R. Varma and S. Mayor, *Nature,* 1998, **394,** 798.

49   A. K. Kenworthy, N. Petranova and M. Edidin, *Mol Biol Cell,* 2000, **11,** 1645.

50   R. Xavier, T. Brennan, Q. Li, C. McCormack and B. Seed, *Immunity,* 1998, **8,** 723.

51   O. Leupin, R. Zaru, T. Laroche, S. Muller and S. Valitutti, *Curr Biol,* 2000, **10,** 277.

52   A. C. Chan, D. M. Desai and A. Weiss, *Annu Rev Immunol,* 1994, **12,** 555.

53   W. Zhang and L. E. Samelson, *Semin Immunol,* 2000, **12,** 35.

54   W. Zhang, J. Sloan-Lancaster, J. Kitchen, R. P. Trible and L. E. Samelson, *Cell,* 1998, **92,** 83.

55   Z. Lou, D. Jevremovic, D. D. Billadeau and P. J. Leibson, *J Exp Med,* 2000, **191,** 347.

56   A. P. Van der Merwe, S. J. Davis, A. S. Shaw and M. L. Dustin, *Semin Immunol,* 2000, **12,** 5.

57   N. Urwyler, P. Eggli and H. U. Keller, *Cell Biol Int,* 2000, **24,** 863.

58   M. M. Al-Alwan, G. Rowden, T. D. Lee and K. A. West, *J Immunol,* 2001, **166,** 1452.

59   M. K. Wild, A. Cambiaggi, M. H. Brown, E. A. Davies, H. Ohno, T. Saito and P. A. Van der Merwe, *J Exp Med,* 1999, **190,** 31.

60   S. Y. Qi, J. T. Groves and A. K. Chakraborty, *Proc Natl Acad Sci U S A,* 2001, **98,** 6548.

61   P. Y. Arnold, D. K. Davidian and M. D. Mannie, *Eur J Immunol,* 1997, **27,** 3198.

62   P. Y. Arnold and M. D. Mannie, *Eur J Immunol,* 1999, **29,** 1363.

63   J. F. Huang, Y. Yang, H. Sepulveda, W. Shi, I. Hwang, P. A. Peterson, M. R. Jackson, J. Sprent and Z. Cai, *Science,* 1999, **286,** 952.

64   D. Hudrisier, J. Riond, H. Mazarguil, J. E. Gairin and E. Joly, *J Immunol,* 2001, **166,** 3645.

65   I. Hwang, J. Huang, H. Kishimoto and e. al., *J Exp Med*, 2000, **191**, 1137.
66   E. Baba, Y. Takahashi, J. Lichtenfeld, R. Tanaka, A. Yoshida, K. Sugamura, N. Yamamoto and Y. Tanaka, *J Immunol*, 2001, **167**, 875.
67   F. D. Batista, D. Iber and M. S. Neuberger, *Nature*, 2001, **411**, 489.
68   M. Vales-Gomez, H. T. Reyburn, R. A. Erskine and J. Strominger, *Proc Natl Acad Sci U S A*, 1998, **95**, 14326.
69   M. Vales-Gomez, R. A. Erskine, M. P. Deacon, J. L. Strominger and H. T. Reyburn, *Proc Natl Acad Sci U S A*, 2001, **98**, 1734.
70   J. J. Peschon, J. L. Slack, P. Reddy, K. L. Stocking, S. W. Sunnarborg, D. C. Lee, W. E. Russell, B. J. Castner, R. S. Johnson, J. N. Fitzner, R. W. Boyce, N. Nelson, C. J. Kozlosky, M. F. Wolfson, C. T. Rauch, D. P. Cerretti, R. J. Paxton, C. J. March and R. A. Black, *Science*, 1998, **282**, 1281.
71   K. Denzer, M. J. Kleijmeer, H. F. Heijnen, W. Stoorvogel and H. J. Geuze, *J Cell Sci*, 2000, **113**, 3365.
72   L. Zitvogel, A. Regnault, A. Lozier, J. Wolfers, C. Flament, D. Tenza, P. Ricciardi-Castagnoli, G. Raposo and S. Amigorena, *Nat Med*, 1998, **4**, 594.
73   G. Raposo, H. W. Nijman, W. Stoorvogel, R. Liejendekker, C. V. Harding, C. J. Melief and H. J. Geuze, *J Exp Med*, 1996, **183**, 1161.
74   W. Sibbet and J. R. Taylor, *J. Lumin*, 1983, **28**, 367.
75   S. Hirayama, Y. Iuchi, F. Tanaka and K. Shobatake, *Chem Phys*, 1990, **144**, 401.
76   R. A. Lampert, S. R. Meech, J. Metcalfe, D. Phillips and A. P. Schaap, *Chem Phys Lett*, 1983, **94**, 137.
77   S. Hirayama and D. Phillips, *J Photochem*, 1980, **12**, 139.
78   M. Kneen, J. Farinas, Y. Li and A. S. Verkman, *Biophys J*, 1998, **74**, 1591.
79   B. Sengupta, J. Guharay and P. K. Sengupta, *Spectrochim Acta A Mol Biomol Spectrosc*, 2000, **56A**, 1433.
80   T. Parasassi, G. Ravagnan, R. M. Rusch and E. Gratton, *Photochem Photobiol*, 1993, **57**, 403.
81   J. B. Birks, *Photophysics of Aromatic Molecules*, Wiley-Interscience, Chichester, 1970.
82   J. R. Lakowicz, *Principles of Fluorescence Spectroscopy*, 2nd edition, Kluwer Academic/Plenum Publishers, New York, 1999.
83   A. D. Scully, A. J. MacRobert, S. Botchway, P. O'Neill, R. B. Ostler and D. Phillips, *J Fluoresc*, 1996, **6**, 119.
84   A. D. Scully, A. J. MacRobert, S. Botchway, P. O'Neill, R. B. Ostler and D. Phillips, *Bioimaging*, 1997, **5**, 9.
85   K. Dowling, M. J. Dayel, M. J. Lever, P. M. W. French, J. D. Hares and A. K. L. Dymoke-Bradshaw, *Opt Lett*, 1998, **23**, 810.
86   K. Dowling, M. J. Dayel, S. C. W. Hyde, P. M. W. French, M. J. Lever, J. D. Hares and A. K. L. Dymoke-Bradshaw, *J Mod Optic*, 1998, **46**, 199.
87   P. I. Bastiaens and A. Squire, *Trends Cell Biol*, 1999, **9**, 48.
88   S. J. Strickler and R. A. Berg, *J Chem Physi*, 1962, **37**, 814.
89   D. V. O'Connor and D. Phillips, *Time-correlated single photon counting*, Academic Press, London, 1984.
90   K. Suhling, D. M. Davis, P. Petrášek, J. Siegel and D. Phillips, *Proc. S.P.I.E.*, 2001, **4259**, 92.
91   C. Dietrich, L. A. Bagatolli, Z. N. Volovyk, N. L. Thompson, M. Levi, K. Jacobson and E. Gratton, *Biophys J*, 2001, **80**, 1417.

PROBING THE STRUCTURE OF VIRAL ION CHANNEL PROTEINS: A COMPUTATIONAL APPROACH.

W.B. Fischer[1]* and M.S.P. Sansom[2]

[1] Department of Biochemistry, Oxford University, South Parks Road, Oxford OX1 3QU, UK
*Correspondence address: wolfgang.fischer@bioch.ox.ac.uk
[2] Laboratory of Molecular Biophysics, University of Oxford, Oxford OX1 3QU, UK

1   INTRODUCTION

There are significant difficulties with the structural characterization of membrane proteins at a molecular level. These difficulties often arise from (i) insufficient quantities of protein available and (ii) high costs of particular chemicals (e.g. specifically labelled amino acids) necessary to obtain spectroscopic data with high structural resolution (e.g. NMR, X-ray). It is expected that these difficulties might be overcome in the future. In the meantime computational methods may play an important role in the construction of reliable models for investigations of the structure and function of this class of proteins. This role can be achieved not only by enabling *in silico* based experiments which are almost impossible to obtain experimentally but also by proposing models prior to their experimental verification. Examples for this will be presented for the viral ion channel proteins NB from influenza B and Vpu from HIV-1.

The genome of both enveloped viruses influenza B and HIV-1 encode for short membrane proteins called NB (100 amino acids)[1] and Vpu (81 amino acids)[2,3], respectively. In similarity with an equivalent short membrane protein from influenza A called M2, which has been found to be involved in viral entry/exit pathway and is conducting protons, it is expected that NB and Vpu have similar functional roles in the life cycle of the viruses. What all these viral membrane proteins have in common is that they span the bilayer once and have to assemble to form ion channels. For NB, ion channel activity has been confirmed for the full length protein expressed in *Escherichia coli* and reconstituted into lipid bilayers.[4] Also a synthetic peptide analogous to the transmembrane (TM) segment of NB under similar conditions as mentioned exhibits channel activity.[5] Full length Vpu and the TM segment alone both show ion conducting properties. Besides these results, 'in vivo' channel formation has not yet been fully proven for these proteins. The cytoplasmic part of Vpu has been found to inhibit the transport of CD4 and other receptor molecules of the infected cell to the cell membrane.

The current knowledge of structural properties of these proteins is based on solution and solid state NMR-, FTIR-, and CD-spectroscopy. NB's topology of the TM segment has been established by CD-spectroscopy to be predominantly helical.[5] Simulations have been done of NB bundles of different size in a fully hydrated lipid bilayer system. For Vpu it is confirmed that the TM segment consists of a helical motif.[6-8] In addition, helical motifs have been identified for the cytoplasmic part of the protein.[9,10] MD simulations have been performed on pentameric bundles consisting of the TM segment of Vpu in a hydrophobic slab representing the lipid bilayer[11] and in a slab of the bilayer mimetic octane.[12]

In this study we compare simulations on the TM segment of NB[5] (Fig. 1) with those performed on TM bundles of Vpu (Fig. 2). Both bundles each consisting of TM segments of 28 amino acids in length are embedded in a hydrated lipid bilayer. The analysis focuses on the structural and functional role of particular amino acids.

2   METHOD

We generated bundles consisting of 4, 5, and 6 TM segments of NB (NB, Lee, with C-38 mutated to Y):

$$\text{IRGS}^{20} \text{ IIITICVSLI}^{30} \text{ VILIVFGYIA}^{40} \text{ KIFI}$$

and Vpu_{6-33} :

$$\text{IVAIV}^{10} \text{ ALVVAIIIAI}^{20} \text{ VVWSIVIIEY}^{30} \text{ RKI}$$

using a SA/MD protocol[13] based on the program Xplor[14]. For NB no starting tilt angle was used, whereas for Vpu we used an initial angle of 5° during the generation process. A detailed description of generating the helical TM segments is given elsewhere.[13,15] In brief, the SA/MD protocol comprises two stages. In Stage 1 the bundles were constructed with parallel, idealized α-helices based on the positions of the Cα-atoms of the peptide unit. All other atoms of the individual amino acids are superimposed. In Stage 2 potential energy functions with the PARAM19 parameter set are introduced into the protocol. Partial charges on side-chain atoms of polar side chains are gradually scaled up (from 0.05 to 0.4 times their full value) during a temperature reduction from 500 to 300 K. We obtain 5 x 5 = 25 structures as a result of the SA/MD protocol. The most symmetric structure in respect to the pseudo five-fold symmetry axis of the bundle was chosen for the simulations.

All bundles were placed within a lipid bilayer of 1-palmitoyl-2-oleoyl-*sn*-glycerol-3-phosphatidyl-choline (POPC) and consequently solvated with water molecules (>30 water molecules (SPC water model[16]) per lipid). Our simulations were done with a system of ca. 20,000 atoms.[17] Simulations were run for 1 ns using GROMACS 1.6 software (http://rugmd0.chem.rug.nl/~gmx/gmx.html) on a 10 processor SGI Origin 2000 machine and a twin cutoff of 1.0 and 1.7 A. For data analysis Gromacs software was used. The structures were visualized with the program MOLSCRIPT.

A   B

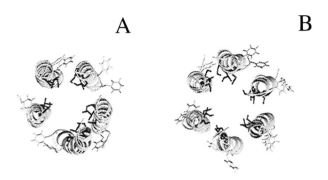

Fig. 1: View from the N to the C terminal end of the pentameric (A) and hexameric (B) bundle of NB. The serine side chains are shown in black and the phenylalanines in light grey. Lipid bilayer and the water molecules are omitted for clarity.

3   RESULTS AND DISCUSSION

All our models are built with the intention that the hydrophilic residues are facing the pore of the water filled channel. This idea is based on experimental findings of the nicotinic acetylcholine receptor where serines and threonines are lining the pore.[18-20] Both NB (Fig. 1) and Vpu (Fig. 2) contain serine residues, as well as threonines in NB only, which all point into the pore. And in both cases this automatically locates the aromatic residues phenylalanine (in NB) and tryptophans (in Vpu) towards the helix / lipid interface. Such a position of the aromatic residues in agreement with findings in other channels such as porins[21] and gramicidin.[22-24] In the case of NB, other hydrophilic residues like arginines (R-18) and lysines (K-41) are located at the helix / lipid interface (data not shown). For Vpu we find a sequence of amino acids (glu-29, tyr-30, arg-31) able to form hydrogen bonds from which the arginines are pointing into the pore (Fig. 2 B and C).

The models adopt a stable structure shown by the leveling off of the root mean square deviation (RMSD), which calculates the positional deviation of each of the subsequent structures from the starting structure (data not shown). Leveling off is achieved after ca. 200 ps and reaches values below 3 nm for each of the bundles. For NB all bundles remain fairly helical. Only single amino acids, especially in the bundle consisting of 4 segments (NB-4) and bundle of 5 segments (NB-5), do not remain in a helical configuration.[5] Val-27 with Φ = -66.9° ± -9.8° and Ψ = -28.1° ± -12.7°, and Ser-28 with Φ = -88.9° ± -15.5° and Ψ = -41.8° ± -10.9° adopt Φ- and Ψ-values which deviate from helical values (Φ = -60°, Ψ = -50°[25]) in one segment of NB-4. For NB-5, residues at the N terminal end in one segment show deviation from ´normal´: Arg-18, which faces the lipid / head group interface (Φ = -120.3° ± -12.0°, Ψ = -61.7° ± -31.5°) and Gly-19 (Φ = -106.5° ± -33.0°, Ψ = -52.2° ± -15.7°). For NB-6 the values for all residues do not deviate from the standard values for a helix. The average kink angle seems to decrease with increasing number of segments (Tab. 1). The average helix tilt angle for all segments adopt values around ca. 6° which indicates the helices tilt slightly compared

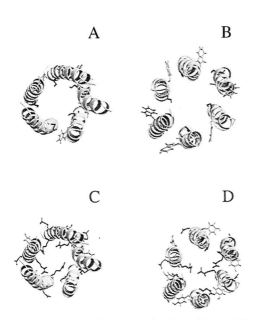

A   B

C   D

Fig. 2: View from the N to the C terminal end of the pentameric (A, C) and hexameric (B, D) bundle of Vpu. In A and B the serines are high lighted in black, the tryptophans in light grey. In C and D black denotes arginines (arg-31), light grey tyrosine (tyr-30), and grey glutamic acid (glu-29).

to their starting structure with no tilt implemented. NB-6 shows the lowest average crossing angle of ca. 4°. All bundles allow for a continuous column of water molecules through their interior with approximately 50 (NB-4), 105 (NB-5), and 145 (NB-6) molecules.

Also for Vpu-4 strong deviations are found for a serine residue (Ser-24: $\Phi = -28.6° \pm -20.4°$, $\Psi = -52.0° \pm -12.2°$). In addition residues Glu-29, Tyr-30, Arg-31, and Lys-32 adopt Φ- and Ψ-values in two segments, which are indicative of an unraveling of the bundle at the C terminal end (for one segment: Glu-29: $\Phi = -53.7° \pm 9.5°$, $\Psi = -30.6° \pm 10.0°$; Tyr-30: $\Phi = -60.9° \pm 11.2°$, $\Psi = -45.7° \pm 11.7°$; Arg-31: $\Phi = -140.2° \pm 20.5°$, $\Psi = 94.4° \pm 13.5°$; Lys-32: $\Phi = -47.7° \pm 15.9°$, $\Psi = -73.8° \pm 18.0°$). These values tend to be less dramatic for Glu-29, Tyr-30, and Arg-31 in Vpu-5 and Vpu-6. The average kink angles are slightly higher for the Vpu bundle (Tab.1). For one segment in Vpu-4 we find a value of ca. 32°. It is interesting to note that the average tilt angles for Vpu-4 and Vpu-5 increase during the simulation from an initial 5° to ca. 15° after 1 ns of simulation. Vpu-6 adopts an almost identical angle than NB-6 of around 6°. The average crossing angle decreases from ca. 20° for Vpu-4 to ca. 5° for Vpu-6. We can only fill the bundles Vpu-5 and Vpu-6 with ca. 92 and ca. 110 waters, respectively. Vpu-4 does not show any space for water molecules within the assembled helices. Also during the simulation no water can access the pore.

As a result for both NB and Vpu the segments seem to straighten with increasing number of segments forming the bundle. During the simulation both NB and Vpu tetrameric

Tab. 1: Kink, tilt, and crossing angles averaged over all the segments in each of the bundles.

Structural Data	NB-4	NB-5	NB-6	Vpu-4	Vpu-5	Vpu-6
Kink Angle (°)	20.6 ± 6.8	13.8 ± 5.9	8.8 ± 4.4	16.9 ± 10.0	19.9 ± 7.8	12.7 ± 7.7
Tilt Angle (°)	4.8 ± 1.3	9.2 ± 2.8	6.2 ± 1.8	15.3 ± 5.0	14.5 ± 3.4	6.0 ± 2.1
Crossing Angle (°)	6.5 ± 2.6	6.6 ± 4.5	3.9 ± 2.0	20.2 ± 7.5	16.7 ± 2.2	4.6 ± 2.3

and pentameric bundles increase their tilt angles by about 7° - 8°. For both of NB´s and Vpu´s hexameric bundles we obtain identical tilt angles of ca. 6° at the end of the simulations. The major sequential differences in both bundles are (i) an increase of β-branched residues from NB (14 residues) to Vpu (17 residues), and (ii) a decrease of hydrophilic residues from 3 (2 serines and one threonine) in NB to one (serine) in Vpu (1). Thus, these differences do not affect the helix tilt angle of the hexamer. The hydrophobic parts of the tetra- and pentameric bundles of Vpu with only one serine allow for the segments to slide around each other adopting higher average crossing angles than their NB counterparts with two serines per segment. It is striking that in both cases for the tetrameric bundle the serines deviate from a regular helical conformation. For Vpu the C terminal end with Glu, Tyr, Arg (EYR-motif) seems to be highly fragile in all bundles in remaining in a helical conformation during the simulation.

4   CONCLUSION

Our simulations indicate that major structural differences between bundles of TM segments are most pronounced if less than 6 segments form the bundle. The reason for this lies in the amino acids used for the TM segments and as a consequence the resulting non-covalent interactions. Hydrophilic residues seem to adopt and support to a lesser degree proper helical conformation. Especially if located at the end of the helices this causes severe deviation from helicity. However, this might reflect the macroscopic behavior in as much as it might form a barrier for the formation of bundles with less than 5 or 6 segments, especially in the case of Vpu. We can expect that the specific properties of an ion channel forming protein can be effective if the bundle size is due to 5 or less segments. Bundles of 6 segments seem to be less sensitive to the amino acid sequence and initial starting structure.

5   ACKNOWLEDGMENT

We acknowledge helpful and stimulating discussion with G. Smith, L. Forrest and P. Tieleman. WBF thanks the EC for a research fellow ship (TMR).

References:

1   M.W. Shaw, R.A. Lamb, B. W. Erickson, D.J. Briedis, P.W. Choppin, *Proc. Natl. Acad. Sci. USA*, 1982, **79**, 6817.
2   K. Strebel, T. Klimkait, M.A. Martin, *Science*, 1988, **241**, 1221.
3   E.A. Cohen, E.F. Terwilliger, J.G. Sodroski, W.A. Haseltine, *Nature*, 1988, **334**, 532.
4   N.A. Sunstrom, L.S. Prekumar, A. Prekumar, G. Ewart, G.B. Cox, P.W. Gage, *J. Membr. Biol.*, 1996, **150**, 127.
5   W.B. Fischer, M. Pitkeathly, B.A. Wallace, L.R. Forrest, G.R. Smith, M.S.P. Sansom, *Biochemistry*, 2000, **39**, 12708.
6   V. Wray, T. Federau, P. Henklein, S. Klabunde, O. Kunert, D. Schomburg, U. Schubert, *Int. J. Peptide Protein Res.*, 1995, **45**, 35.
7   V. Wray, R. Kinder, T. Federau, P. Henklein, B. Bechinger, U. Schubert, *Biochemistry*, 1999, **38**, 5272.
8   A. Kukol, I.T. Arkin, *Biophys. J.*, 1999, **77**, 1594.
9   T. Federau, U. Schubert, J. Floßdorf, P. Henklein, D. Schomburg, V. Wray, *Int. J. Peptide Protein Res.*, 1996, **47**, 297.
10   D. Willbold, S. Hoffmann, P. Rösch, *Eur. J. Biochem.*, 1997, **245**, 581.
11   A.L. Grice, I.D. Kerr, M.S.P. Sansom, *FEBS Lett.*, 1997, **405**, 299.
12   P.B. Moore, Q. Zhong, T. Husslein, M.L. Klein, *FEBS Lett.*, 1998, **431**, 143.
13   I.D. Kerr, R. Sankararamakrishnan, O.S. Smart, M.S.P. Sansom, *Biophys. J.*, 1994, **67**, 1501.
14   A.T. Brünger, *X-PLOR Version 3.1. A System for X-ray Crystallography and NMR*; Yale University Press: New Haven, Ct., 1992.
15   F. Cordes, A. Kukol, L.R. Forrest, I.T. Arkin, M.S.P. Sansom, W.B. Fischer, *Biochim. Biophys. Acta*, 2001, **1512**, 291.
16   H.J.C. Berendsen, J.R. Grigera, T.P. Straatsma, *J. Phys. Chem.*, 1987, **91**, 6269.
17   Fischer, W. B.; Forrest, L. R.; Smith, G. R.; Sansom, M. S. P. *Biopolymers* **2000**, *53*, 529
18   J.P. Changeux, J.I. Galzi, A. Devillers-Thiéry, D. Bertrand, *Quart. Rev. Biophys.*, 1992, **25**, 395.
19   H. Lester, *Ann. Rev. Biophys. Biomol. Struct.*, 1992, **21**, 267.
20   F. Hucho, V.I. Tsetlin, J. Machold, *Eur. J. Biochem.*, 1996, **239**, 539.
21   A. Hirsch, T. Wacker, J. Weckesser, K. Diederichs, W. Wolfram, *Proteins: Struc. Func. Genet.*, 1995, **23**, 282.
22   A.M. O'Connell, R.E. Koeppe II, O.S. Andersen, *Science*, 1990, **250**, 1256.
23   N.D. Lazo, W. Hu, T.A. Cross, *J. Chem. Soc., Chem. Commun.*, 1992, 1529.
24   W. Hu, N.D. Lazo, T.A. Cross, *Biochemistry*, 1995, **34**, 14138.
25   C. Branden, J. Tooze, *Introduction to Protein Structure. Garland, New York.*, 1991.

THE IMPACT OF H$_2$O$_2$ ON THE STRUCTURE OF CATALASES BY MOLECULAR MODELLING METHODS

Susana G. Kalko, José Ll. Gelpí, and Modesto Orozco

Departament de Bioquímica i Biologia Molecular, Universitat de Barcelona. C/ Martí i Franquès 1. Barcelona 08028. Spain

1 INTRODUCTION

Catalases are responsible for the reduction of reactive oxygen species such as small peroxides (including hydrogen peroxide), which are dismutated into water (or alcohol) and oxygen. The speed of this reaction is very high, reaching in some cases the diffusion limit. Malfunction of catalases may lead to severe effects, among them increased susceptibility to thermal injury,[1] inflammation[2] and accelerated ageing.[3]

Crystallographic studies show that the active site of catalase is deeply buried in the interior of the protein, at the end of a long and narrow channel connecting the exterior surface with the heme group.[4] Other smaller channels were suggested as possible product release pathways,[4,5] but no clear experimental evidence has been provided to confirm their existence in active catalases.

In a previous study[6] the substrate recognition mechanisms of peroxisomal catalase from *Saccharomyces cerevisiae* were explained by using crystallographic and theoretical data. Our calculations suggested that water was a competitive inhibitor of the enzyme, blocking the access of hydrogen peroxide to the active site. Furthermore, theoretical calculations gave support to the putative role of secondary channels[6] like those suggested by Fita and coworkers based on crystallographic data.[4,5]

Catalases are designed by evolution to work in environments having a very large concentration of H$_2$O$_2$, such as those existing at the peroxisomes. The impact of the large concentration of hydrogen peroxide on the structure and reactive properties of catalases is unclear due to difficulties in obtaining crystals in such environments. However, this effect might be crucial for the physiological role of the enzyme. In this paper we will use *state of the art* molecular dynamics simulations to gain insight into the impact of a high concentration of H$_2$O$_2$ on catalase structure and ligand binding properties.

2 METHODS AND RESULTS

2.1 Molecular Dynamics

Simulations in pure water and a hydrogen peroxide/water mixture were performed starting from the crystallographic structure of the *Saccharomyces cerevisiae* peroxisomal catalase solved at 2.8 Å resolution.[4] The active form is an homotetramer but due to the size of the system a reduced model was used in our simulations (see ref. 6 for details). The optimised structure in water was used to generate a starting model for the protein in a c.a. 27% mixture of hydrogen peroxide and water. AMBER-95[7] and TIP3P[8] force-fields were used to describe protein atoms and water, respectively. Bonded parameters for hydrogen peroxide were determined using the PAPQMD strategy[9] and HF/6-31G(d) calculations as reference. Charges were determined using RESP[10] and HF/6-31G(d) wave functions.

2.1.1 Analysis of the root mean square deviation (RMSD). The all atoms RMSDs between the flexible part of the protein and the crystal structure were small in both trajectories, around 1.0 Å in pure water and 1.5 Å in the mixture. The greater RMSD for the simulation in the mixture agrees with the fact that the crystal environment is closer to pure water than to the 27% H_2O_2/H_2O mixture. The all atoms RMSD between the MD-averaged structure and the trajectory along a simulation can be used to obtain a measure of the flexibility of the protein in both solvents. This analysis was extended (see Table 1) by inspection of the fluctuations of four regions of special interest: g1, the internal part of the main channel (A66-A73, A109-A126 and A148-A156 residues); g2, the external part of that channel (A157-A189 residues); g3, a channel connecting the heme with the interface between monomers (residues D59-D63); and g4, a channel connecting the heme with the NADP(H) binding position (residues A142-A147).[6] It becomes evident that g2 is the most sensitive region to the solvent composition, while g4 has a very stable conformation, either in pure water or in the H_2O_2/H_2O mixture.

Table 1   *RMSD between the MD-averaged structures and the corresponding pure water and mixture trajectories. Standard deviations in parenthesis.*

reference	g1/Å	g2/Å	g3/Å	g4/Å	total/Å
Pure water	0.7(0.1)	1.1(0.1)	1.2(0.3)	0.5(0.3)	1.0(0.1)
Mixture	1.0(0.1)	1.7(0.3)	1.0(0.2)	0.5(0.1)	1.2(0.1)

The all atom RMSD between the average structure in pure water and the trajectory of the H_2O_2/H_2O mixture provides us with a measure of the influence of solvent composition on the structure (see Figure 1). Moderate structural deviations are found during this trajectory, the largest of them located at the g2 region. The g4 region shows the smallest RMSD values, but a noticeable transition occurs around 1 ns due to the "flipping" of the ASN143 side chain in the H_2O_2 solution. It is worth noting that this residue was suggested to participate in a "back-door" mechanism of this enzyme.[4] No other major changes were found in the structure related to the change of solvent indicating that the main structure of catalase is insensitive to changes in the concentration of H_2O_2 (within the range that occurs in the peroxisome).

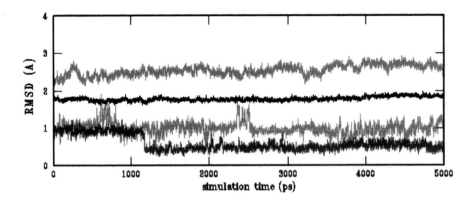

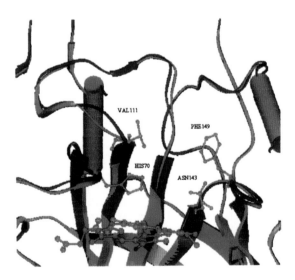

Figure 1   *RMSD plot between MD trajectory of mixture simulation and MD average structure of pure water simulation in regions g1 (black), g2 (red), g3 (green) and g4 (blue). RMSD for total mobile residues (data not shown) are similar to g1 values.*

2.1.2   Analysis of secondary structure.   The secondary structure of the protein is highly conserved irrespective of the composition of the solvent (see Fig. 2). A detailed analysis of dihedral angles in the g1 and g2 regions of the main chain indicates a slightly larger flexibility (around 20-30% larger) in the simulation in the mixture solution compared with the simulation in pure water. This finding agrees with the RMSD analysis noted above, suggesting these regions as the most sensitive to changes in solvent composition.

Figure 2   *Secondary structure representation for MD averaged structure of the main channel for mixture simulation (red) and pure water simulation (blue). Relevant residues and heme group are explicitly represented with balls and sticks.*

2.1.3 Analysis of key distances in the mouth of the main channel. The principal effect of the presence of H_2O_2 in terms of heme accessibility is a slight widening of the main channel. Thus, residues VAL111 and PHE149 define the narrowest point of the main channel in water (shortest heavy atom-heavy atom distance around 6 Å). On the contrary, in the H_2O_2/H_2O mixture the side chain of PHE149 is more flexible, showing a *flip-flop* transition, which opens the channel at this point. Therefore, PRO124 and PHE159 define then the narrowest point of the channel (shortest heavy atom-heavy atom distance around 6.8 Å). In summary, the accessibility to the heme site is higher in the presence of a large concentration of peroxide in the solution. This fact might increase the catalytic efficiency of the enzyme under conditions of high hydrogen peroxide concentration.

2.1.4 Analysis of protein-protein hydrogen bonds in channels. VAL111 has a key role in defining the channel accessibility, as demonstrated by site-directed mutagenesis,[11] which makes it interesting to study the pattern of hydrogen bonds (HB). The occupancy degree of HB for this residue, together with those involving the essential HIS70 and its close neighbour SER109 are shown in Table 2. The stability of these important protein-protein hydrogen bonds is very large in the case of pure water simulation, whereas in the mixture solvent the stability is reduced. This fact may be related to a higher number of hydrogen bonds between protein polar atoms and the hydrogen peroxide molecule, as suggested in previous calculations.[6]

Table 2   *Occupancy degree of HB during the simulation.*

Partners	Pure water (%)	Mixture (%)
O VAL111-NH HIS70	96.2	77.4
NH VAL111-O PRO124	97.3	92.6
* O SER109-N GLY126	99.4	86.2
* OG SER109-ND1 HIS70	98.4	0

* *HB in crystal structure*

2.1.5 Analysis of preferential solvation. The total and relative number of solvent molecules at less than 3 Å distance of the protein atoms were calculated along the trajectory and are shown in Table 3. The quotient between the solvation number (SN) and the total number of solvent molecules was calculated for all residues -fist raw of Table 3- whereas for residues classified as charged, polar, aromatic, exposed[12] or buried,[12] the relative value was referred to the SN of all residues. The analysis show that in all the cases hydrogen peroxide is more closely associated to the protein than is water, suggesting that H_2O_2 is a better solvent than water for catalase. Interestingly, there are not differences between the solvation of water or hydrogen peroxide in exposed residues, but on the contrary, important differences are found between solvent-protein interactions for buried residues, since these residues are more accessible to hydrogen peroxide than to water. Finally, it is worth noting that water has approximately the same level of contact to residues in both simulations, indicating that the intrinsic ability of water to interact with different types of residues does not vary in the presence of hydrogen peroxide.

Table 3   *Average number of molecules in the first solvation shells (SN). Error limits correspond to standard deviations from average values. Bold numbers are described in the text.*

Residues	WATER in PURE WATER (Total:3754)	WATER in MIXTURE (Total:2096)	PEROXIDE in MIXTURE (Total:560)
ALL	854 ± 18 **0.23**	503 ± 17 **0.24**	179 ± 8 **0.32**
CHARGED	437 ± 10 **0.51**	263 ± 11 **0.52**	98 ± 6 **0.54**
POLAR	320 ± 9 **0.38**	189 ± 9 **0.38**	75 ± 6 **0.42**
AROMATIC	139 ± 7 **0.16**	84 ± 7 **0.17**	32 ± 5 **0.18**
EXPOSED	778 ± 17 **0.91**	465 ± 16 **0.92**	167 ± 9 **0.92**
BURIED	111 ± 5 **0.13**	63 ± 4 **0.13**	34 ± 3 **0.19**

2.2 Classical Molecular Interaction Potential (CMIP)

CMIP calculations[13] were carried out using a neutral TIP3P[8] O as a probe, and the MD-averaged structures in pure and mixture solution. The grid was positioned in the centre of the main channel, and a spacing of 1 Å was used. This type of calculation allowed us to obtain a picture of the variation of the binding site accessibility in presence of large concentration of hydrogen peroxide. CMIP plots in Fig. 3 clearly define the shape of the main channel. Remarkably, in the structure derived from mixture simulation there is a continuous spine of solvation, whereas in the pure water case the spine is discontinuous, at the VAL111 position. A branching of the main channel is observable in the mixture simulation, going from the heme site to the NADP(H) binding site, supporting previous hypotheses on the role of the secondary channel for the release of reaction products.[4,5]

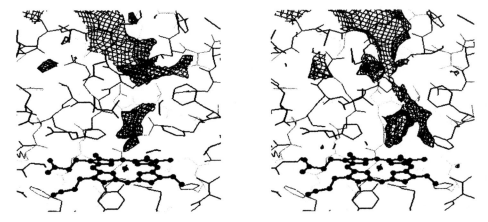

Figure 3   *CMIP plots corresponding to 0 Kcal/mol interaction energy between a neutral TIP3P O and MD average structure of pure aqueous solution (left) and mixture solution (right) around the active site of catalase.*

3 CONCLUSIONS

The structure of catalase is very similar in the crystal, in pure aqueous solution, and in the presence of c.a. 27% of hydrogen peroxide. The stability of the structure in very different environments, including those with a very large concentration of H_2O_2 should be important from a biological point of view, since it allows the enzyme to be active under these conditions. The enzyme shows a significant widening of its channels in the presence of peroxide. This suggests a solvent-induced conformational change, which might increase solute accessibility to the active site under conditions of large oxidative stress. In summary, the results presented here indicate that hydrogen peroxide does not lead to dramatic effects on catalase structure, but its presence in the medium promotes conformational changes of the enzyme for an easier arrival of the ligand to the active site and the posterior clearance of the reaction products.

References

1   J.A. Leff, *Inflammation*, 1993, **17**, 199.
2   B. Halliwell and O.I. Aruoma, *FEBS Lett.*, 1991, **281**, 9.
3   J. Taub, J.F. Lau, C. Ma, J.H. Hahn, R. Hoque, J. Rothblatt and M. Chalfie, *Nature*, 1999, **399**, 162.
4   M.J. Maté, M. Zamocky, L.M. Nykyri, C. Herzog, P.M. Alzari, C. Betzel, F. Koller and I. Fita, *J. Mol. Biol.*, 1999, **286**, 135.
5   M.S. Servinc, M.J. Maté, J. Switala, I. Fita and P. Loewen, *Protein Sci.*, 1999, **8**, 490.
6   S. G. Kalko, J. Ll. Gelpí, I. Fita and M. Orozco, *J.Am.Chem.Soc.*, 2001, **123**, 9665.
7   W.D. Cornell, P. Cieplak, C.I. Bayly, I.R. Gould, K. Merz, D.M. Ferguson, D.C. Spellmeyer, T. Fox, J.W. Caldwell and P.A. Kollman, *J.Am.Chem.Soc.*,1995, **117**, 11946.
8   W.L. Jorgensen, J. Chandrasekhar, J. Madura, R.W. Impey, and M.L Klein, *J.Chem.Phys*, 1983, **79**, 926.
9   C. Alemán, E. I. Canela, R. Franco and M. Orozco, *J.Comp.Chem.*, 1991, **12**, 664.

10 C.I. Bayly, P. Cieplak, W.D Cornell and P.A. Kollman, *J.Phys.Chem.*, 1993, **97**, 10269.
11 M. Zamocky, C. Herzog, L.M Nykyri, and F. Koller, *FEBS Lett.*, 1995, **367**, 241.
12 S.J. Hubbard, S.F. Campbell and J. Thornton, *J. Mol. Biol.*, 1991, **220**, 507.
13 J.Ll. Gelpí, S.G. Kalko, X. Barril, J. Cirera, X. de la Cruz, F.J. Luque and M. Orozco, *Proteins*, 2001, **45**, 428.

ENTROPY IN THE ALIGNMENT AND DIMERIZATION OF CLASS C G-PROTEIN COUPLED RECEPTORS

Mark K Dean[1], Christopher Higgs[1], Richard E Smith[1], Paul D Scott[2], Robert P Bywater[3], Trevor J Howe[4] and Christopher A Reynolds[1]

[1] Department of Biological Sciences, Central Campus, University of Essex, Wivenhoe Park, Colchester, Essex, CO4 3SQ, United Kingdom
[2] Department of Computer Science, University of Essex, Wivenhoe Park, Colchester, Essex, CO4 3SQ, United Kingdom
[3] Biostructure Department, Novo Nordisk A/S, Novo Nordisk Park DK-2760 Måløv, Denmark,
[4] Novartis, Horsham Research Centre, Wimblehurst Road, Horsham, West Sussex, RH12 5AB, United Kingdom

1 INTRODUCTION

Class C G-protein coupled receptors (GPCRs), which include the metabotropic glutamate, calcium sensing and GABA$_B$ receptors, have played an important role in elucidating the link between G-protein coupled receptor dimerization and function. Here we present a sequence-based study on the alignment and dimerization of class C GPCRs.

The observation of disulfide-linked metabotropic glutamate receptor dimers gave rise to one of the earlier reports of GPCR dimers[1]; this was soon followed by a report of disulfide-linked calcium-sensing receptors[2]. The most significant link between G-protein coupled receptor (GPCR) dimerization and function however, came with the observation that GABA$_B$ receptor subtypes R1 and R2 were not functional unless co-expressed[3-6]. It was suggested that the GABA$_B$ dimer was held together by coiled-coil interactions between the C-termini[6]; these articles[3] also introduced the idea of heterodimerization. Structural evidence for class C receptor dimers has been provided by crystal structures of the N-terminal domain of the metabotropic glutamate receptor, in both closed and open dimeric forms (with and without agonist), implying that the agonist induced activation involves large relative movement between the two parts of the N-terminal domain[7]; we suggest that the function of this movement is to bring the transmembrane domains together.

Within the class A GPCRs, Devi has shown that heterodimerization of the κ and μ opioid receptors could give rise to novel pharmacology not seen in the κ or μ systems alone[8] and these ideas have also arisen in other articles describing GPCR heterodimerization[9-11]. However, the current interest in G-protein coupled receptor dimerization probably arose from the work of Maggio et al. Maggio constructed chimeric muscarinic-adrenergic receptors in which the N-termius through to intracellular loop 3 was taken from the muscarinic M3 receptor and helix 6 through to the C-terminus was taken from the α$_2$-

adrenergic receptor. The receptors were inactive: they did not bind ligand or activate the G-protein. The alternative adrenergic-muscarinic chimeras were similarly inactive but when co-expressed, the chimeras bound both adrenergic and muscarinic ligands and activated the G-protein. This functional rescue on coexpression was interpreted as "cross-talk" by the authors[12]. Equally important was the work of Hebert et al. who showed using co-immunoprecipitation that transmembrane helix 6 of the β_2-adrenergic receptor inhibited both dimerization and activation[13], implying that the dimerization is driven by interactions between the transmembrane domains. Our theoretical work on receptor dimerization has provided an interpretation of Maggio's results in terms of domain swapping, with helices 5 and 6 nominally forming the receptor dimer interface[14,15].

Studies using fluorescence resonance energy transfer (FRET) and bioluminescence resonance energy transfer (BRET) have given further evidence for GPCR dimerization[16,17], but generally, the link between dimerization and function is not understood. We note that parallel dose-response curves for signalling and ligand-induced dimerization have been observed in some cases[13,16] but not in others[18,19]. However, recent studies have shown that heterodimerization can affect the G-protein coupling preferences of the receptors[20] and this provides evidence that the GPCR dimer is relevant to activation.

Approaches to predicting protein-protein interactions from sequence are based primarily on correlated mutation analysis[21] or the evolutionary trace (ET) method[22-24]. Both methods are essentially data-mining approaches for determining highly correlated patterns of change within a multiple sequence alignment[15,23-31] and both have been usefully applied to GPCRs and their associated G-proteins[15,23-31]. Here we apply a related entropy-based method to the problem of G-protein coupled receptor dimerization in the class C receptors. Initially, we have used the entropy analysis to assist in the alignment of class C sequences with the class A opsin receptor sequences; this enabled the entropy analysis to be mapped onto the rhodopsin X-ray crystal structure. As a control, we have also studied the protein interface and substrate binding site in the ras-rasgap X-ray crystal structure.

2 METHODS

2.1 Determination of functionally important residues

The basic assumptions of the ET method, as applied to a multiple sequence alignment of a protein family are:
- the family retains its fold throughout the series of proteins
- the location of the functional sites is conserved within the family
- these functional sites have lower mutation rates than the rest of the protein
- the lower mutation rate is punctuated by mutation events that cause divergence.

Clearly these assumptions will not apply rigorously because certain structure-stabilising residues may also have low mutation rates while certain binding sites may display convergent evolution. Similarly, certain family members may display subtle variations in the fold. Nevertheless, they do provide a working hypothesis. In practical implementations of the ET method, conserved-in-class residues are identified for positions associated with successively more branched regions of the dendrogram (derived from the multiple sequence alignment) and plotted onto a structure until the residues no longer cluster but rather are scattered in a seemingly random manner over the protein surface.

These ET assumptions apply equally well to the entropy analysis used here, where the entropy (in an information sense) at a given position j is given by equation (1)

$$S_j = \log\left(\frac{N!\,P!}{(P-V_j)!\prod\limits_{i=1}^{V} M_i!}\right) \times 100/S_{MAX} \tag{1}$$

where N is the number of sequences (24 for class C, 112 for rhodopsin), P is the number of amino acid residues (21, including gaps), V is the variability and M_i is the number of amino acids of type i at position j. Here the entropy was normalized so that the maximum value was 100. To determine functional sites, the entropy values were sorted into 20 bins of increasing relative entropy and residues from successive bins were plotted until the residues no longer appeared to cluster, as in previous applications of the ET method. There are two conceptual advantages of this method over the ET method. Firstly, a dendrogram is not required and so discussions on the most appropriate tree (e.g. neighbour joining[32], UP-GMA[33]) are not relevant. Secondly, while a sequencing error prevents a residue from being identified as conserved-in-class in the ET method, the error merely increases the entropy but does not necessarily prevent the residue from being identified in a higher bin.

2.2 Alignment of class A and class C receptor sequences

The class A GPCRs can be readily aligned[34,35] and for this family, the X-ray structure of rhodopsin provides a suitable structural model[36,37]. However, the similarity between class A receptors and those of class B and class C is too low to obtain a multiple sequence alignment using conventional methods. Elsewhere an alignment between the class A and class B receptors has been determined by assuming that the *positions* of the functionally important residues are conserved[38] even if their *identity* is not conserved; the positions of the functionally important residues were identified as having a conservation level of 70% or more. The alignment then reduced to finding the alignment with the maximum coincidence of functionally important residues subject to the constraint of aligning the internal residues, as determined using PERSCAN[39]. Here we have followed a similar procedure but have used a relative entropy cut-off of 30% or less to identify the functionally important residues. The GABA_B sequences were aligned to the other class C sequences using essentially the same method.

3 RESULTS AND DISCUSSION

3.1 The rhodopsin-class C alignment

The entropy-based alignment of the 7 transmembrane helices of the rhodopsin family and the class C family are shown in Table 1. Several conserved motifs of the class A receptors are picked out by the entropy analysis, including the GN on helix 1, the DRY on helix 3, the CWXXP on TM6, the charged residues at the intracellular end of helix 6 and the NPXXY motif of helix 7. In each case apart from helix 7, at least one of the corresponding positions in the class C receptor family is identified by the entropy analysis. There is a reasonable degree of overlap between the low entropy residues of the rhodopsin family and the low entropy residues of the class C receptors, and these matches are denoted by an X in Table 1.

The extent to which we should expect the low entropy residue positions to align depends on the similarity between the structure function relationships, and this similarity essentially arises from the common seven transmembrane helical structure and a common G-protein coupling mechanism. However, there are notable differences that will be reflected in low entropy residues occurring in different regions of the structure. Thus, for the class C receptors, the ligand binding site is within the N-terminus[7] but for the class A opsin receptors it resides with the helical bundle[36]. Moreover, the role of the closed N-terminus structure is probably to favour dimerization[1] but other regions, e.g. helix 6, may fulfil this role within the class A receptors[13]. While these variations on the basic theme could underlie some differences in the co-location of low entropy residues,

Table 1. *The alignment of class A opsin family with class C family GPCRs. The alignment of low entropy residue positions within the transmembrane (TM) regions 1-7, denoted in bold, e.g. A, is indicated by "X". Identities, high similarities and medium similarities and predicted internal residues are denoted "|", ":", "." and by underlining, e.g. A or A respectively.*

TM1	AVLLAALYSLLFLVGLLGNLLVILVILR |::.:... |..|: |.X : X: . AIAPVFFACLGILATLFVIVTFVRYNDT
TM2	RTPTNYFLLNLAVADLLVALTLPPFAL :. X. .:XX .: | . :|X...| ASGRELSYILLTGIFLCYCITFLMIAK
TM3	ALCKLVGFLDVLNGTASIFSLTAIAIDRYLAICHP |:X | ...| . :|X X XX .X. XX AICSLRRLFLGLGFAISYSALLTKTNRIARIFEQK
TM4	RRAILVIALVWVLSLLLSLPPLFGW X::: . X. :X||. :..|.. ASQLVITFSLISVQLLGVVIWLVVE
TM5	VIYSSICGFLLPLLVMLFCYGRILRATQKA .| || ... .|:|: |X. X: x . SDLSLICSLGYSGLLMVTCTVYAFKTRGVP
TM6	KAERKAAKMLVIIVGVFLLCWLPFFIVALL |. XX. XX ||XX:X:.X:..X. | KFIGFTMYTTCIIWLAFIPIFFGTAQSAEK
TM7	FLITLWLAYLNSCLNPIIYAFLNK . ::|X: X:X || ..X...: X ISVSLSASVALGCLFMPKVYIILF

the *optimal match* of low entropy residue positions does give rise to an alignment, as shown in table 1. In addition to the alignment of low entropy residue positions, the few identities, high similarities and medium similarities are denoted by "|", ":" and "." respectively. Moreover, the optimal entropy alignment is consistent with the alignment of internal residues as predicted by PERSCAN. These internal residues do not fully coincide with the

Figure 1. *Conserved and low entropy residues (grey) plotted onto the structure of ras. (a) The interface with rasgap, (b) The binding site for GTPγS.*

internal residues in the rhodopsin crystal structure but are used in preference to the crystal structure data for two reasons. Firstly, no crystal structure is available for the transmembrane domain of the class C receptors and secondly, the rhodopsin crystal structure is for the inactive structure and while the class C inactive structure may differ slightly, the PERSCAN results are equally applicable to both the active and inactive structures. Thus, the alignment enables a model of the transmembrane domain of the class C receptor to be constructed and the entropy analysis to be applied to the aligned class C receptor sequences.

3.2 Functional sites predicted by entropy analysis: ras

The entropy analysis has been applied to a seed alignment, obtained from the Pfam database[40] (accession number PF00071) of 61 sequences of the ras family. Residues from the first 9/10 entropy bins, i.e. up to a relative entropy of 45/50%, were plotted onto the structure of the ras protein (a small GTPase, taken from the ras-rasgap complex[41], pdb code 1WQ1), as shown in figure 1. Figure 1a shows that the residues in the vicinity of the interface with rasgap, a protein that modifies the catalytic properties of ras, are clearly identified by the entropy analysis. In addition, the GDP binding site is also identified, as shown in figure 1b. Although figure 1b also shows residues from the first 9/10 bins, most of the residues in contact with GDP identified by the entropy were identified in the first 4 bins. The observation that substrate binding sites are identified at a lower entropy than protein interfaces appears to be a common trend across many proteins[42]. The observation that entropy analysis works well for the protein-protein interface between ras and rasgap implies that it could also be used to identify putative protein interfaces in the class C receptors.

3.3 Functional sites predicted by entropy analysis: class C GPCRs

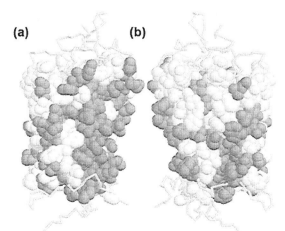

Figure 2. *(a) Low entropy residues for the class C receptors plotted onto the 3D structure of rhodopsin; helices 5 and 6 form the centre of the figure, with helix 7 to the right. (b) As for (a), but helices 2 and 3 form the centre of the figure.*

Figure 2a shows the low entropy residues plotted onto the external face of transmembrane helices 5 and 6. Studies on the class A receptors, particularly those of Maggio[12,43] and Hebert[13] but also those of Ng[44] have shown that the transmembrane helices play a crucial role in receptor dimerization, and this has been supported by computational studies[15;28]. These results are in marked contrast to the initial studies on class C receptor dimerization. Firstly, Romano et al showed that the glutamate receptor dimer formed through disulfide bonds between the N-termini[1]; similar results were obtained for the calcium sensing receptor[2]. These results were further clarified by the crystallization of the glutamate N-terminus dimer in both an open and a closed form.[7] Secondly, the initial studies on the GABA_B heterodimer[6], involving yeast two hybrid studies, suggested that it is held together by coiled coil interactions between the C-termini. Subsequent studies have indicated that the coiled coil interaction is an endoplasmic reticulum retention motif and that removal of the coiled coil does not prevent dimer formation[45]. Moreover, modelling studies have shown that the C-terminus of the GABA$_B$ receptors is sufficiently long for the transmembrane domains to form a 5,6-dimer simultaneously with a coiled coil interaction[27,46]. Here in figure 2a we have shown that the class C receptors display the appropriate functionality on helices 5 and 6, as predicted by entropy analysis, to dimerize through their transmembrane domains. Naturally, the prediction of a functional site on helices 5 and 6 does not identify the interacting partners. Thus, while it seems reasonable to suggest that these signify the regions involved in homodimerization, the same regions are also involved in heterodimerization[47] but a more in depth analysis would be required to identify which heterodimer partners would be permitted by particular sequences.

Elsewhere we have performed an evolutionary trace analysis and reported similar results[25]. However, the entropy analysis shows a stronger signal on helix 7, visible in figure 2a. This region displays some interesting structural features in the vicinity of the conserved class A NPXXY motif (which is not present in class C receptors). Weinstein has shown from NMR studies on peptides, from molecular modelling and from structural database searches that NP does not readily occur in α-helical regions but that it does have a strong tendency to form β-turns[48,49]. The 2D and 3D structural data on the other hand is unequivocal that this region of helix 7 is relatively straight in the inactive receptor (at least in comparison to Weinstein's proposed structures)[36,37,50-52], though the X-ray data shows that this key region has a 3$_{10}$-helical structure rather than an α-helical structure, as predicted by Gouldson[14]. (Distortions elsewhere in TM7 were earlier predicted by Pappin and Findlay[53].) These conflicting observations would be resolved if the activation process involved conformational changes, and indeed the importance of residues in this region may underlie the cluster of low entropy residues on helix 7 shown in figure 2a.

The entropy analysis identifies a functional site on helices 2 and 3, as shown in figure 2b. A wide range of effects resulting from mutating external residues on helices 2 and 3 have been recorded, as listed in reference[25] but generally the function, if any, of these external residues is unknown for both the class A and class C receptors. Preliminary evidence however suggests that residues in this region may be involved in binding RAMPs[54,55].

It is interesting to compare the functional site predicted by entropy analysis, the evolutionary trace method and by correlated mutation analysis. Correlated mutation analysis has been extremely useful for identifying specific functional residues, particularly those involved in selective binding, for both ligand binding and subtype specific dimerization[14,15,27,29,56,57]. However, in our hands the evolutionary trace method is superior for identifying the spatial extent of the binding site[27,28]. The entropy analysis generally gives similar results to the evolutionary trace method[42]. Here, entropy gives excellent results for Ras (figure 1) but the uncertain nature of biological sequence data dictates that this may not always be the case. Thus, while entropy analysis identifies a putative dimerization interface on helices 5 and 6 of the class C receptors, the evolutionary trace method appears to give superior results[25]. We note that the evolutionary trace and entropy analyses give slightly different information, thus while the dimerization interface is not identified so clearly by entropy analysis, residues on helix 7 are identified more readily.

4 CONCLUSIONS

We have presented several applications of entropy analysis for identifying functional residues from multiple sequence alignments. Entropy analysis is related to correlated mutation analysis and the evolutionary trace method but has the advantages of not requiring a dendrogram and not requiring residues to be 100% conserved-in-class. The entropy measures conservation, so it can be used to create alignments between sequence families that share low levels of similarity by aligning positions of low entropy. Here the assumption is that while identity of amino acids is not conserved, the functional positions are conserved. Using this method we have aligned the class A and class C GPCRs. The method has been applied to a seed alignment of the ras family. The GTP binding site is identified at relatively low entropy and the rasgap binding site has been identified at medium entropy values. The method also identifies potential functionality amongst the external resides of the class C GPCRs. A putative dimerization site is identified on helices 5 and 6, but not as clearly as with the evolutionary trace method. This dimerization site appears to extend onto helix 7, and this extension may be related to requirements for flexibility in this region. Additional potential functionality has been observed on helices 2 and 3. In the class B receptors this site may be associated with RAMP (Receptor Activity Modifying Protein) binding but this region has no known function in class C receptors.

Acknowledgements. We wish to acknowledge Novo Nordisk (CH), Novartis (RES) and the BBSRC (MKD, RES) for funding.

References

1   C. Romano, Yang, W. L., O'Malley, K. L. *J. Biol. Chem.* 1996, **271** , 28612.
2   M. Bai, Trivedi, S., Brown, E. M. *J. Biol. Chem.* 1998, **273**, 23605.
3   H. Mohler, Fritschy, J. M. *Trends Pharmacol. Sci.* 1999, **20**, 87.
4   K. Kaupmann, Malitschek, B., Schuler, V., Heid, J., Froestl, W., Beck, P., Mosbacher, J., Bischoff, S., Kulik, A., Shigemoto, R., Karschin, A., Bettler, B. *Nature* 1998, **396**, 683.
5   K. A. Jones, Borowsky, B., Tamm, J. A., Craig, D. A., Durkin, M. M., Dai, M., Yao, W. J., Johnson, M., Gunwaldsen, C., Huang, L. Y., Tang, C., Shen, Q., Salon, J. A., Morse, K., Laz, T., Smith, K. E., Nagarathnam, D., Noble, S. A., Branchek, T. A., Gerald, C. *Nature* 1998, **396**, 674.
6   J. H. White; Wise, A.; Main, M. J.; Green, A.; Fraser, N. J.; Disney, G. H.; Barnes, A. A.; Emson, P.; Foord, S. M.; Marshall, F. H. *Nature* 1998, **396**, 679.
7   N. Kunishima; Simada, Y.; Tsuji, Y.; Sato, T.; Yamamoto, M.; Kumasaka, T.; Nakanishi, S.; Jingami, H.; Morikawa, K. *Nature* 2000, **407**, 971.
8   B. A. Jordan; Devi, L. A. *Nature* 1999, **399**, 697.
9   M. Rocheville; Lange, D. C.; Kumar, U.; Patel, S. C.; Patel, R. C.; Patel, Y. C. *Science* 2000, **288**, 154.
10 M. Mellado; Rodriguez-Frade, J. M.; Vila-Coro, A. J.; Fernandez, S.; Martin, D. A.; Jones, D. R.; Toran, J. L.; Martinez, A. *EMBO J* 2001, **20**, 2497.
11 B. A. Jordan; Trapaidze, N.; Gomes, I.; Nivarthi, R.; Devi, L. A. *Proc. Natl. Acad. Sci. U.S.A* 2001, **98**, 343.
12 R. Maggio, Vogel, Z., Wess, J. *Proc. Natl. Acad. Sci. U.S.A* 1993, **90**, 3103.
13 T. E. Hebert, Moffett, S., Morello, J. P., Loisel, T. P., Bichet, D. G., Barret, C., Bouvier, M. *J. Biol. Chem.* 1996, **271**, 16384.
14 P. R. Gouldson, Snell, C. R., Reynolds, C. A. *J. Med. Chem.* 1997, **40**, 3871.
15 P. R. Gouldson, Snell, C. R., Bywater, R. P., Higgs, C., Reynolds, C. A. *Protein Eng* 1998, **11**, 1181.
16 M. Rocheville, Lange, D. C., Kumar, U., Sasi, R., Patel, R. C., Patel, Y. C. *J Biol. Chem* 2000, **275**, 7862.

17 S. Angers, Salahpour, A., Joly, E., Hilairet, S., Chelsky, D., Dennis, M., Bouvier, M. *Proc. Natl. Acad. Sci. U.S.A* 2000, **97**, 3684.
18 S. R. George, Lee, S. P., Varghese, G., Zeman, P. R., Seeman, P., Ng, G. Y., O'Dowd, B. F. *J. Biol. Chem.* 1998, **273**, 30244.
19 S. Cvejic, Devi, L. A. *J.Biol.Chem.* 1997, **272**, 26959.
20 S. AbdAlla, Lother, H., Quitterer, U. *Nature* 2000, **407**, 94.
21 F. Pazos, Helmer-Citterich, M., Ausiello, G., Valencia, A. *J. Mol. Biol.* 1997, **271**, 511.
22 O. Lichtarge, Yamamoto, K. R., Cohen, F. E. *J. Mol. Biol.* 1997, **274**, 325.
23 O. Lichtarge, Bourne, H. R., Cohen, F. E. *Proc. Natl. Acad. Sci. U.S.A* 1996, **93**, 7507.
24 O. Lichtarge, Bourne, H. R., Cohen, F. E. *J. Mol. Biol.* 1996, **257**, 342.
25 M. K. Dean, Higgs, C., Smith, R. E., Bywater, R. P., Snell, C. R., Scott, P. D., Upton, G. J. C., Howe, T. J., Reynolds, C. A. *J. Med. Chem.*, 2001, **44**, 4595.
26 W. Kuipers, Oliveira, L., Vriend, G., Ijzerman, A. P. *Receptors. Channels* 1997, **5**, 159.
27 P. R. Gouldson, Higgs, C., Smith, R. E., Dean, M. K., Gkoutos, G. V., Reynolds, C. A. *Neuropsychopharmacology* 2000, **23**, S60-S77.
28 G. V. Gkoutos, Higgs, C., Bywater, R. P., Gouldson, P. R., Reynolds, C. A. *Int. J. Quant. Chem. Biophys Quarterly* 1999, **74**, 371.
29 P. R. Gouldson, Dean, M. K., Snell, C. R., Bywater, R. P., Gkoutos, G., Reynolds, C. A. *Protein Eng,* 2001, **14**, 759.
30 M. E. Sowa, He, W., Slep, K. C., Kercher, M. A., Lichtarge, O., Wensel, T. G. *Nat. Struct. Biol.* 2001, **8**, 234.
31 M. E. Sowa, He, W., Wensel, T. G., Lichtarge, O. *Proc. Natl. Acad. Sci. U.S.A* 2000, **97**, 1483.
32 N. Saitou, Nei, M. *Mol. Biol. Evol.* 1987, **4**, 406.
33 S. Sneath *Numerical Taxonomy* 1973.
34 F. Horn, Weare, J., Beukers, M. W., Horsch, S., Bairoch, A., Chen, W., Edvardsen, O., Campagne, F., Vriend, G. *Nucleic Acids Res.* 1998, **26**, 275.
35 Vriend, G. GPCRDB: Information system for G protein-coupled receptors (GPCRs), http://www.gpcr.org/7tm/. 2000.
36 K. Palczewski, Kumasaka, T., Hori, T., Behnke, C. A., Motoshima, H., Fox, B. A., Le, T., I, Teller, D. C., Okada, T., Stenkamp, R. E., Yamamoto, M., Miyano, M. *Science* 2000, **289**, 739.
37 D. C. Teller, Okada, T., Behnke, C. A., Palczewski, K., Stenkamp, R. E. *Biochemistry* 2001, **40**, 7761.
38 T. M. Frimurer, Bywater, R. P. *Proteins* 1999, **35**, 375.
39 D. Donnelly, Overington, J. P., Blundell, T. L. *Protein Eng* 1994, **7**, 645.
40 A. Bateman, Birney, E., Durbin, R., Eddy, S. R., Finn, R. D., Sonnhammer, E. L. *Nucleic Acids Res.* 1999, **27**, 260.
41 K. Scheffzek, Ahmadian, M. R., Kabsch, W., Wiesmuller, L., Lautwein, A., Schmitz, F., Wittinghofer, A. *Science* 1997, **277**, 333.
42 M. K. Dean, Smith, R. E., Scott, P. D., Upton, G. J. C., Reynolds, C. A. *to be submitted.* 2001.
43 R. Maggio, Barbier, P., Fornai, F., Corsini, G. U. *J.Biol.Chem.* 1996, **271**, 31055.
44 G. Y. Ng, O'Dowd, B. F., Lee, S. P., Chung, H. T., Brann, M. R., Seeman, P., George, S. R. *Biochem. Biophys. Res. Commun.* 1996, **227**, 200.

45 Calver, A. R., Robbins, M. J., Medhurst, A. D., Hirst, W. D., Pangalos, and M.N. Effects of GABAB receptor subunit chimeras on receptor localisation and ligand binding, European Neuroscience 2000, Brighton, UK. 2000.
46 Higgs, C and Reynolds, C. A. Modelling G-protein coupled receptors, in Theoretical Biochemistry, ed. Eriksson, L., Elsevier, 2001, Amsterdam, pp341-376.
47 F. Ciruela, Escriche, M., Burgueno, J., Angulo, E., Casado, V., Soloviev, M. M., Canela, E. I., Mallol, J., Chan, W. Y., Lluis, C., McIlhinney, R. A., Franco, R. *J Biol.Chem* 2001, **276**, 18345.
48 K. Konvicka, Guarnieri, F., Ballesteros, J. A., Weinstein, H. *Biophys J* 1998, **75**, 601.
49 D. Fu, Ballesteros, J. A., Weinstein, H., Chen, J., Javitch, J. A. *Biochemistry* 1996, **35**, 11278.
50 A. Krebs, Villa, C., Edwards, P. C., Schertler, G. F. *J.Mol.Biol.* 1998, **282**, 991.
51 J. M. Baldwin, Schertler, G. F., Unger, V. M. *J.Mol.Biol.* 1997, **272**, 144.
52 V. M. Unger, Hargrave, P. A., Baldwin, J. M., Schertler, G. F. *Nature* 1997, **389**, 203.
53 J. B. Findlay, Pappin, D. J. *Biochem.J.* 1986, **238**, 625.
54 N. Tilakaratne, Christopoulos, G., Zumpe, E. T., Foord, S. M., Sexton, P. M. *J Pharmacol. Exp. Ther.* 2000, **294**, 61.
55 E. T. Zumpe, Tilakaratne, N., Fraser, N. J., Christopoulos, G., Foord, S. M., Sexton, P. M. *Biochem.Biophys Res.Commun.* 2000, **267**, 368.
56 P. R. Gouldson, Bywater, R. P., Reynolds, C. A. *Biochem.Soc.Trans.* 1997, **25**, 434S.
57 P. R. Gouldson, Bywater, R. P., Reynolds, C. A. *Biochem.Soc.Trans.* 1997, **25**, 529S.

ELECTROSTATIC STABILITY OF WILD TYPE AND MUTANT TRANSTHYRETIN OLIGOMERS.

S. Skoulakis and J.M. Goodfellow.

Department of Crystallography, Birkbeck College, University of London, Malet Street, London WC1E 7HX, UK

1 INTRODUCTION

Transthyretin (TTR), formerly know as prealbuim[1], is a protein that binds thyroid hormones and transports retinal binding protein that carries retinol. Many structures of TTR and of its mutants have been determined[2]. The physiologically active form of wild type TTR is a homotetrameric plasma protein in which the monomer is composed by 127 amino acid residues (Figure 1). The residues in the monomeric unit are in two four-stranded beta-sheets, which form a beta-sandwich structure. In general, all mutant structures crystallise in the same space group as the wild type, and display only minor changes compared with the wild type. TTR, both wild type and mutant, is one of at least 20 human proteins that under suitable conditions has been shown to form amyloids[3,4].

Deposits of wild-type TTR amyloids accumulate in the heart, in a disease called senile systemic amyloidosis (SSA) with the age of onset at about 80. A related disease, familial amyloidotic polyneuropathy (FAP), occurs at a much younger age, in some cases during the second decade of life, and affected individuals are found to have a mutation of TTR. So far 73 point mutations are associated with FAP and many of these destabilise the molecule leading to the formation of fibrils.

One of the models that has been proposed, for the formation of amyloids, is the conformational change hypothesis[5]. First, by lowering the pH, the tetrameric structure of the protein is disrupted and disassociates into a monomeric structure, which is structurally different from the monomer at normal pH. These altered monomers can associate and eventually form amyloid fibres. The mutations shift the tetramer-monomer equilibrium in favour of the structurally altered monomer, leading to a relative loss of stability of the mutated proteins relative to the wild type.

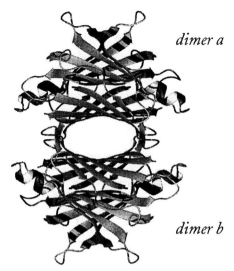

dimer a

dimer b

Figure1 *Diagram of the tetramer of wild type transthyretin (TTR) taken form crystal structure wt2, plotted using Molmol (version 2.1k.1.0)[9].*

The pH-dependent effects in proteins are mainly electrostatic in nature and originate from changes in the protonation states of acidic and basic residues[6,7,8]. Some of the titratable residues in a protein exhibit different behaviour, compared with that of the same residue in isolation, due to their interactions with other residues in the protein, and due to the altered interactions with the solvent.

In order to test the conformational change hypothesis we calculate the protonation state of all titratable residues. We find that some residues undergo a large change in their protonation state upon dimer and tetramer formation. In order to understand further the role of these residues in the binding process, we calculate their contribution to the electrostatic free energy of binding.

2 METHODS

Initial models are constructed using the GROMACS package[10]. The atomic coordinates are taken from the RCSB Protein Data Bank for the wild type transthyretin (1tta) and (1f41) at 1.7Å, and 1.5Å resolution respectively and the mutants M119T (1bze) at 1.8Å and V30M (1ttc) at 1.7Å resolution. Only the coordinate files of the dimers are provided and the tetramers are generated using the CCP4 suite of programs. As explained in Hörnberg et al. (2000)[2], the placement of the first 9 and the last 2 residues is not correct in the above pdb files due to the increased flexibility of these segments. We exclude these segments from all the structures in our pK$_a$ calculations and electrostatic calculations, and since these are exposed to the solvent, we do not expect their pK$_a$ values to be significantly perturbed. The positions of hydrogen atoms, that are added to the structures, are minimised using 50 cycles of steepest descent followed by conjugate gradient with a force tolerance of 100 kJ/mol/nm. During this minimisation procedure, all the atoms except those of hydrogen are frozen.

The electrostatic potential can be found by solving the linearized form of the Poisson-Boltzmann equation (PB)[11,12,13], as follows:

$$\nabla [\varepsilon(r) \nabla \phi(r)] - k^2 \varepsilon(r) \phi(r) = -4\pi \rho(r) \tag{1}$$

where $\phi(r)$ is the electrostatic potential at position r, $\rho(r)$ is the charge density and $\varepsilon(r)$ is the dielectric constant as a function of position, and k^2 is the modified Debye-Huckel parameter, which accounts for counterion screening in the solvent. Since water is more easily polarized by an electric field than the protein at least two values for ε must be used. In the following we use value 80 for the water and 20 for the protein, and repeat the calculations with value 4 for the protein.

We can define the pK_a value of a residue i on the basis of protonation probability $\langle q(i) \rangle$, of this residue, from the equation[7]:

$$pK_a(i) = pH + (1/2.303) \ln[\langle q(i) \rangle / (1 - \langle q(i) \rangle)]. \tag{2}$$

The pH value at which the protonation probability is 0.5 is named as $pK_{1/2}$ and is often used to describe the titration behaviour of the residues. A large change in $pK_{1/2}$ indicates a large change in the titration curves, upon oligomer formation. It can be proved[6,8], that the change in the protonation curves upon oligomer formation is related to the pH dependent electrostatic free energy contribution according to the following formula

$$\Delta G_{bind} (pH) = = -2.303RT \int\limits_{pH}^{\infty} \Delta \langle Q(pH') \rangle \; dpH', \tag{3}$$

where, $\Delta \langle Q(pH) \rangle$ is the pH-dependent mean charge difference between two states of the system and is given by the following equation

$$\Delta \langle Q \rangle_{bind} = \langle Q \rangle_{oli} - n \times \langle Q \rangle_{mon}, \tag{4}$$

where n is the number of monomers involved in the binding process, $\langle Q \rangle_{oli}$ and $\langle Q \rangle_{mon}$ are the mean charges for the oligomer and monomer respectively and are given by the sum of the mean charges of all the residues,

$$\langle Q \rangle = \sum_{i=1}^{N} \langle q(i) \rangle. \tag{5}$$

This implies that the residues with the largest shift in the $pK_{1/2}$ values, are the most important in the pH related binding process.

To understand further the contribution of specific residues, or networks of residues, to the energetics of binding, we calculate the electrostatic free energy difference that results from simultaneously mutating the residues or network of residues to uncharged ones. The calculated free energy corresponds to the reversible work associated with the process of inserting all charges, to each side chain of the network in the oligomeric form of the protein, relative to the monomer. The total electrostatic free energy ΔGtot, can be decomposed into three distinct terms[14,15] (Figure 2).

$$\Delta Gtot = \Delta Gdslv + \Delta Gbrd + \Delta Gprot \qquad (6)$$

ΔGdslv is due to desolvating the network of residues in the dimer or the tetramer when all the other charges are 0, the "bridging" term ΔGbrd is the electrostatic free energy of interaction of the residues embedded in a uncharged protein, and ΔGprot is the difference in the interaction, of the network with the charges of the protein, between the dimer or tetramer and the monomer.

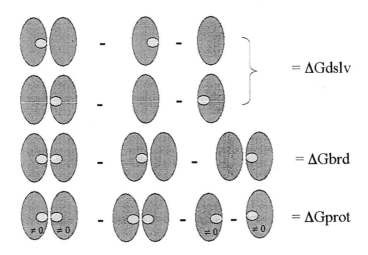

Figure 2 *Decomposition of the ΔGtot for the dimer. The network of charges is shown in light grey. The Poisson Boltzman equation is solved in every case shown. Only the charges of the network are included in the calculations of ΔGdsl and, ΔGbrd. All the charges are included to the structures' snapshots noted by (≠0).*

3 RESULTS

Initially we undertook calculations to determine the titration curves of all the residues in the monomer, dimer and tetramer of wild type and mutant conformations. Using MEAD[16] and Monti[17], we solve the Poisson-Boltzman equation and we calculate the titration behaviour of the residues. The titration curves of the residues in the wild type are similar to those of the mutants, which is consistent with the close structural similarity of the conformations. The results of the $pK_{1/2}$ calculations for the monomers of the non-amyloidogenic mutant M119T (mut_noa), the amyloidogenic mutant V30M (mut_a), and the two structures for the wild type (wt1), (wt2), are shown in Table 1. The values are very similar for all structures. The calculations are performed with dielectric constant $\varepsilon = 20$[18].

Table 1 *pK$_{1/2}$ values for TTR monomers.*

Residues	mut_noa	wt1	wt2	mut_a
Arg21	11.1	11.6	11.4	11.4
Arg34	11.4	11.6	11.4	11.7
Arg103	12.0	11.9	12.0	12.0
Arg104	11.4	11.6	11.4	11.4
Lys15	9.6	9.5	9.5	9.6
Lys35	9.6	9.8	9.8	9.3
Lys48	10.2	10.1	9.3	9.9
Lys70	9.3	9.2	9.3	9.3
Lys76	9.9	9.8	9.6	9.6
Lys80	10.2	10.1	9.9	9.9
Tyr69	9.3	10.4	9.9	10.2
Tyr78	8.7	9.2	8.7	9.3
Tyr105	9.3	9.2	8.7	9.3
Tyr114	9.6	9.5	9.6	9.6
Tyr116	9.3	9.2	9.6	9.3
Cys10	8.1	7.7	7.5	7.8
His31	5.1	5.0	5.1	5.1
His56	4.8	4.1	4.8	5.1
His88	3.9	4.1	3.9	3.9
His90	3.6	4.1	3.8	3.9
Glu42	2.4	3.2	3.0	2.7
Glu51	3.6	3.5	3.6	3.6
Glu54	2.1	1.7	1.8	1.8
Glu61	3.3	2.9	3.0	3.3
Glu62	3.3	3.5	3.3	3.6
Glu63	3.3	2.9	3.0	3.0
Glu66	3.6	3.8	3.6	3.9
Glu72	0.9	0.5	1.2	0.6
Glu89	2.1	2.0	2.1	2.1
Glu92	0.3	0.5	0.3	0.3
Asp18	-0.8	-0.1	-0.8	-0.4
Asp38	3.0	2.6	2.7	2.7
Asp39	2.4	2.6	2.4	2.4
Asp74	0.9	0.2	0.3	0.9
Asp99	1.8	2.0	2.1	1.8

Most of the residues have reasonable pK$_{1/2}$ values. His88, His90, Glu42, Glu54, Glu72, Glu92, Asp18, Asp74, Asp99, have lower values than expected. Due to the long range of the table interactions and the number of titratable residues, it is not easy, in general, to attribute the anomalous pK$_{1/2}$ value to the proximity of specific residues. However, the interactions between the residues can be analysed from visual inspection of the crystal structures and from the matrix of interactions (as output from MEAD). We find that there is a network of strongly interacting residues, consisting of Glu72, His88, Glu92, His90, Tyr116. The distances between the closest atoms are shown in Table 2. The closest atoms are the ones that are strongly interacting as we confirm from the output of the matrix of interactions. For the other residues, Glu42 is influenced by Arg34, Glu54 by Lys15, Asp18 by Tyr78, and Asp99 by Tyr105.

Table 2 *Close contacts within the monomers.*

	Distances(Å)			
Closest non-hydrogen atoms	mut_noa	wt1	wt2	mut_a
Glu72(OE1)-His90(NE2)	4.12	3.84	4.55	4.29
His90(NE2)-Glu92(OE2)	5.90	4.26	5.61	4.26
His90(ND1)-Glu92(OE1)	5.60	4.57	3.29	4.65
His90(ND1)-Glu92(OE2)	4.19	2.57	5.83	3.39
Glu92(OE1)-Tyr116(OH)	3.44	3.74	5.27	7.60
His88(NE2)-Tyr116(CE2)	3.84	3.70	6.17	3.90
His88(NE2)-Tyr116(OH)	5.20	4.95	3.99	5.29
His88(ND1)-Tyr116(CE2)	3.81	3.92	5.29	3.93

Upon dimer formation, the pK_a of some the residues changes and the results for the largest changes are shown in the Table 3. These changes cannot easily be explained by the approach of the other monomer, for example, by formation of salt bridges. The only extra hydrogen bond that is formed involving titratable residues is between OG1 (Thr96) and OE2 (Glu89), in all structures studied.

Table 3 $\Delta pK_{1/2}$ *for dimer formation. $\Delta pK_{1/2}$ values $((pK_{1/2}(dimer) - pK_{1/2}(monomer))$ for the residues with the largest changes in values i.e. $|\Delta pK_{1/2}| > 0.6 \cong 0.8$ kcal/mol at 300 K. Values for the residues of only one monomer are shown. Due to symmetry, we find similar behaviour for the other monomer. The values are calculated with dielectric constant $\varepsilon = 20$. These residues noted by (#) titrate at a pH less than –4 in the dimer.*

	mut_noa	wt1	wt2	mut_a
Lys35	-0.8	-1.0	-0.8	-0.9
Lys70	-1.9	-2.1	-2.4	-2.5
Lys76	-1.1	-1.2	-1.0	-1.6
Tyr105	-1.1	-1.3	-1.6	-1.1
Tyr114	-1.0	-1.2	-1.1	-1.2
Tyr116	-1.7	-1.6	0.0	-1.5
His31	-0.9	-0.9	-0.8	-0.7
His88	-5.3	-5.3	-5.7	-5.7
His90	-7.6	-7.4	-6.9	<-7.6[#]
Glu66	-0.4	-0.4	-0.6	-0.7
Glu72	-2.3	-2.8	-2.7	-2.6
Glu89	-2.5	-2.5	-2.8	-3.9
Glu92	<-5.3[#]	<-5.3[#]	<-4.5[#]	<-4.3[#]
Asp18	-0.8	-0.4	-0.5	-0.6
Asp39	-1.8	-2.0	-1.4	-2.0
Asp74	-0.9	-0.7	-0.6	-0.9
Asp99	-2.4	-2.1	-2.2	-1.8

Notable is the close proximity of Glu92 in one monomer to the Glu92 and the OH atom of Tyr116 table the other monomer. These belong to the network of strongly interacting intramolecular residues, that become buried upon dimer formation, and interact intermolecularly with the same network of the other monomer, (Table 4).

On dimer association to form the tetramer, a similar pattern of behaviour is observed. The values of $pK_{1/2}$, shown in Table 5, decrease but not to the same extent as upon dimer formation. This is not surprising because the dimer-dimer interface is populated by hydrophobic residues. This is in contrast to the monomer-monomer interface, which is occupied mainly by charged and polar residues. Once more the interactions between residues are the most important factor leading to the low $pK_{1/2}$ values. The most affected

Table 4 *New contacts upon dimer formation. Upon dimer formation the above titratable groups come in close proximity resulting in low $pK_{1/2}$ values.*

monomer a	monomer b	Distances(Å)			
		mut_noa	wt1	wt2	mut_a
Glu92(O)	Tyr116(OH)	2.98	3.00	2.76	2.88
Glu92(OE1)	Glu92(OE1)	4.53	4.14	4.11	2.63
Glu92(OE1)	Glu92(OE2)	6.03	6.02	3.29	3.89
Glu92(OE2)	Glu92(OE1)	5.81	5.07	4.46	4.81
Tyr116(OH)	Tyr116(OH)	4.03	4.79	4.25	4.31

residues are the ones closest to the interface between the dimers, as expected, even though the only titratable residues that come closer than 5Å to the dimer, due to tetramer formation, are Tyr114 and Arg21 from the third and fourth monomer of the second dimer. The appearance of the second dimer changes the strength of the interactions already present in the first dimer, and this causes the shift of the $pK_{1/2}$ values.

Table 5 *$\Delta pK_{1/2}$ for tetramer formation. $\Delta pK_{1/2}$ values $((pK_{1/2}(tetramer) - pK_{1/2}(dimer))$ for the residues with the largest changes i.e. $|\Delta pK_{1/2}| > 0.6 \cong 0.8$ kcal/mol at 300 K. The values for the residues of only one monomer, (a), are shown. Due to symmetry we find similar behaviour for the others. The dielectric constant used is $\varepsilon = 20$.*

monomer a	mut_noa	wt1	wt2	mut_a
Lys15	-2.5	-2.1	-1.8	-2.3
Tyr78	-1.2	-1.2	-1.0	-1.2
Tyr116	-2.4	-1.7	-2.9	-3.0
His56	-0.8	-1.0	-1.1	-1.0
His88	<-2.6	<-2.6	<-2.4	<-2.2
Glu54	-2.6	-2.5	-2.5	-2.7
Asp18	<-2.4	<-2.8	<-3.4	<-3.0
Asp74	-0.7	-0.1	-0.5	-0.5

From the above calculations we find that $\Delta \langle Q \rangle_{bind}$ is negative, implying that ΔG_{bind} is increasing as we lower the pH (equation 3). This is consistent with the experimental results[3,4,5], proving the instability of the oligomer structure as the pH is lowered. Moreover the importance of the network of Glu72, His88, Glu92, His90, and Tyr116, in the pH dependent binding process, makes the investigation of its role in the binding process in general worthwhile.

We calculate the contribution of this network (Nt1), in the electrostatic free energy of dimer and tetramer formation. We also repeat the calculations for a subset (Nt2) of this network composed by Glu92, His90, and Tyr116. In both cases we find that the network formation is unfavourable, mainly due to the desolvation term, which is large since the network is located at the interface between the monomers.

Table 6 *Eelectrostatic free energy contribution, for tetramer, $\Delta\Delta Gtot(t\text{-}m)$, (dimer, $\Delta\Delta Gtot(d\text{-}m)$) formation, from the monomers, for subsets of residues, namely Nt1: Glu72, His88, Glu92, His90, and Tyr116 and Nt2: Glu92, His90, and Tyr116. All energies are in kcal/mol.*

$\Delta\Delta Gtot(d\text{-}m)$		ΔGprot		ΔGbrd		ΔGdslv		ΔGtot	
		ε=4	ε=20	ε=4	ε=20	ε=4	ε=20	ε=4	ε=20
wt1	Nt1	-11.7	-4.1	7.2	3.6	24.5	3.6	20.0	3.1
	Nt2	-10.2	-2.0	5.6	1.8	22.2	4.1	17.6	3.9
wt2	Nt1	-10.7	-4.6	7.3	1.9	20.7	3.0	17.3	0.3
	Nt2	-8.8	-4.0	5.1	1.7	20.8	2.8	17.1	0.5
mut_noa	Nt1	-11.6	-3.8	7.0	3.3	27.4	4.4	22.8	3.9
	Nt2	-7.9	-1.5	5.3	1.6	24.1	3.5	21.5	3.6
mut_a	Nt1	-10.8	-4.0	7.8	3.9	17.2	3.2	14.2	3.1
	Nt2	-7.0	-1.4	6.0	3.9	13.6	3.2	12.6	5.7

$\Delta\Delta Gtot(t\text{-}m)$		ΔGprot		ΔGbrd		ΔGdslv		ΔGtot	
		ε=4	ε=20	ε=4	ε=20	ε=4	ε=20	ε=4	ε=20
wt1	Nt1	-13.2	-4.3	7.3	3.6	24.8	5.3	18.9	4.6
	Nt2	-11.2	-2.3	5.7	1.8	22.5	4.1	17.0	3.6
wt2	Nt1	-12.3	-4.1	7.7	3.4	21.3	5.5	16.7	4.8
	Nt2	-10.3	-2.2	5.5	1.7	20.1	4.0	15.3	3.5
mut_noa	Nt1	-13.7	-4.3	7.0	3.3	26.5	4.3	19.8	3.3
	Nt2	-8.9	-1.7	5.2	1.8	23.7	3.4	20.0	3.5
mut_a	Nt1	-12.9	-4.4	7.6	3.9	16.1	3.1	10.8	2.6
	Nt2	-6.8	-1.4	5.6	2.2	12.4	2.1	11.2	2.9

Our results (Table 6) are robust with respect to changes in the dielectric constant. When we use ε = 20, ΔGtot is less unfavourable, as expected from the smoothing role of a higher value of ε, but still consistent with general view of electrostatic strain in this network. The results are very similar for the network in the dimer or tetramer, indicating that the monomer-monomer interface is the one contributing the most to the electrostatic interactions.

Table 7 *Electrostatic free energy of binding.*

		ε = 4				ε = 20			
		mut_noa	wt1	wt2	mut_a	mut_noa	wt1	wt2	mut_a
	d-m	53.5	47.9	35.0	45.0	9.6	9.4	8.2	9.5
ΔGelect	t-d	33.7	48.9	32.1	30.2	10.2	12.0	8.9	10.2
	t-m	140.7	144.7	102.1	120.2	29.4	30.8	25.3	29.2

We also calculate the total electrostatic free energy of binding for all the structures, (Table 7). The table energy found is positive in all cases and with all dielectrics used. This means that the electrostatic interactions oppose the oligomer formation. Even though we can predict the trend of the free energies of binding the calculations are not sensitive enough to discriminate the mutant from the wild type structure, which is not surprising considering the high accuracy required for such discrimination. Our calculations show that the reduced stability of the oligomer when we lower the pH is not due to the loss of favourable interactions upon binding of H^+, but a 'bad' situation becoming 'worse' as the electrostatic interactions are unfavourable even at low pH.

4 CONCLUSION

We predict the loss of stability of oligomeric TTR, when the pH is lowered. The importance of the network of residues, Glu72, His88, Glu92, His90, and Tyr116, is indicated. We also find that the contribution of the above residues in the electrostatic free energy of binding is unfavourable and the binding of H^+ increases this contribution, leading to the formation of monomeric structures.

Acknowledgements

We would like to thank the Wellcome Trust for a training fellowship in mathematical biology to SS and the BBSRC for some computer resources. The work was carried out within the Bloomsbury Structural Biology Centre.

References

1   C.C.F. Blake, M.J. Geisow, S.J. Oatley, B. Rérat, and C. Rérat, *J. Mol. Biol.*, 1978, **121**, 339.
2   A.Hörnberg, T. Eneqvist, A. Olofsson, E. Lundgren, and A. E. Sauer-Eriksson, *J. Mol. Biol.*, 2000, **302**, 649.
3   J.W. Kelly, *Structure*, 1997, **5**, 595.
4   A.M. Damas, M.S. Saraiva, *Journal of Structural Biology*, 2000, **130**, 290.
5   J.W. Kelly, *Current Opinion in Structural Biology*, 1998, **8**, 101.
6   A.S. Yang, B. Honig, *J. Mol. Biol.*, 1993, **231**, 459.
7   G. M. Ullmann, E.W. Knapp, *Eur.Biophys. J.*, 1999, **28**, 533.
8   M. Schaefer, M. Sommer, M. Karplus, *J. Phys. Chem.*, 1996, **101**, 16663.
9   R. Koradi, M. Billeter, K. Wüthrich, *J. Mol. Graphics*, 1996, **14**, 51.
10 H.J.C. Berendsen, D. van der Spoel, R. van Drunen, *Comp. Phys. Comm.*, 1995, **91**, 43.
11 J. Warwicker, H.C. Watson, *J. Mol. Biol.*, 1982, **157**, 671.
12 M.K. Gilson, B. Honig, *Proteins Struct. Funct. Genet.*, 1988, **4**, 7.
13 B. Honig, A. Nicholls, *Science*, 1995, **268**, 1144.
14 V. Lounnas, R.C. Wade, *Biochemistry*, 1997, **30**, 5402.
15 Z.S. Hendsch, B. Tidor, *Prot. Sci.*, 1994, **3**, 211.
16 D. Bashford, K. Gerwert, *J. Mol. Biol.*, 1992, **224**, 473.
17 P.Beroza, D.R. Fredkin. M.Y. Okamura, G. Feher, *Proc. Natl. Acad. Sci. USA*, 1990, **87**, 5804.
18 J. Antosiewicz, J, J.A. McCammon, M.K. Gilson, *Biochemistry*, 1996, **35**, 7819.

SIMULATIONS OF HUMAN LYSOZYME: CONFORMATIONS TRIGGERING AMYLOIDOSIS IN I56T MUTANT

G. Moraitakis[1] and J.M. Goodfellow[1]

[1]School of Crystallography, Birkbeck College, University of London, Malet Street, London WC1E 7HX, UK

1 INTRODUCTION

Human Lysozyme is a 130-residue protein found in secretions (e.g. saliva, sweat and mucus) and more generally in leukocytes and kidneys. It is an enzyme which hydrolyses preferentially the β-1,4 glucosidic linkages between N-acetylmuramic acid and N-acetylglucosamine which occur in the mucopeptide cell wall structure of certain micro-organisms.[1] The wild type human lysozyme has been crystallised and its structure elucidated by Artymiuk and Blake[2] at 1.5Å resolution (Figure 1A). Its native structure consists of two domains: an α domain that has four α-helices (A-D) and one 3_{10} helix, and a β domain, which consists mainly of an antiparallel β-sheet and a long loop. The active site is located in the cleft that is formed between these two domains. The protein contains 4 disulphide bonds of which two are located in the α-domain, one in the long loop of the β-domain, and one that connects the two domains.

There are two known natural mutations of the human lysozyme: I56T and D67H.[3] They both cause autosomal dominant hereditary *non-neuropathic systemic amyloidosis*. This is a condition whereby there is tissue deposition in viscera and other body cavities, of normally soluble autologous proteins as insoluble fibrils called *amyloids*. The core of these fibrils structure consists of β-sheet with the strands perpendicular to the long axis of the fibre.[4] Amyloids can be formed from proteins of diverse sequence, fold and function and are known to lead to serious medical conditions such as Alzheimer's disease and spongiform encephalopathies.[5] Although the mechanism of lysozyme fibrilogenesis is not very clear, experimental evidence[6] suggests that it is related to changes in stability and tendency of aggregation due to the mutations. More specifically, Booth et al.[7] proposed that, during folding, a partially folded transient population of the two human lysozyme amyloidogenic proteins, which lacks global cooperativity, undergoes structural transformation (a helix-to-sheet transition) and creates the first template, the "seed", for further protein deposition and fibril formation. In a recent paper,[8] Morozova-Roche et al. show that the presence of these "seeds" for both wild and the two mutants of human lysozyme facilitates the formation of fibrils.

We focus on the behaviour of the I56T variant. The crystallised structure of this mutant[7] at 1.8 Å resolution, is quite similar to that of the wild type (0.12 Å difference) and

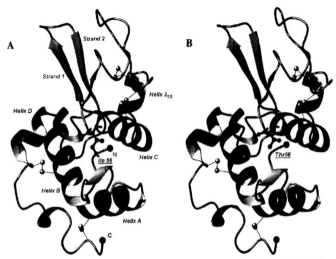

Figure 1   *The crystal structures of the wild type (A) and the I56T (B) human lysozyme and the mutation site, Ile56, located in the interface of the α and β domains, are shown as ball and sticks. The 4 disulphide bonds are indicated as ball and sticks. The sulphur atoms of these bonds as well as the N and C termini atoms are shown as spheres.*

is shown in figure 1B. There are very small changes near the mutation site whereby the introduction of a charged group (hydroxyl) and the removal of a methyl group lead to the N-terminal approach more Thr56 and result to minor changes at the C-terminal of helix C.

In order to understand the behaviour of the partially folded structures that may trigger amyloid formation, we explore the conformations adopted during the induced unfolding of the wild type and mutants at high temperatures (500K) using molecular dynamics (MD) simulations. The temperature 'denaturation' of proteins using MD is considered as one of the most straightforward computational 'experiments'.[9] The great advantage of unfolding simulations is that they allow the investigation of the conformational properties at every point along the unfolding pathway.[10] There are many examples of their use[11] and these have led to detailed insight of experimental results of protein unfolding[12,13] and the study of the folding pathway.[14] Lysozyme, from hen egg-white has been extensively investigated by means of unfolding simulations with the aim to investigate several issues such as the stability and folding of the protein.[15,16,17,18,19] The human and hen forms of lysozyme have 60% sequence similarity but very similar 3D structure.

In our study, we attempt to examine and compare the structure and properties of the partially folded intermediates obtained from the induced unfolding of the wild type and the I56T form.

2 METHODS

2.1 Model and Simulation Details

The crystal structures of the wild type (1REX) and the mutant (1LOZ) lysozyme are the starting point of the simulations. Hydrogen atoms are added and the final system includes solvent molecules which are represented by the SPC216 model.[20] All atoms are explicitly represented. The solvated model is contained in a rectangular box (70x70x70 Å) using periodic boundaries conditions.[21] The building of the protein models and all their simulations are carried out with the *GROMACS* suit of programs.[22] The GROMOS 96 force field is used to describe the atomic interactions.[23] For the correct treatment of long-range electrostatics we make use of the particle mesh Ewald summation algorithm.[24] The high frequency degrees of freedom from the covalent bonds of hydrogen atoms are constrained using the LINCS algorithm[25] and thus the time step is increased from 1 to 2 fs. The system is coupled to an external temperature bath with a separate bath for the solvent and the solute. The regulation of the pressure is achieved by means of a "pressure" bath. Data on the trajectories are saved every 0.2ps. For these simulations, we use an in-house multi-processor Origin 2000 with 4 processors. The total CPU time for all simulations was about 48 days.

In order to minimise the initial system, we use a combination of the *conjugate gradients* and *steepest descent* methods in which after every 50 steps of conjugate gradients, one step of steepest descent is performed. The minimisation is terminated when the overall force of the system is 100N or after 5000 steps. This protocol is first used to minimise hydrogen atom and water molecules positions and then extended to the whole system. To further optimise the arrangement of the solvent around the protein and alleviate high energy regions (hot-spots), the water molecules only, are assigned initial velocities from a Gaussian distribution generated from a random seed and then warmed-up from 50K to 300K over 10ps of MD. This is followed by equilibration of the water molecules at 300K for 30ps. Then the whole system (including the previously constrained protein) is assigned initial velocities from a Gaussian distribution generated by a random seed for 50K, warmed up to 300K for 10ps and finally equilibrated at this temperature for 1000ps. The 1000ps trajectory at 300K serves as a control simulation. The last conformation of the system obtained from this trajectory is used as the starting structure for the unfolding simulation in which the system is warmed up to 500K for 10 ps and is subsequently maintained at this temperature. To enhance the sampling of the unfolding pathway, long multiple trajectories are performed by assigning velocities generated from different seed numbers. Three 5000ps high temperature (500K) and three 1000ps control (300K) simulations are performed for each lysozyme form.

2.2 Analysis of Molecular Dynamics

In the analysis, the root mean square deviation (RMSD), solvent accessible surface area and secondary structure content plots are calculated using the analysis programs provided by GROMACS.[26] The secondary structure analysis is carried out using the Kabsch and Sander algorithm[27] incorporated in their DSSP program. The inter-residue distances and the clustering analysis is performed with software written in-house. The percentages of secondary content per region across all trajectories are derived as follows: First the percentage in each of the four secondary structure conformation (helix, strand, turn, coil) contained in the specified regions is calculated for each structure sampled from the multiple unfolding trajectories of each lysozyme form. Then, these percentages are obtained across all multiple trajectories (Table 1).

The number of pairwise distances of all residues within 8Å (closest atoms distance) is calculated for each conformation in the trajectory and for the crystal structure. For the latter, all these residue contacts are referred to as *native contacts* as opposed to the *non-native contacts* of pairs of residues that are situated within the 8Å cut-off but are not present in the native structure (crystal structure).

The distance matrix used in the clustering, is constructed from the column vectors of a properties' matrix. Here the properties' matrix of the conformations observed during the trajectories contains four rows corresponding to four properties: (1) the number of native residue contacts, (2) the number of non-native residue contacts, (3) the number of residues in secondary structure elements (α-helix, β-sheet, β-turn) and (4) the number of residues in random coils. The reason that those properties have been chosen is because it has been observed from our plots that they change with time during the unfolding unlike other properties like Solvent Accessible Surface area (SASA), radius of Gyration (R_g) and number of H-bonds which oscillate around their initial value. Thus, the latter cannot discriminate easily between conformations obtained in different times during the unfolding).

The Mahalanobis distance[28] is used for the construction of the distance matrix. The elements of this matrix are calculated by the following formula:

$$D_{ij}^2 = (x_i - x_j)^T C^{-1} (x_i - x_j) \qquad (1)$$

Here, C is the covariance matrix and x is a column vector of the properties' matrix. The use of Mahalanobis distance removes several of the limitations of Euclidian distances used by others for classifications of MD simulations[29] as it automatically accounts for the scaling of the coordinate axes and also it corrects for correlation between the different features (e.g. residue contacts and secondary structure content).[30]

We use hierarchical clustering to group the conformations by transforming a set of data points with a given measurement for dissimilarity (the distance matrix) into a sequence of nested partitions (a dendrogram). We use agglomerative methods to built the dendrogram. These involve starting with each data point as a single cluster and subsequently merge these together. In particular we make use of the *group average distance* hierarchical algorithm: At each step, we seek the shortest distance of a pair of clusters in the distance matrix and merge them together. A new distance matrix is created in which the new distances of all the clusters from the newly created matrix are based on the average distance of their members.[31,32] This process is continued until no distance in the distance matrix occurs below a set cutoff. We have selected this cutoff to be the point in the process at which a cluster formed contains more than 20% of the conformations. Our results using this clustering algorithm have been verified against other statistical packages such as S-PLUS.[33]

3 RESULTS AND DISCUSSION

Three 5ns trajectories at high temperature are performed for the wild type (WT 1, WT 2, WT 3) and three for the mutant (I56TH 1, I56T 2, I56T 3) in order to increase the sampling of the unfolding conformational space.[29] Clustering techniques are used to probe for the most populated average conformations. Analysis of these conformations is carried out to ascertain the structural features of any "seeds" that have the potential to lead to fibre formation.

3.1 Analysis of the individual trajectories

At 300K we observe that the root mean square deviation (RMSD) based on the Cα atoms plateaus around 0.2 nm from the crystal structure for all trajectories (Figure 2) and thus agrees with previous results on the equilibration of a native protein structure at this

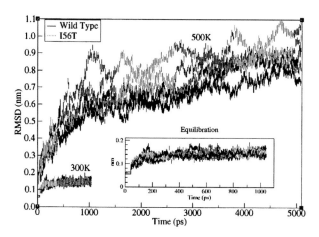

Figure 2 *The plot of root mean square deviation (RMSD) (in nm) for Cα atoms as a function of time (ps) with respect to the crystal conformation. Results are shown for unfolding (500K), control and equilibration trajectories (300K).*

temperature by Kazmirski et al.[29] At 500K, we find that all simulations have similar rates of increase in RMSD for the first 1000ps. However, after this point, the I56T mutant trajectories exhibit a larger increase in RMSD compared to that of the wild type trajectories. The values of the RMSD near 5000 ps seem to converge for all trajectories with the exception of one that reaches about 1.1nm after 5ns.

The root mean square fluctuations of the atoms across all the simulations of the mutant and the wild type form have been calculated for the Cα atoms (Figure 3). We observe that during unfolding, I56T tends to have slightly higher fluctuations in most of the regions. This difference is most notable in the β domain and, in particular, in the loop region of this domain between residues 68 and 75. The mutation site is however slightly more rigid in the mutant during the simulations than in the wild type. The higher flexibility of parts of the β-domain in I56T implies that it is less stable than that of the wild type, which is in agree-

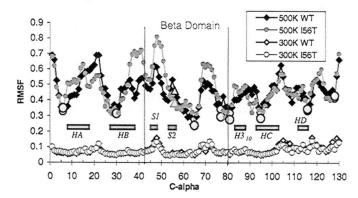

Figure 3 Root mean square fluctuations (RMSF) of the Cα atoms during the unfolding and 300K simulations. The triangles represent the Cα atom at residue 56. The atoms of the residues involved in disulphide bonds are also shown (larger circles). Regions of β-strand and α-helix are shown.

ment with the work of Booth et al.[7] where the calculated B-factors for this domain are increased. We also observe some increased fluctuations in the loop region just after helix B in the mutant conformations.

Inter-residue distances change dramatically during the unfolding simulations with residues originally distant in the native structure now being found in close proximity and vice versa. Residue contacts can be classified as *native contacts* (those found in the native structure) and *non-native contacts* which are formed during the simulation but not found in the native structure. For the unfolding simulations at 500K, we have observed that the trajectories have a similar rate of loss of native contacts and formation of non-native contacts for the first 1000ps. Subsequently, the mutant simulations show a very slight increase in loss of native contacts and a similar increase formation of non-native with all trajectories converging by 5000 ps. For the control simulations at 300K, we observe that the number of inter-residue contacts remains steady fairly constant. Thus, results from these contact plots are consistent with the analysis of RMSD.

The major impact of unfolding can be shown from plots of inter-residue contact distances, coloured according to the fraction of conformations sampled exhibiting each contact. A similar pattern in native contact loss is observed for the wild type and the mutant. Some differences are observed in the area of the beta domain and in particular in the contact region of residue 56 and the C-terminal of helix B. Helix C shows very high contacts retention for the mutant. In the non-native plots (figure 4). We observe that new relatively strong contacts (50-60%) appear in the beta domain of the wild type. These are near the native contacts suggesting that this domain is more resistant. A relatively stronger contact (30-50%) forms between Thr56 and helix 3_{10}. This is more clearly shown in figure 5 where we plot the percentage of the total conformations sampled carrying each of the contacts that residue 56 makes with the rest of the aminoacids in the polypeptide chain. We observe that Thr56 has indeed particularly strong contacts with the residues 80-90 (helix 3_{10}) and residues 59-61 (part of the beta domain loop). In contrast Ile56 is in a lot less conformations connected to this helix. However it shows some very loose contacts with the C-terminal residues (110-129) as opposed to its mutant.

The solvent accessible surface area (SASA) does not change greatly with time in any trajectory at 500K, for wild type or mutant. During the unfolding of a protein, it is expected that more residues become accessible to solvent. Although a slight increase is observed for all the trajectories between 0 and 1000 ps in the SASA of hydrophobic residues, most of the time it is around the average value of 40 nm^2. The *total* SASA increases slightly in the first 500ps and then acquires a steady value. We noted also a small "jump" near 4500 ps that is common for all the trajectories. These results for SASA are consistent with the analysis of the radius of gyration against time, which similarly shows no major change with time during the simulations.

The reason for the lack of overall change of the radius of gyration and SASA may be the presence of four disulphide bonds. In the classical force field, which is used in MD simulations, the disulphide bonds are kept intact irrespective of the increased temperature conditions. In a separate set of simulations (data not shown) in which the disulphide bonds of the two forms are reduced, it is observed that at high temperature both SASA and the radius of gyration increase after 1.5 ns. At the same time, the percentage of secondary structure decreases faster than in the normal simulations. One effect of the disulphide bonding is that the residues involved in this are less flexible, as illustrated from our root mean square fluctuation plots (Figure 3), where the Cα of these residues have the lowest fluctuations. This is consistence with the crystallographically derived B-factors for the structure.

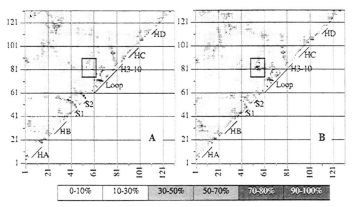

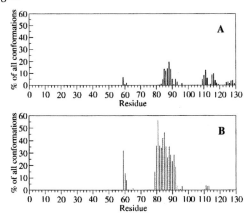

Figure 4 *Maps of non-native contacts of (A) the wild type and (B) the mutant across all conformations sampled. Both axes represent residue number. The labels on the diagonal correspond to regions of the molecule. The cut-off for a pair of residues to be in contact is 8Å between their Cα atoms. Colour shade is based on percentage of conformations sampled have each specific contact and is shown. The area of contact between residue 56 and helix 3_{10} is shown in the rectangle.*

Figure 5 *Distribution of contacts that Ile56 (A) and Thr56 (B) make with the rest of the residues across all the conformations sampled.*

For all the trajectories, we find that the secondary structure profile changes significantly during the simulation. Initially, most residues are in a specific secondary structural conformation but the number of such residues drops gradually and reaches a plateau after about 4ns. At the same time, random coil conformations become gradually predominant. Analysis of the time course of secondary structure shows that (a) the original β-strands are destroyed before the helices, (b) the helical elements are eventually replaced by coils, bends and turns and then (c) transient beta strands and bridges appear in many areas across the polypeptide chain.

The changes of the secondary structure content during the unfolding in separate regions of the lysozyme are depicted in Table 1. As the regions selected are entirely comprised of one of the four conformations in the native structure, we can have a rough estimation of (a) how much of the original conformation is retained during the unfolding and (b) what other conformations are acquired during the simulation. Our first observation is that helical regions original secondary conformation is converted mostly to coils and turns during the simulations. However, there is a wide difference in the percentage of each helical region undergoing this conversion between the mutant and the wild type. Helix A in the latter has helical components for 50% of the conformations sampled while only one third of the structures in the mutant are helical in this region. In the same way, helix 3_{10} and helix D are a lot more retained in the wild type than in the mutant. In contrast helix B in the mutant tends to be helical in twice as many structures as the wild type. Similarly helix C retains helical components mostly in the mutant conformations. In very few cases there is partial conversion from helix to strand and even less from strand to helix. Generally we observe that in most of the cases the wild type structures sampled tend to have higher percentages of the original secondary structure conformation than the mutant with the sole exception of helix B. Conversely, the turn and coil content is higher for the mutant in all the regions.

The hydrogen bonding across the protein, during unfolding, has also been examined, although in part it is related to the inter-residue contacts analysis described previously. The total number of H-bonds is initially about 100 and then, after 500ps, it drops to around an average of 75 (within the range of 60 to 90) for the rest of the trajectory. This applies to all of the trajectories at 500K. The explanation for this observation may be related to the changes in the secondary structure; the native structure is rich in helices and strands, both of which involve many hydrogen bonds. During the unfolding, we have observed that these elements are converted to coils (no hydrogen bonds) and turns (contain smaller hydrogen bonding network).

3.2 Clustering of the trajectories' conformations and analysis

More informative analysis of multiple trajectories can be obtained by accumulating all snapshots from each trajectory and clustering into those with similar properties. This is particularly useful since we cannot discriminate which trajectory is more "significant" than the others.[34] All the conformations generated from the three trajectories of each lysozyme form, are clustered (using an hierarchical clustering procedure) to generate average structures ranked by the population of the cluster they represent. This clustering is based on residue contacts and secondary structure content (as described in methods). The resulting conformations are shown in Figure 6.

Table 1   *Secondary structure content in lysozyme during the unfolding simulations.*

Regions (residues)	Helix (%)[a]		Strand (%)[a]		Turn (%)[a]		Coil(%)[a]	
	WT	I56T	WT	I56T	WT	I56T	WT	I56T
Helix A (5-14)	54	31	1	4	27	30	19	34
Helix B (25-36)	23	43	4	6	37	30	36	20
Strand 1 (43-46)	1	1	20	17	21	21	59	61
Strand 2 (51-54)	0	1	35	20	16	28	49	50
Loop (60-78)	4	3	11	9	50	51	35	37
Helix 3_{10} (81-85)	23	8	2	2	44	50	31	40
Helix C (90-100)	40	57	2	3	35	29	23	9
Helix D (110-115)	29	13	6	2	44	43	21	41

[a] The percentage is calculated from accumulative sums of secondary structure within given regions across all conformations sampled during the trajectories

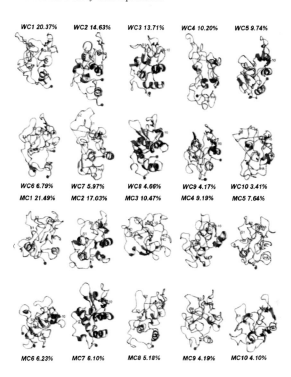

Figure 6   *The average structures of the ten most populated clusters are shown. The mutation site (residue 67) is shown as ball and sticks. The N- and C-terminal are shown as green and blue spheres respectively. The disulphide links are also shown as spheres. The clusters for the WT are denoted as WC1, WC2 etc. while the ones from the mutant as MC1, MC2 etc. The models have been obtained after least square fitting each of these conformations to the respective initial models (crystal structure + hydrogens). Thus, the orientation and shape of the structures are relative to the ones shown in Figure 1. The dotted lines show the distance between Ile56 and helix 3_{10} (orange) and the distance between Ile56 and helix B (green).*

Analysis of the wild type trajectories leads to 21 clusters. The most populated, WC1, contains 20.4% of the total population, followed by two clusters with the order of 14% and two more of 10%. Five more clusters ranging between 3-7% are found and the rest are below 2%. For the I56T mutant, where 22 clusters are assigned using the same cutoff criteria as the WT, the distribution is slightly different. The largest group contains 21.5% and followed by another big cluster of about 17% of the total population. Two clusters of about 10% exist and six more groups ranging from 4-7% are assigned. There is one group of about 2% and two of about 1% and the rest have populations ranging between 0.54% and 0.07%. A small proportion of the conformations of both lysozyme forms are grouped in clusters (named WC8 and MC7) that both contain elements from the initial stages of the unfolding trajectory.

The average structural features of these clusters are given in Table 2. The most populated conformations for both the WT and the mutant have lost most of their secondary structure content and the original native contacts. It appears though, that the WT clusters are usually populated by conformations retaining more native contacts (i.e. a higher fraction of the initial conformation) than its mutant counterpart. Additionally, the populated mutant clusters contain conformations usually with greater non-native contacts. In many of the clusters there are conformations that contain a high proportion of residues in random coils. However, the WT clusters appear to contain a higher percentage of secondary structure elements (helices and sheets) than the mutant.

The major impact of unfolding can be shown from plots of inter-residue contact distances. These contact maps for the average structures of the two most populated clusters and the crystal structures are plotted in a 2D colour map of all-against-all residue distances (Figure 7). In the most populated cluster of the mutant (MC1), the beta domain (especially in the region of the beta strands) exhibits a very different set of contacts compared with those in crystal structure and with those in other clusters from the mutant trajectories.

Specifically, in the MC1 cluster, the region around Ile56, located in the interface between the α and β domain, loses contact with helix B but has stronger contacts with helix 3_{10} compared to the wild type. Those two contacts are present in the crystal structures and 300K simulations of both lysozyme forms and they exist as well in the average structure of the most populated clusters of the wild type. The distribution of these distances in the conformational space sampled with the simulations is shown in Figure 8. We observe that at 500K, for the Ile56-helix B contact, the mutant exhibits a distribution lying further away

Table 2 *Average structural properties of the (A) Wild Type and (B) the I56T mutant clusters given as the numbers of residues in a particular secondary structure conformation.*

A. WILD TYPE

Cluster	Population (%)	Native Contacts (% retained)	Non-Native Contacts	No. residues in helix or sheet or turn[a]	No. residues in Coil[a]
WC1	20.37	140 (38)	240	31	52
WC2	14.63	182 (49)	175	55	42
WC3	13.71	191 (52)	173	51	48
WC4	10.20	117 (32)	267	39	56
WC5	9.74	164 (44)	237	50	45
WC6	6.79	116 (31)	238	32	61
WC7	5.97	107 (29)	247	33	54
WC8	4.66	270 (73)	94	78	28
WC9	4.17	148 (40)	265	34	54
WC10	3.41	108 (19)	283	30	49
Crystal[#]	-	370 (100)	0	89	19

B I56T MUTANT

Cluster	Population (%)	Native Contacts (% retained)	Non-Native Contacts	No. residues in helix or sheet or turn[a]	No. residues in Random Coil[a]
MC1	21.49	131 (36)	245	32	54
MC2	17.03	183 (50)	189	56	42
MC3	10.47	108 (29)	263	40	52
MC4	9.19	125 (34)	287	35	57
MC5	7.64	127 (34)	288	30	49
MC6	6.23	116 (31)	299	42	49
MC7	6.10	273 (74)	98	70	34
MC8	5.18	174 (47)	212	51	39
MC9	4.89	149 (40)	250	25	57
MC10	4.10	113 (30)	222	31	54
Crystal[b]	-	357 (100)	0	93	18

[a] The rest of the 130 residues are in β-bends and β-bridges.
[b] For comparison, the structural features of the respective crystal structures are shown.

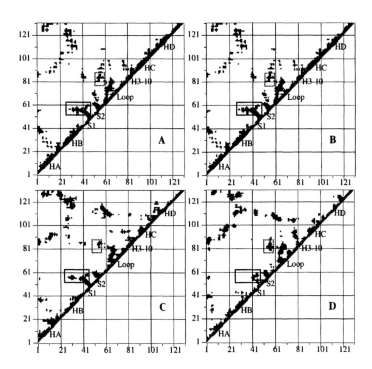

Figure 7 *Contact maps of (A) the WT crystal structure, (B) the mutant crystal structure, (C) the WC1 cluster and (D) the MC1 cluster. Both axes represent residue number. The labels on the diagonal correspond to regions of the molecule. The cut-off for a pair of residues to be in contact is 8Å between their Cα atoms. The residue 56 contacts with helix B (rectangle) and helix 3₁₀ (dashed rectangle) are shown.*

(1.3 nm) from the distributions of the 300K simulation and the wild type 500K simulations (around 0.9 nm). The latter however have a small population of conformations where this distance ranges between 1 and 2.2 nm. On the contrary we observe that I56T mutant conformations have in most of the time shorter distance between Thr56 and helix 3_{10}. A comparison of the means and standard deviations of the plots are shown in Table 3. The mean and the standard deviations of the distributions of the unfolding simulations have been compared using standard statistical methods (t-test and F-test) and found that their difference is significant (i.e it does not stem from random data differences).

Table 3 *Mean and standard deviation of the distance distribution (in nm) of Ile56-Helix B and Ile56-Helix 3_{10} contacts.*

Simulation	Ile56 – HelixB (nm)		Ile56 – Helix 3_{10} (nm)	
	WT	I56T	WT	I56T
Crystal	0.90	0.91	0.99	1.01
Control	0.91 ± 0.04	0.83 ± 0.04	0.94 ± 0.05	0.99 ± 0.04
Unfolding	0.89 ± 0.35	1.23 ± 0.26	1.34 ± 0.42	0.82 ± 0.28

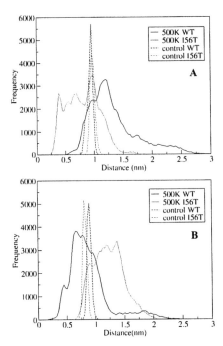

Figure 8   *Distribution of the distances between (A) residue 56 (Ile) to helix 3_{10} and (B) between residue 56 (Ile) and Helix B in the sampled conformations of the unfolding and control trajectories. The distances at the native structure are shown as black (wild type) and red (mutant) rectangles just above the x-axis.*

The tendency of Thr56 to approach helix 3_{10} (residues 81-85) might be related to some favourable hydrogen bonding occurring in the area. In fact from the contact maps of figure 7D and also from the distribution of residue contacts in figure 5, it is clear that Thr56 has a high tendency to be in contact distance (less than 8Å) with residues 80 and 82. These two are serines which implies that hydrogen bonding may be involved between them and Thr56 thus giving an explanation for the persistence of the latter to migrate towards helix 3_{10}.

3.3 Comparison with experimental data and other simulations

Comparisons with experimental data can be made by consideration of the RMSD and residue contacts against time. Thus, we observe that at 500K the RMSD of the mutant increases at a higher rate than the RMSD of the wild type after 1000ps. These results are in agreement with data from circular dichroism[7] and stopped-flow fluorescence experiments,[35] which show that both natural mutant forms of lysozyme are less thermostable than the wild type.

In high temperature (498K) unfolding MD simulations carried out in hen egg-white lysozyme,[18] the partially folded intermediates identified are showing increased non-native structure. This is consistent with our findings for the most populated clusters of conformations sampled during the high temperature simulations whereby non-native contacts are increased across the polypeptide chain. Another common feature is that the radius of gyration of the intermediates increases slightly (~10%) in comparison to the 300K simulations. These intermediate states show a slight change in non-polar solvent accessible surface area (SASA) in comparison to the crystal structure and again this is consistent with our findings

During induced unfolding at 500K, both wild type and mutant lysozyme demonstrate increasing loss of their original secondary structure. The results from the analysis of secondary structure content against time show that most of the original secondary structure elements are lost within the first 2.5 ns. An interesting feature is that the beta domain is destabilised first and then the alpha domain follows. This is in agreement with previous unfolding simulations of hen egg-white lysozyme at 500K.[18] Also data from refolding experiments (using stopped-flow amide hydrogen exchange and mass spectroscopy by Miranker et al.[36,37] show that about 80% of the refolding molecules have their amide hydrogen atoms in the α-domain protected before those in the β-domain. Pepys et al.[3] also showed that the beta domain is more unstable than the alpha domain.

The destabilised secondary structure in both wild type and mutant lysozyme forms gives rise to random transient β-turns across the whole polypeptide chain. β- turns do not generally constitute stable structural elements[38] which explains their transient nature in our simulations. However based on both experimental evidence[39,40] and also lattice simulations,[41] it is argued that β-turns may play a role in the initial stages of the formation of helices and sheets during folding and thus they appear to direct folding pathways while tending to adopt conformations that minimise the "local" conformational free energy of the residues in the turn.[42] In our results (Table 1) we observe that the sampled conformations of the mutant, contain slightly higher percentage of turns than the WT in most of the regions. Using Fourier Transform Infra-Red spectroscopy data[7] propose that the partly folded forms of the lysozyme variants associate through the unstable beta domain to form the initial seed for the generation of amyloids. From Table 1 we observe that for this beta domain (strand 1, strand 2 and loop) the partially unfolded intermediates of the mutant have higher beta turn content, which, may be an indication of the relative instability in this region in the mutant lysozyme compared to the wild type.

Booth and collaborators[7] suggest that the motion in the β-domain due to the mutation may not be the direct cause for amyloidogenicity. Rather, changes in the interface region of the two domains, transmitted from the disturbed β-sheet may be the actual cause. In particular Ile56, found to have increased *B*-factors for both amyloidogenic mutants of human lysozyme, could play an important role in the mutant's amyloidogenic properties. In our unfolding simulations, the fluctuation for this residue is higher in the variant and thus in agreement with these experimental results.

4 CONCLUSIONS

Our results show that the unfolding of the mutant appears to occur faster than the wild type and we observe that it takes longer for the wild type to lose its native structure, under our artificial higher temperature conditions. In addition to that, in the conformational space sampled, it seems that the wild type retains higher content of its original secondary structure in most of the regions relatively to the mutant. Both wild type and mutant simulations include conversions of the original secondary structure conformations to coils and beta turns. These β-turns, however, seem to be more predominant in the beta domain of the mutant, a region which has been suggested to be involved in fibrilogenesis.[3,7] Also in the beta domain we observe higher atomic fluctuations in I56T than the wild type implying that this region is more flexible and thus reinforcing the previous data. However, Thr56 appears to be more rigid than Ile56 with the former tending towards helix 3_{10}. Turns, are thought to occur in abundance prior to formation of helices and sheets during folding[41] and thus the overall higher content in turns in the sampled unfolding conformational area of the mutant may be an indication that it unfolds faster than the wild type.

The clustering analysis reinforces the points mentioned above. The most populated clusters of the wild type retain near half of their native contacts compared with one third for I56T mutant. The lack of retention of native contacts is a useful measure of the degree of unfolding. So, it appears that most of the conformations of the mutant are further away from their native structure than those of the wild type and thus, the mutant protein is more susceptible to changes under unfolding conditions. Interestingly, the most populated structures of the mutant have their Thr56 near helix 3_{10} unlike the wild type ones. Some favourable hydrogen bonding with Ser80 and Ser82 may be related to this observation. The restrain due to this hydrogen bonding to the motion of Thr56, a pivotal residue in the interface of the two lysozyme domains, could result in destabilisation of the structure of the molecule that has been suggested is related to amyloidogenesis.[6,7]

In our unfolding simulations we attempt to identify features that distinguish the partially unfolded conformations of the WT and I56T human lysozyme. Although there is a vast diversity in the conformational features sampled for both the forms, the different distances of the Ile56 and two regions of the alpha domain seems to differentiate a number of conformations of the mutant from the wild type. The significance of this finding cannot directly be linked at this time to amyloidogenesis. However, it does demonstrate clearly that one of the effects of the mutation at residue 56 is the resulting distortion of the important region at the interface of the two domains.

Acknowledgements

This work has been carried out within the BBSRC sponsored Bloomsbury Centre for Structural Biology. We thank the BBSRC for computer hardware and the EPSRC for a studentship to GM.

References

1   D.M. Chipman and N.Sharon, *Science*, 1969, **165**, 454.

2.   P.J. Artymiuk and C.C.F. Blake, *J. Mol. Biol.*, 1981, **152**, 737.

3.   M.B. Pepys, P.N. Hawkins, D.R. Booth, D.M. Viguishin, G.A. Tennent, A.K. Soutar, N. Totty, O. Nguyen, C.C.F. Blake, C.J. Terry, T.G. Feest, A.M. Zalin and J.J. Hsuan, *Nature*, 1993, **206**, 553.

4.   M.B. Pepys, in *Amyloidosis: The Oxford Textbook of Medicine*, 3rd Edn., ed. D.J. Weatherall, J.G.G. Ledingham and D.A. Warell, Oxford University Press, Oxford, 1996, vol. 2, p. 1512.

5.   J.W. Kelly, *Curr. Opin. Struct. Biol.*, 1998, **8**, 101.

6.   J. Funahashi, K. Takano, K. Ogasahara, Y. Yamagata and K. Yutani, *J. Biochem.*, 1996, **120**, 1216.

7.   D.R. Booth, M. Sunde, V. Bellotti, C.V. Robinson, W.L. Hutchinson, P.E. Fraser, P/N Hawkins, C.M. Dobson, S.E. Radford, C.C.F. Blake, and M.B. Pepys, *Nature*, 1997, **385**, 787.

8.   L.A. Morozova-Roche, J. Zurdo, A. Spencer, W. Noppe, V. Receveur, D.B. Archer, M. Joniau and C.M. Dobson, *Journal of Structural Biology*, 2000, **130**, 339.

9.   C.L. Brooks III, *Curr. Op. Struct. Biol.*, 1998, **8**, 222.

10. A. Li, V. Daggett, *Proc. Natl. Acad Sci. USA.*, 1994, **91**, 10430.

11. V. Daggett, *Curr. Op. Struct. Biol.*, 2000, **10**, 160.

12. A. Li, and V. Daggett, *J. Mol. Biol.*, 1998, **275**, 677.

13. D.O.V. Alonso and V. Daggett, *Biophysics*, 2000, **97**, 133.
14. M. Karplus and A. Sali, *Curr. Opin. Struct. Biol.* 1995, **5**, 58.
15. A.E. Mark and W.F. van Gunsteren, *Biochemistry*, 1992, **31**, 7745.
16. P.H. Hunenberger, A.E. Mark and W.F. van Gunsteren, *Nature: Struct. Funct. Genet.*, 1995, **21**, 196.
17. M.A. Williams, J.M. Thornton and J.M. Goodfellow, *Protein Eng.*, 1997, **10**, 895.
18. S.L. Kazmirski and V. Daggett, *J. Mol. Biol.*, 1998, **284**, 793.
19. B. Gilquin, C. Guilbert and D. Perahia, *Proteins*, 2000, **41**, 58.
20. H.J.C. Berendsen, J.P.M. Postma, W.F. van Gunsteren and J. Hermans, in *Intermolecular Forces*. Dordrecht, Reidel. 1981.
21. M.P. Allen and D.J. Tildesley, in *Computer Simulations of Liquids*, Clarendon, Oxford, 1989.
22. H.J.C. Berendsen, D. van der Spoel and R. van Drunen, *Comp. Phys. Comm.*, 1995, **91**, 43.
23. W.F. van Gunsteren, S.R. Billeter, A.A. Eising, P.H. Hünenberger, P. Krüger, A.E. Mark, W.R.O. Scott I.G. Tironi in *Biomolecular Simulation: The GROMOS96 manual and user guide*, Hochschulverlag AG an der ETH Zurich, 1996.
24. T. Darden, D. York and L. Pedersen, *J. Chem. Phys.*, 1993, **98**, 10089.
25. B. Hess, H. Bekker, H.J.C. Berendsen and J.G.E.M. Fraaije, *J. Comp. Chem.*, 1997, **18**, 1463.
26. H.J.C. Berendsen, P.M. Postmaj, W.F. van Gusteren, A. Di Nola and J.R. Haak, *J. Chem .Phys.* 1984, **81**, 3684.
27. W. Kabsch and C. Sander, *Biopolymers*, 1983, **22**, 2577.
28. P.C. Mahalanobis, *Journal of the Asiatic Society of Benagal*, 1983, **26**, 541.
29. S.L., Kazmirski, A. Li and V. Daggett, *J. Mol. Biol.*,1999, **290**, 283.
30. M. Otto, in *Chemometrics: statistics and computer application in analytical chemistry*, Weinheim, Cambridge, 1998.
31. B.S. Everitt, in *Cluster Analysis*, London, E. Arnold New York Halsted, 1995.
32. H. Späth,. in *Cluster analysis algorithms for data reduction and classification of objects*, Chichester Ellis Horwood New York Chichester, 1980.
33. W.N. Venables and B.D. Ripley, in *Modern Applied Statistics with S-PLUS.* 3rd Edn., Springer, 1999.
34. G. A. Worth, F. Nardi and R.C. Wade, *J. of Phys. Chem. B*, 1998, **102**, 6260.
35. D. Canet, , M. Sunde, A.M. Last, A. Miranker, A. Spencer, C.V. Robinson and C.M. Dobson, *Biochemistry*,1999, **38**, 6419.
36. A. Miranker, S.E. Radford, M. Karplus and C.M. Dobson, *Nature*,1991, **349**, 633.
37. A. Miranker, C.V. Robinson, S.E. Radford, R.T. Alpin and C.M. Dobson, *Science*, 1993, **262**, 896.
38. D.J. Tobias, S.F. Sneddon and C.L.Brooks III, *J. Mol. Biol.*, 1990, **216**, 783.
39. S.S. Zimmerman and H.A. Scheraga, *Proc. Natl Acad. Sci. USA.*, 1977, **74**, 4126.
40. H.J. Dyson, J.R. Sayre, G. Merutka, H.C. Shin, R.A. Lerner and P.E. Wright, *J. Mol. Biol.*, 1992, **226**, 819.
41. J. Skolnick and A. Kolinski, *J. Mol. Biol.*, 1991, **221**, 499.
42. A. Yang, B. Hitz and B. Honig, *J. Mol. Biol.*, 1996, **259**, 873.

COLLECTIVE EXCITATION DYNAMICS IN MOLECULAR AGGREGATES: EXCITON RELAXATION, SELF-TRAPPING AND POLARON FORMATION.

M. Dahlbom, W. Beenken, V. Sundström, T. Pullerits

Department of Chemical Physics, Lund University, Box 124, S-221 00 Lund, Sweden

1 INTRODUCTION

The excited state properties and dynamics in various types of condensed systems are of great interest to the community of chemical physics. In particular, a diverse range of molecular systems where collective excited states occur has in recent time been under intense study due to their cooperative electronic behavior. Systems, such as conjugated polymers, fluorescent proteins and biological light-harvesting systems have attracted a great interest since ultra-short laser pulses became available and a large number of studies have been performed both theoretically and experimentally, see review[1] and references therein.

In these types of systems, the inter-molecular interactions can lead to delocalization of excited states, denoted exciton states (excitons). Molecular excitons were introduced by Frenkel[2] early in the last century and further studied by Davydov[3]. The properties and ultra-fast dynamics of cooperative excited state behavior in molecular aggregates have been studied extensively over the last decade. For example, a number of studies have utilized different techniques in order to characterize the exciton delocalization in bacterial photosynthetic antennas[4,5]. It has been suggested that an optical excitation in the tightly coupled peripheral antenna of purple bacteria spans a considerable size of the aggregate and relaxes within a few hundreds of femtoseconds to an equilibrium size of just a few chromophores[5]. Delocalized excitations can also be trapped at lattice sites due to the displacement of the excited state potential and the localization of the wave function as a result of the excitation-phonon coupling. Peierls[6] and Frenkel[7] introduced the possibility of exciton self-trapping as early as in the 1930's. Excitons that interact strongly with the nuclear degrees of freedom are denoted polarons. Due to this coupling they distort their surrounding to a considerable degree. Polaron formation has, for example, been suggested to occur in photosynthetic antenna aggregates at low temperatures [8]. Polaron with their sensitivity to nuclear motions can also be seen as possible non-destructive spectroscopic probes of protein and/or membrane dynamics.

The aim of this paper is to study the collective excitations coupled to the molecular vibrations in the real-space representation. We present a model that treats the molecular vibrations classically via the Langevin equation whereas the electronic system is treated quantum mechanically[9]. The excitation dynamics can henceforth be studied as a function of the molecular vibrations. It will be shown that in the presented model, following an optical excitation, the excitons are trapped at low energy sites (self-trapping) due to the population enhancement of the displacement factor introduced via a mean field correction. The degree of excitation migration and self-trapping depends on temperature, exciton-phonon couplings (Huang-Rhys factors) and static distribution of site energies[10,11]. Special interest is directed towards the real-space motion of the excitation on the femtosecond time-scale and polaron formation. In order to distinguish between the two processes, we studied the wave function mobility and localization as a function of the vibrational frequency, displacement, temperature and intermolecular interaction strength. Besides the general considerations of real-space excitation dynamics in molecular aggregates on the femtosecond time-scale there is also the question of large scale energy funnelling in photosynthetic light-harvesting pigment-protein systems. In order to make the transfer between the B850 aggregate and the LH1 antenna complex more efficient in purple bacteria, the excitation in the B850 ring should be localized at the junction where the distance to the LH1 is the closest. Polivka et. al.[8] has attributed the appearance of a red-shifted feature in the transient absorption spectra at cryogenic temperatures to polaron formation. They propose that the polaron may be formed in inter-ring connection points. We will not address this problem specifically in the current work, but rather address the global aspects of the polaron formation as a viable process in photosynthetic light-harvesting systems.

2 THEORY

According to the Born-Oppenheimer approximation we separate the electronic degrees of freedom from the nuclear motion. The electronic system will be treated quantum mechanically in the molecular exciton picture[2,3], while for the nuclear motion a classical approach will be used. Furthermore, the nuclear system will be separated into explicit and bath modes. The explicit modes are assumed to be coupled the electronic system via the adiabatic potential energy surface. To take into account the nonadiabatic coupling of the electronic system to the explicit modes, we use Tully's surface-hopping method[9]. The other modes will be described by use a Caldeira-Leggett[10] type of bath resulting in statistically fluctuating external forces driving and damping the explicit modes.

2.1 Frenkel Excitons

According to the Born-Oppenheimer approximation[12] the molecular electronic Hamiltonian is given by

$$H_{el} = \sum_i \frac{\hbar^2}{2m_i} \nabla_i^2 - \sum_{i,n} \frac{e^2 Z_n}{|r_i - R_n|} + \sum_{i,i\neq j} \frac{e^2}{|r_i - r'_j|}, \tag{1}$$

where r_i and r'_j assign electronic coordinates and R_n the nuclear ones with index n denotes the nuclear coordinate. $\hbar^2 \nabla_i^2$ represents the electron momentum and m_i the corresponding mass. The nuclear charges are denoted as Z_n. In what follows, it has been assumed that the multi-electron problem is solved for the single molecule to obtain the single molecule energies as h_j, and the excitation creation and annihilation operators $B_j^\dagger$ and B_j, respectively, where j denotes the molecular site[13]. The creation and annihilation operators are given by

$$B_j^\dagger = \sum_{a>b} b_{jab} a_a^\dagger a_b \,, \tag{2}$$

where $a_a^\dagger$ and a_a are the creation and annihilation operators for electrons in the molecular orbitals $|a\rangle$ and $|b\rangle$, and b_{jab} are the corresponding expansion coefficients. Due to the adiabatic approach the single molecule energies h_j as well as the operators $B_j^\dagger$ and B_j depend parametrically on the nuclear coordinates R_n. The electronic Hamiltonian can be written in the excitonic representation as

$$H_{el}(R_n) = \sum_j h_j(R_n) B_j^\dagger B_j + \sum_{j,i \neq j} J_{ji}(R_n) B_j^\dagger B_i \,, \tag{3}$$

where J_{ji} describes the excitation transfer matrix elements between different molecules (depending on nuclear coordinates R_n as well). For sake of simplicity we will restrict our consideration to molecules that are well described as two level systems. Introducing the molecular ground and excited states as $|\phi_j^{(g)}\rangle$ and $|\phi_j^{(e)}\rangle$, respectively, we can expand the exciton eigenstates as

$$|\alpha\rangle = \sum_j c_{\alpha j}(R_n) |\phi_j^{(e)}\rangle \prod_{i \neq j} |\phi_i^{(g)}\rangle. \tag{4}$$

These states can be obtained by solving the eigenvalue equation

$$\langle \alpha | H_{el}(R_n) | \beta \rangle = U_\alpha(R_n) \delta_{\alpha\beta} \tag{5}$$

The eigenstates $|\alpha\rangle$, as well as the expansion coefficients $c_{\alpha j}(R_n) = \langle \alpha | \phi_j^{(e)} \rangle$, are dependent on the nuclear coordinates R_n. The eigenvalues $U_\alpha(R_n)$ yield the adiabatic potential energy surfaces for the nuclear motions as

$$U_\alpha(R_n) = \sum_j \left(h_j^{(g)}(R_n)(1 - c_{\alpha j}^2(R_n)) + h_j^{(e)}(R_n) c_{\alpha j}^2(R_n) + \sum_{i \neq j} J_{ij}(R_n) c_{\alpha i}(R_n) c_{\alpha j}(R_n) \right), \tag{6}$$

depending on the nuclear coordinates R_n via $h_j^{(g)}(R_n)$, $h_j^{(e)}(R_n)$, $J_{ij}(R_n)$ and $c_{\alpha j}(R_n)$.

After transformation of the nuclear coordinates R_n to molecular normal coordinates $q_{\mu,j}$ as performed in the next section, for an aggregate consisting of two-level system molecules one can write down the diagonal matrix elements of the general electronic Hamiltonian in the site representation as

$$H_{jj}^{(el)}(t) = E_j - \sum_{\mu}^{M} \omega_{\mu,j} d_{\mu,j} q_{\mu,j}. \tag{7}$$

Here μ denotes the M different explicit modes with frequency $\omega_{\mu,j}$ belonging to the j'th molecule. $d_{\mu,j}$ is the displacement between the minima of the excited and ground state potential energy surfaces, and E_j the energy gap between excited and ground state for $q_{\mu,j} = 0$, means minimum of the ground state potential surface. This Hamiltonian is taken in the Frank-Condon approximation and both, the ground and excited state potential energy

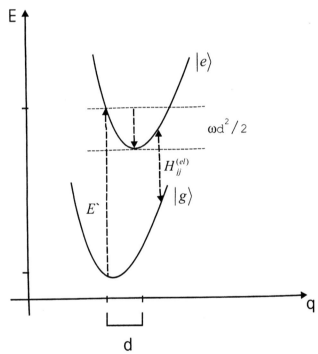

Figure 1. The two level schemes used for the individual molecules where the energy of the direct Frank-Condon transition from the ground state is E' and the energy gap is $H_{jj}^{(el)}$.

surfaces are taken to have same slope means frequencies $\omega_{\mu,j}$.

In the present work, we restrict ourselves to one explicit mode per molecular site with a unique frequency ω and displacement d. If also the energy gap is site independent, Eq. (7) simplifies to

$$H_{jj}^{(el)}(t) = E + \hbar\omega d\, q_j. \tag{8}$$

2.2 Explicit Modes

In the gas-phase the molecular nuclear system can be described by three kinds of motion: translation, rotation and vibration. In the condensed phase the surrounding matrix hinders the first two. Nevertheless, it is useful to distinguish between inter-molecular motion, which means that the whole molecule changes its position relative to the neighbors; and intra-molecular motion, which means that the nuclei of a single molecule change their position and orientation in respect to each other. Inter-molecular motion will mainly affect the interaction matrix $J_{ij}(R_n)$, while intra-molecular motion has also an effect on the site energies $h_j^{(g)}(R_n)$ and $h_j^{(e)}(R_n)$. In what follows, we will neglect the inter-molecular motion and focus on the intra-molecular motion. Expanding the ground state potential energy surface $h_j^{(g)}(R_n)$ of site j around its minimum at $R_n^{(j)}$ one can write

$$h_j^{(g)}(R_n) = h_j^{(g)}(R_n^{(j)}) + \sum_{n,m} \frac{\partial^2 h_j^{(g)}}{\partial R_n \partial R_m}(R_n - R_n^{(j)})(R_m - R_m^{(j)}) + \dots . \tag{9}$$

By transformation to principal axis we obtain a multidimensional harmonic oscillator potential in the normal mode representation, given by

$$h_j^{(g)}(q_{j\xi}) = h_j^{(g)}(0) + \sum_\xi \hbar\omega_{j\xi}q_{j\xi}^2 \tag{10}$$

using dimensionless coordinates $q_{j\xi}$. The excited state potential $h_j^{(e)}(R_n)$ of site j can be expanded using the same coordinates as

$$h_j^{(e)}(q_{j\xi}) = h_j^{(e)}(0) + \frac{1}{2}\sum_{\xi,\xi'} \hbar\omega_{j\xi'}W_{\xi\xi'}(q_{j\xi} - d_{j\xi})(q_{j\xi'} - d_{j\xi'}), \tag{11}$$

where $W_{\xi\xi'}$ take into account a different curvature and $d_{j\xi}$ a different minimum of the potential for the excited state. For sake of simplicity, the curvature of excited as well as ground state potential surface shall be unique too, i.e. $W_{\xi\xi'} = \delta_{\xi\xi'}$. Now one can deduce from Eqs.(6), (10) and (11) the excitonic potential surface as

$$U_\alpha(...q_j...) = E + \frac{1}{2}\sum_j \hbar\omega_j q_j^2 - \sum_j \hbar\omega_j q_j d_j c_{\alpha j}^2 + \sum_{i,j} J_{ij} c_{\alpha i} c_{\alpha j}, \tag{12}$$

One has to note that the $c_{\alpha j}$ still depends parametrically on q_j. The potential surface given in Eq. (12) will be used to determine the intrinsic forces for the nuclear motion with respect to the site mode j as

$$F_{\alpha j}(...q_j...) = -\frac{1}{\hbar}\left\langle \frac{\partial U_\alpha(...q_j...)}{\partial q_j} \right\rangle = -\omega_j(q_j - d_j c_{\alpha j}^2) \tag{13}$$

Thus Eq. (13) leads us to Hook's force law. This is the core of our theoretical model.

2.3 Bath Modes

The complete Hamiltonian for the nuclear motion on the potential surface $U_\alpha(...q_j...)$ is given by

$$\langle \alpha | H_{nuc} | \alpha \rangle = \frac{1}{2}\sum_j \hbar\omega_j p_j^2 + U_\alpha(...q_j...) + \frac{1}{2}\sum_\xi \hbar\omega_\xi(p_\xi^2 + q_\xi^2) + \sum_{j,\xi} K_{j\xi} q_j q_\xi. \tag{14}$$

Here the index ξ assigns the bath modes, while the index j assigns the explicit site modes as above. Following Caldeira-Leggett[10], the bath modes are assumed to be harmonic oscillators and the coupling between bath and explicit modes to be bilinear in the coordinates q_j and q_ξ with coupling parameter $K_{j\xi}$. The dimensionless momentum $p_{j,\xi}$ are related to the coordinates $q_{j,\xi}$ by the canonical equations

$$\dot{q}_{j,\xi} = \frac{1}{\hbar}\left\langle \frac{\partial H_{nuc}}{\partial p_{j,\xi}} \right\rangle = \omega_{j,\xi} p_{j,\xi} \tag{15}$$

$$\dot{p}_{j,\xi} = -\frac{1}{\hbar}\left\langle \frac{\partial H_{nuc}}{\partial q_{j,\xi}} \right\rangle. \tag{16}$$

Resolving these canonical equations for the explicit mode and treating the bath modes statistically (see ref.10 for details), Eq. (16) results in the equation of dissipative nuclear motion

$$\dot{p}_j + \int_{-\infty}^t \gamma_j(t-t')p_j(t')dt' = F_{\alpha j}(...q_j...) + f_j(t), \tag{17}$$

where the intrinsic forces $F_{\alpha j}(...q_j...)$ are given by Eq. **(13)**. The stochastical bath forces $f_j(t)$ are connected to the dissipation, which is given by $\gamma_j(\tau)$, according to the dissipation-fluctuation theorem as

$$\langle f_j(t) f_j(t-\tau) \rangle = \gamma_j(\tau) \langle p_j^2 \rangle \tag{18}$$

For a thermalized bath the momentum variance $\langle p_j^2 \rangle$ is given by

$$\langle p_j^2 \rangle = \frac{1}{\exp[\hbar \omega_j / k_B T] - 1} + \frac{1}{2} \tag{19}$$

If one neglects the memory of the bath, i.e. invokes the Markov approximation, Eq.(17) reduces to the Langevin equation[12]

$$\dot{p}_j(t) + \gamma_j p_j(t) = F_{\alpha j}(...q_j...) + f_j(t) , \tag{20}$$

where the integral $\int_{-\infty}^{t} \gamma_j(t-t') p_j(t') dt'$ is substituted by the damping term $\gamma_j p_j$. The dissipation-fluctuation theorem time-averaged gives the variance of external forces $f_j(t)$ as

$$\langle f_j^2 \rangle = \frac{2\gamma_j}{\tau^*} \left(\frac{1}{\exp[\hbar \omega_j / k_B T]} + \frac{1}{2} \right), \tag{21}$$

where τ^* is the correlation time of the fluctuations[1].

3 NONADIABATICITY AND THE SURFACE HOPPING METHOD

In general, the time-dependent one-exciton wave function $|\Psi(t)\rangle$ can be expanded in the adiabatic basis set $|\alpha\rangle$ and propagated by

$$|\Psi(t)\rangle = \sum_{\alpha} C_\alpha(t) \exp\left[-\frac{i}{\hbar} \int_0^t U_\alpha(\tau) d\tau \right] |\alpha\rangle \tag{22}$$

---

[1] For the numerical calculations we will use random $f_j(t_n)$ and set τ^* equal to the time-step $\Delta t = t_n - t_{n-1}$ of the propagation. This is justified because the value of $f_j(t)$ will not change within the time interval $[t_n, t_n + \Delta t[$, which means that $f_j(t)$ is fully correlated.

where $C_\alpha(t)$ are time-dependent expansion coefficients. One has to note that the eigen-functions $|\alpha\rangle$ as well as the eigen-energies $U_\alpha(t)$ are time-dependent due to the parametric dependency of the eigen-value problem Eq.(5) on the nuclear coordinates. These, now represented by normal coordinates q_j are propagating in time according Eq.(20). From Eqs.(4) and (15) one can deduce the identity

$$\hbar\omega\sum_j\langle\alpha|\frac{\partial}{\partial q_j}|\beta\rangle p_j = \sum_j c_{\alpha j}\frac{\partial c_{\beta j}}{\partial t} \tag{23}$$

for the nonadiabatic coupling between two excitonic eigenstates $|\alpha\rangle$ and $|\beta\rangle$. The left side of the equality means that for a finite nuclear motion, i.e. $p_j \neq 0$, the adiabatic excitonic states are mixed by the nonadiabaticity of the electronic Hamiltonian beyond Born-Oppenheimer approximation[12]. On the right side the same coupling is represented by the time evolution of the coefficients $c_{\alpha j}$. The latter can be determined without explicit knowledge of their dependency on the coordinate normal coordinates q_j by diagonalisation of the excitonic Hamiltonian in the site representation, with diagonal elements H_{jj} as given in Eq.(8) and off-diagonal elements as $H_{ij} = J_{ij}$.

3.1 Surface Hopping Probability

Instead of a fully coherent propagation of the exciton, in this work we will apply the surface hopping method introduced by Tully[9]. Therefore, the time-dependent Schrödinger equation for the exciton wave function, given as

$$\frac{\partial}{\partial t}|\Psi\rangle = H(t)|\Psi\rangle, \tag{24}$$

is transformed into a system of differential equations for the coefficients $C_\alpha(t) = \langle\alpha|\Psi\rangle$, yielding as

$$\frac{dC_\alpha(t)}{dt} = -\sum_\beta k_{\alpha\beta}(t)C_\beta(t) \tag{25}$$

using transport coefficients given by

$$k_{\alpha\beta}(t) = \sum_j\left(c_{j\alpha}\frac{\partial c_{j\beta}}{\partial t}\right)\exp\left[i\int_0^t \omega_{\alpha\beta}(\tau)d\tau\right], \tag{26}$$

where $U_\alpha(\tau) - U_\beta(\tau) = \hbar \omega_{\alpha\beta}(\tau)$. Since one can write down the probability to jump from a exciton eigen-state $|\alpha\rangle$ to another $|\beta\rangle$ within a time interval $[t_0, t]$ as

$$P(\alpha \to \beta) = \int_{t_0}^{t} \left[k_{\beta\alpha}(\tau)\rho_{\beta\alpha}(\tau) - k_{\alpha\beta}(\tau)\rho_{\alpha\beta}(\tau) \right] d\tau \tag{27}$$

using the density matrix elements $\rho_{\alpha\beta}(\tau) = C_\alpha^*(\tau)C_\beta(\tau)$, the relation $k_{\alpha\beta}^*(\tau) = -k_{\beta\alpha}(\tau)$, Eq. (25) and where t_0 is the time of the last hop. According to Eqs.(26) and (23) the transport coefficients $k_{\alpha\beta}(\tau)$ are mainly determined by the nonadiabatic coupling. With other words, $P(\alpha \to \beta)$ gives the probability for the nonadiabatic hopping between adiabatic potential surfaces. Between the jumps, the exciton propagation is performed coherently on the adiabatic potential surfaces due to the exponential term in Eq.(26).

3.2 The Surface Hopping Algorithm

In the surface hopping method, two sets of states are simultaneously followed, the reference and the auxiliary states. The latter are propagated using Schrödinger equation as discussed above and used for calculating the surface hopping probabilities. The populations on the auxiliary states are propagated coherently. The populations on the reference states are always 0 or 1; it will only change if the surface hopping occurs. In figure 2 below, the reference states are shown verses time and the thick black solid line corresponds to the populated state at that given point in time. The surface-hopping Algorithm can be defined by following steps:

(i) The system is initialized, as the electronic Hamiltonian at time zero is generated using a random distributed set of coordinates and momenta, $\{q_i(0), p_i(0)\}$, corresponding to the given system temperature. Solving the Schrödinger equation for this configuration one obtains the initial exciton energies and wave functions and the initial population can be created in a predefined manor. Note that at time zero the reference states and the auxiliary states are the same, which is not true for the following time steps.

(ii) The explicit mode oscillators are then propagated a time step Δt using the Langevin equation, Eq.(20), and the new electronic Hamiltonian is generated. The eigen-value problem is solved generating new states and energies. The populations on the reference states are always integer, but the populations on the auxiliary states are propagated according to Eq.(24).

(iii) Based on the auxiliary states the hopping probability for the reference configuration is calculated using Eq.(27). A random number ξ is pulled from a uniform distribution $[0,1]$ and the population is moved from state $|\alpha\rangle$ to $|\beta\rangle$ if the criterion

$$\sum_{\beta' \neq \alpha, \beta'=0}^{\beta-1} P(\alpha \to \beta') < \xi < \sum_{\beta' \neq \alpha, \beta'=0}^{\beta} P(\alpha \to \beta') \tag{28}$$

is fulfilled. Otherwise the decision is taken not to jump. Then one continues with step (ii).

3.2 Localization and Motion of Excitons

As discussed in the Introduction, the states in a molecular aggregate with a sufficiently large intermolecular interaction will be more or less delocalized and one usually denotes them as exciton states. If, however, the excitons couples strongly to the (intra- or inter-) molecular vibrations, localization may occur, this means polaron formation. In our model the polaron formation results from a feedback between excitonic and nuclear Hamiltonian, as follows: Since the exciton states are delocalized to some degree by dipole-dipole interaction, each individual molecular site will contribute only a certain amount to the exciton wave-functions according to Eq.(4). However, the displacement d is weighted with these contributions to give the internal forces in Eq.(13). For occupation of the lowest excitonic state, these forces will drive the molecular site that contributes most to the exciton wave-function to a lower molecular site energy. Consequently, this site will give a still higher contribution to the lowest excitonic state than before. This results in a self-amplifying increase of the exciton localization. Likewise the mobility of the exciton decreases - eventually up to self-trapping - since it becomes harder for the excitation to slip out of the trap to a neighboring site. Nevertheless due to Eq. (20) at finite temperature the external forces from the thermal bath may statistically kick the system out of the self-trapping configuration. On the other hand, for small displacements localized polarons will not be formed. In this case only the center of the delocalized excitonic wave function may fluctuate thermally, due to the kicking bath forces.

To study the localization quantitatively we use the customary inverse participation ratio, defined as

$$L_{ipr}^{-1}(t) = \frac{1}{Z} \sum_{j,\alpha} c_{\alpha j}^4 \rho_{\alpha\alpha}(t) \tag{29}$$

where the $c_{j\alpha}(t)$ are the parametrically time dependent expansion coefficient of the exciton eigen-state $|\alpha\rangle$. The diagoal density matrix element $\rho_{\alpha\alpha}(t)$ is just the population of this eigen-state and results from Eq.(25). Z represents the normalization factor[5]. In order to analyze the mobility of the exciton we begin by introducing the exciton (polaron) position operator $\hat{Q} = \sum_j j B_j^\dagger B_j$, following Mak et. al.[14], to obtain

$$Q(t) = \sum_{\alpha,j} j c_{\alpha j}^2(t) \rho_{\alpha\alpha}(t) . \tag{30}$$

This quantity gives information on which site is the center of gravity of a localized as well as a delocalized excitation. However, it can trace the motion of an exciton only for single

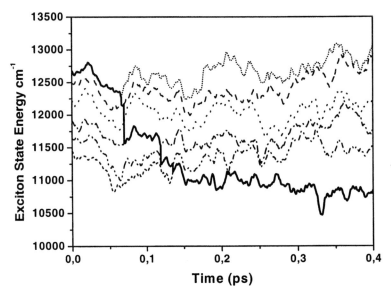

Figure 2. The energies for the electronic states overlaid with the current reference state (thick solid line). The Huang-Rhys factor is 1.125 and the temperature 0K.

Monte-Carlo trajectory. For sampling over a huge amount of trajectories on homogeneous aggregates it yields only a constant averaged value. In order to remedy this situation, we define the velocity of an exciton as

$$I(t) = \left| \frac{dQ(t)}{dt} \right| \tag{31}$$

which will be not canceled by the Monte-Carlo averaging. If the velocity has a high value, the exciton moves rapidly through the system, probably on a random walk. However, if the value of $I(t)$ is small the interpretation is ambiguous: either the exciton is trapped, e.g. by "small" polaron formation, or it is delocalized in such a degree, that its motion does not result in a significant change of the site populations. The second kind of interpretation can be only excluded, if both, $I(t)$ and $L_{ipr}(t)$ are small, means the exciton is trapped and localized on mainly one site.

4 RESULTS AND DISCUSSION

The model system considered here is a linear aggregate of six identical two-level systems with the molecular optical transition at 12400 cm^{-1} and a value of the nearest neighbor intermolecular interaction set to 342 cm^{-1} for all sites. We restrict ourselves to a single explicit intra-molecular uncorrelated vibrational mode per molecular site with a unique

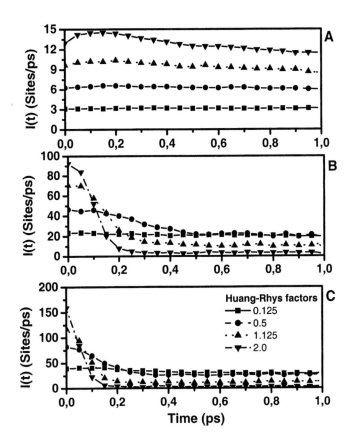

Figure 3. The exciton (polaron) velocity as a function of time, where panel A displays ω_{vib} equal to 250 cm^{-1}, B 684 cm^{-1}, C 900 cm^{-1}. The symbols are displayed in the legend of panel C.

vibrational frequency and displacement d.

The electronic level dynamics is displayed in figure 2. The excitonic states are displayed using different line styles; the thick solid line represents the currently populated reference level. In the displayed case the population has relaxed to the lowest exciton state in a few hundred femtoseconds. On the longer time scale the broadening of the energetic band gap can be seen.

For detailed analyzing of the polaron formation we start to figure out the influence of the displacement d (expressed as Huang-Rhys factors $S = 0.5d^2$) on the localization ($L_{ipr}(t)$) and the exciton velocity ($I(t)$) of the exciton. Figures 3 and 4, where the Huang-Rhys factor is varied between 0.125 and 2.0. shows the strong dependence of both quantities on the mode frequency and displacement (Huang-Rhys factors). The influence of the displacement on I(t) and L_{ipr} is easy to understand since it controls the molecular

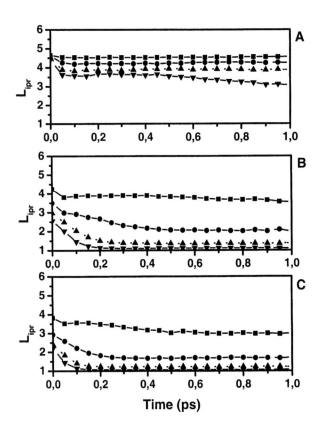

Figure 4. The inverse participation ratio displayed as a function of time. The symbols and parameters are the same as in figure 3.

energy gap according to Eq.(8). Due to the feedback channel in the Langevin equation, the exciton-phonon coupling increasingly influences the localization.

We know from the dimer that the inter-level transfer rates are maximized if the vibrational frequency is tuned to twice the inter-molecular interaction energy. In this case the vibrational frequency matches exactly the energy gap between the levels and a resonance occurs. We tuned the inter-molecular vibrational frequencies from 250 cm^{-1} as shown in Panel A, 642 cm^{-1} Panel B up to 900 cm^{-1} as in Panel C. There is a fast initial rise followed by a monotonic decrease in all kinetics in figure 3, panels A through C. Even though the time-scales change, the dynamics behaves similarly in all three cases, i.e. the velocity decrease as the vibrational frequency increase.

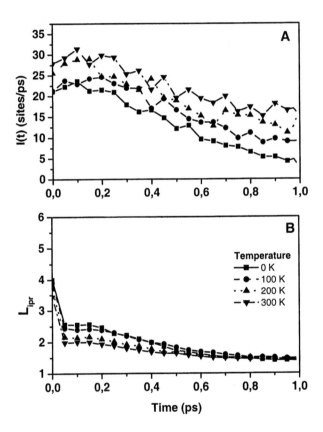

Figure 5. The time-dependence exciton velocity and inverse participation ratio displayed for four different temperatures. Intermolecular coupling V= 100 cm^{-1}, ω=200cm^{-1} and the Huang-Rhys factor is 1.125.

In figure 4, there is a fast initial decrease of L_{ipr} followed by a slower phase present in all the three cases. From the frequency dependence, we conclude that the polaron formation occurs faster at higher vibrational frequencies. Further, it is clear that even for very low temperatures, were the vibrations are almost negligible, polaron formation depends on the ratio of the inter-molecular interaction and the Huang-Rhys factor multiplied with the vibrational frequency. Stated differently, polaron formation does not occur until the

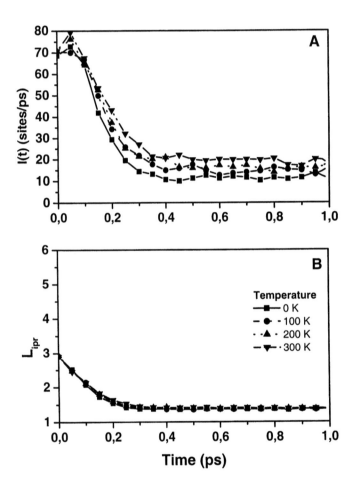

Figure 6. The exciton velocity and inverse participation ratio, with a intermolecular coupling $V= 342$ cm^{-1}, $\omega=684cm^{-1}$ in the hexamer model system, as a function of temperature. The Huang-Rhys factor is 1.125.

coupling between the nuclear and electronic degrees of freedom become sufficiently strong to overcome the delocalization of the intermolecular dipolar coupling.

Next we turn our attention to the temperature dependence for the inverse participation ratio (L_{ipr}) and the exciton velocity $(I(t))$. The important parameter for the temperature dependence is the ratio of the vibrational quantum and the thermal energy ω_{vib}/kT. We performed two sets of simulations; the first with V=100 cm^{-1} and ω_{vib}=200 cm^{-1}. The temperature was varied from 0 to 300 K. The second set was done exactly in the same way, but for V=342 cm^{-1} and ω_{vib}=684 cm^{-1}. Figure 5 displays the first case with panels of both, the exciton velocity (see panel A) and the inverse participation ratio (panel B) and in Figure 6 corresponding data is displayed for the second scenario. In Figure 5, the ratio between vibrational and thermal energy is close to unity, and temperature dependence is clearly visible. The velocity $(I(t))$ at time zero is a factor of 1.5 larger for 300K than for 0K, but after 1ps the velocity at 300K is nearly twice as large as at 0K. In panel B, the inverse participation ratio (L_{ipr}) is displayed as a function of time and temperature. Here the temperature dependence is slightly, but still clearly visible below 400fs, indicating the importance of the ratio $k_B T/\hbar\omega$ mentioned above. In Figure 6, the ratio between the intermolecular vibrational frequency and thermal energy is much larger than unity. There is weak temperature dependence visible in panel A. The final velocity for the different temperatures lies between 10 and 20 sites/ps.

For all temperatures the inverse participation ratio is between 1 and 2 molecules. However, the time dependence of the velocity I(t) and the inverse participation ratio L_{ipr} is qualitatively similar for all temperatures. The apparent difference between the panels in figure 5 and 6 is deceiving. Due to the lower interaction energies and vibrational frequencies only the time scale is shifted. If one takes this into consideration the two panels show very similar time dependence.

Finally we want to turn the attention to the exciton band gap of the molecular aggregate. The polaron formation will involve a lowering of the energy of a specific site due to the feedback channel introduced in Eq.(13), hence causing the lowest exciton energy to decrease and widening the exciton band. The band gap is here defined by the energy difference between the lowest and highest exciton states and is displayed in the two panels in Figure 7. Panel A shows the averaged exciton energies and panel B the time-dependent band gap. In figure 7 two effects can be seen when the exciton-phonon coupling is increased. The increase of the band gap due to the increase of the molecular site fluctuations, i.e. the increase of the slope in panel A and the initial rise in panel B. Secondly, the polaron formation due to the decrease of the lowest exciton state energy so that it energetically deviates from the rest of the equally spaced exciton energy levels. The first effect arises from the displacement dependent Hamiltonian in Eq.(8) which causes the energy fluctuations of the excitonic states to become larger with increasing Huang-Rhys factors. The decrease of the lowest exciton energy is a result of the polaron formation, i.e. the energetic trap is formed by the interaction of the localized exciton and the molecular vibrations on the occupied site. The second effect indicates that the polaron formation process has come into balance with the delocalizing effect of the inter-molecular interactions. Further, it is clear that after a few hundred femtoseconds the band gap stays constant for all Huang-Rhys factors. This is caused by the exciton-vibrational coupling, since an increase of Huang-Rhys factor does not only lower the energy on the populated site but does also increase the amplitude of the molecular energy fluctuation.

To summarize, we have proposed a model, based on the surface-hopping method in the real-space (site) representation, capable of calculating the excitation dynamics in molecular aggregates where the molecular vibrations are incorporated on a real-time basis. We have

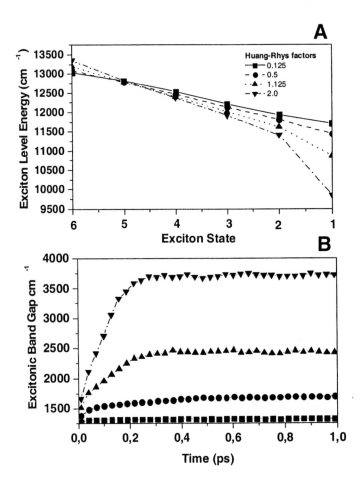

Figure 7. Panel A displayes the exciton energies as a function of state number for four different realizations of the Huang-Rhys factor. Panel B shows the time dependent exciton band gap for the same Huang-Rhys factors as panel A. The parameters are the same as in figure 4.

shown that this model can capture such phenomena as polaron formation and self-trapping. The dependency on temperature, exciton-phonon coupling, phonon frequency and intermolecular interaction strength were investigated.

Reference:

1. V. Sundström; T. Pullerits; R. van Grondelle, *J. Chem. Phys. B* 1999, **103**, 2327.
2. J. I. Frenkel, *Phys. Rev.* 1931, **37**, 17.
3. A. S. Davydov, *Theory of Molecular Excitons*, Plenum Press: New York, 1971.
4. M. Dahlbom; T. Minami; V. Chernyak; T. Pullerits; V. Sundström; S. Mukamel, *J. Chem. Phys. B.* 2000, **104**, 3976.
5. M. Dahlbom; T. Pullerits; S. Mukamel; V. Sundström, *J. Phys. Chem. B.* 2001, **105**, 5515.
6. R. Peierls, *Ann. Phys.* 1932, **13**, 905.
7. J. I. Frenkel, *Phys. Z* 1936, **9**, 158.
8. T. Polivka; T. Pullerits; J.L. Herek; V. Sundström, *J. Chem. Phys. B* 2000, **104**, 1088.
9. J. C. Tully, *J. Chem. Phys.* 1990, **93**, 1061.
10. U. Weiss, *Quantum Dissipative Systems*; World Scientific: Singapore, 1999.
11. S. Mukamel, *Principles of Nonlinear Optical Spectroscopy*; Oxford University Press: New York, 1995.
12. I.B. Bersuker; V.Z., Polinger *Vibronic Interactions in Molecules and Crystals*; Springer Verlag: Berlin, 1998.
13. F. C. Spano, *Phys. Rev. Lett.* 1991, **67**, 3424.
14. C. Mak; R. Egger, *Phys. Rev. E.* 1994, **49**, 1997.

SURPRISING ELECTRO-MAGNETIC PROPERTIES OF CLOSE PACKED ORGANIZED ORGANIC LAYERS- MAGNETIZATION OF CHIRAL MONOLAYERS OF POLYPEPTYDE

Itai Carmeli[1], Viera Skakalova[1], Ron Naaman[1*], Zeev Vager[2]

[1]Department of Chemical Physics
[2]Department of Particle Physics
Weizmann Institute, Rehovot 76100, Israel

1 INTRODUCTION

Close-packed, organized organic layers are the focus of substantial studies in recent years, due to their abilities to modify electronic properties of substrates[1,2], metals or semiconductors[3,4], and to serve as elements in modern optical[5] and electronic devices,[6,7,8,9] light emitting diodes[10], solar cells[11], sensors,[12] etc. Closed packed molecular system are also the main building blocks in biological membranes.

It is usually assumed that the electronic properties of the adsorbed molecules are similar to that of the isolated molecule or of the molecule embedded in an isotropic medium. The weak coupling between the molecules in a monolayer seems to support this notion. This is taken as a justification to use molecular based calculations for predicting the properties of the monolayer.[13,14]

In what follows we present theoretical and experimental results that point to the fact that this assumption is generally not justified and that properties of molecules can vary significantly upon adsorption to a close packed layer. This observation may be of importance in understanding physical properties of natural membranes. In the present work, by studying well-characterized monolayers of polyalanine we are able to obtain an insight on the details of a mechanism that may account for the previously observed magnetic behavior of biological membranes[15].

It is known that adsorbed layers can affect the work function of the substrates by acting as a dipole layer in which a force is exerted on the electrons while passing through it. Hence, the energy required to remove an electron from the substrate (that is, the work function) depends on the layer's dipole density.

When atoms are adsorbed on the surface, the dipole layer arises from either charge transfer between the substrate and the adsorbate layer or an induced polarization of the atom. It has been known for quite some time that for adsorbed atoms the size of the dipole of the layer is limited due to the dipole-dipole interaction between the adsorbed species. Clearly the atomic adsorbates are modified and their electronic structure differs from that of the isolated atoms[16]. The dipole properties result directly from the adsorption process.

The situation is different when molecules are the adsorbates since the free molecules may carry their own dipole moment. A simple picture is that the properties of the dipole layer result solely from the molecular properties. Indeed, several studies[17,18] find correlations between the dipole moment of the isolated molecule and the dipole density of the adsorbed layer. Furthermore, this approach looks as if this is a good way to achieve a very high dipole density. The rationalization is as follows. Strong chemical binding between the adsorbed molecule and the substrate provide enough energy to overcome the dipole-dipole repulsion, keeping negative the total change in the free energy of the adsorption process.

However, it has been recognized already that when molecules are assembled into a dense packed layer, their electronic structure varies. The surface potential theory of thin films on water or metal substrates is probably best described by Taylor and Bays[19]. They used a perturbative approach to deal with the effects caused by the layer formation. Similar approaches have been taken in more recent publications that show that indeed the dielectric constant of an organized molecular monolayer varies with the layer density.[20,21]

The theoretical approach taken by Taylor and Bays[19] deals with an effective molecular dipole M which is modified from the free molecular dipole μ by the self consistent electric field in which the dipoles are immersed. The modification occurs through the Stark effect on the free molecular states and is calculated by first order perturbation theory as

$$\vec{M} = \vec{\mu} + \alpha \vec{E} \qquad (1)$$

where α is the electric molecular polarizability.

A relative permeability (dielectric constant), ε, is defined by

$$\varepsilon = \frac{\mu}{M} \qquad (2)$$

The condition for the first order perturbation to be valid is

$$|\varepsilon - 1| << 1 \qquad (3)$$

It is shown here that for packing of molecules with substantial dipoles into a monolayer, the quantity $|\varepsilon|$, defined by Eq. 2, is always much larger than one. Furthermore, the effective dipoles within monolayers diminish drastically in a non-pertubative manner. Most importantly, as in adsorbed atoms, the electronic states of the adsorbed molecules differ considerably from the states of the free molecule. This electronic rearrangement by adsorption to a monolayer, must be accompanied by a current and plays a substantial role in defining the properties of the adsorbed layer. It alter significantly the electronic properties of the molecule imbedded in the monolayer, as compared to the isolated molecule or a molecule in an isotropic medium. Following this part, it will be shown that when the molecules are homochiral, a splitting between their spin states occurs which may result in a magnetic layer. This is due to the electronic rearrangement and the electric potential on the layer.

2 THEORETICAL CONSIDERATIONS

2.1 Substantial charge rearrangements

We start by following the electrostatic formulation as in ref. 19, except that ε will be treated empirically. Denote by q the effective charge separation in the molecule before adsorption and w the effective distance of the charge separation, such that $\mu = qw$ is the electric dipole of the unadsorbed molecule. Idealize the layer as an electrostatic bilayer with smoothly distributed opposite charges on each side of the adsorbed layer. The displacement vector D of a bilayer is confined to the inner volume of the layer and it is zero elsewhere. Therefore, by Gauss law (esu), it is given by

$$D = \frac{4\pi q}{\sigma} = \varepsilon E \tag{4}$$

where σ is the effective area occupied by a single adsorbed molecule and E is the physical electric field, also confined within the layer. A typical bilayer potential jump of

$$V = wE = \frac{4\pi\mu}{\varepsilon\sigma} \tag{5}$$

occurs along the width w of the layer. This potential jump is compared with a fictitious bilayer of effective electric dipole moments M such that

$$V = \frac{4\pi M}{\sigma} \tag{6}$$

Thus the effective electric dipole moment of an adsorbed molecule is

$$M = \frac{\mu}{\varepsilon} \tag{7}$$

which is the same as Eq. 2 with no reference to first order perturbation theory.
The conversion of Eqs. 4 and 6 to practical units (Debye, Å, and eV) results in

$$E(Volts/\text{Å}) = \frac{38M(Debye)}{v(\text{Å}^3)} \tag{8}$$

$$V(Volts) = \frac{38M(Debye)}{\sigma(\text{Å}^2)} \tag{9}$$

where $v = w\sigma$ is the effective molecular volume.

Already here, the validity of first order pertubation theory for ε is suspect, since for $\varepsilon \approx 1$ (Eq. (3)) unrealistically large electric fields are obtained.
The real obstacle for reaching large electric fields in thin films (as well as bulk material) is the dielectric breakdown. For example, highest breakdown electric field in a very thin film of polypropylene[22] is

$$E_{DB} = 0.06 \, (V/\text{Å}) \tag{10}$$

Using Eq. 8 with this limitation, an upper bound is found for the effective dipole in monolayers

$$M(\text{Debye}) \leq 1.6 \cdot 10^{-3} \, v \, (\text{Å}^3) \tag{11}$$

The upper bound for the dipole moment obtained from this consideration, is not related to the initial electric dipole moment of the molecule. As is pointed out in ref. [23], the breakdown mechanism may be related to the wave-like character of the electrons when laterally confined and therefore related to a universal phenomenon. In general, this upper bound is much smaller than the unadsorbed molecular dipole moment, hence by Eq. 7, $|\varepsilon| \gg 1$. Figure 1 presents the calculated effective dielectric constant and the effective electric field within a layer as a function of the dipole moment of the adsorbed molecules. It is assumed that each molecule occupies an area of 40 Å^2 and that the thickness of the layer is of 20 Å. Since the electric field cannot exceed the breakdown limit, the effective dielectric constant of the layer must increase linearly with the free molecule dipole moment.

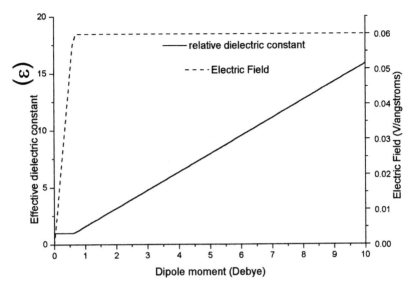

Figure 1: *The relative dielectric constant (solid line) and the electric field within a monolayer (dashed line) as a function of the dipole moment of the molecules, assuming that each molecule occupies an area of 20Å² and its length is 20 Å.*

2.2 Chiral thin films

As a result of the above discussion, it is clear that intensive charge rearrangement occurs within the molecules of the layer. The rearrangement happens with the aid of the surroundings. This surrounding can be a substrate, in the case of a layer adsorbed on solid, or a solution in the case of Langmuir films or natural membranes. A formal way to express charge rearrangement is by strongly mixing the ground state either with excited electronic states of the molecule itself or of the surroundings. This process allows a finite probability that the molecules will be left with unpaired electrons.

In most molecular layers, the electric field is just below the breakdown field (due to previously existing dipole moments). This symmetry breaking huge field cannot split the spin states of the electrons, unless another intrinsic symmetry breaking element exists in the layer. Specifically, consider a state of a homochiral molecular layer and its mirror image. Though the Hamiltonian is invariant under parity transformation, the states are characterized by a handedness c, which changes sign upon a mirror reflection. Thus, the handedness c is an isoscalar quantity. For simplicity, choose a mirror plane parallel to the field E (normal to the surface of the layer). Such a transformation does not change E but does change c. Therefore, one can define another operator $B=cE$ which is an axial vector and can interact with the spin of electrons in the same way as angular momentum does in atoms or as a magnetic field does in a macroscopic system. Different handedness are expressed by $c=1$ or $c=-1$ and their corresponding B points either parallel or anti-parallel to E.

The upper bound for the electric potential, given in Eq. 10, corresponds to a value of B, which is equivalent to a magnetic field of up to 20,000 gauss. This value is larger than the fields found in common ferrites.

From the above considerations one can conclude that when chiral molecules form a close packed layer, due to the charge rearrangement, the layer may gain magnetic properties. This effect may be of relevance in biological membranes where homochiral molecules with relatively large dipole moment are packed together.

3 EXPERIMANTAL

Self-assembled monolayers of either L or D polyalanine polypeptides in the form of α helices were prepared on a gold substrate and the orientation of the molecules was monitored by IR spectroscopy. Monolayers of either L or D polyalanine polypeptides were prepared on glass slides coated with 100 nm thick annealed gold film. By connecting a sulfide group either at the C- or N- terminal of the peptide, the dipole moment of the attached molecules is pointing either away from the substrate or towards the substrate. Three types of films were investigated, two were of poly L-alanine and poly D-alanine, both connected to the surface at the C-terminal (referred to as LC and DC respectively), and one type of poly D-alanine connected to the surface at the N-terminal (referred to as DN).

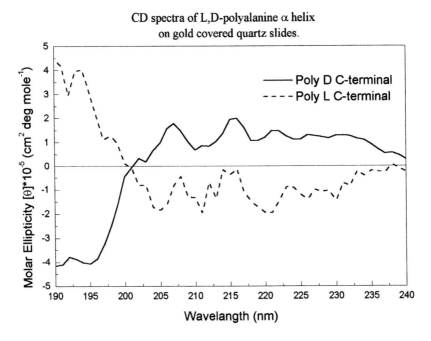

Figure 2: *Circular Dichroism (CD) spectra of monolayers of L and D polyalanine attached through the C-terminal to a 10 nm thick gold film coated quartz slide.*

Polypeptide lengths between sixteen and twenty-two amino acid units were used. The structure of the films was determined by their FTIR spectra[24]. The handedness of the adsorbed films was verified by circular dichroism absorption as shown in Figure 2, and their thickness was measured by ellipsommetry. Characterizations by AFM measurements were carried out as well. All the results reported here were obtained at room temperature from closed packed layers.

IR spectra of LC and DC monolayers were measured at magnetic field strengths of 0, ±900, and ±4500 Gauss applied perpendicular to the layer. Henceforth, "North" indicates magnetic field lines starting at the North Pole and penetrating through the gold surface to the monolayer and vice versa for "South".

For the electron transmission studies, the samples were inserted into an ultrahigh vacuum chamber at $<10^{-8}$ mbar. The polarized photoelectrons are ejected from the substrate by applying a laser beam at 248 nm using a $\lambda/4$ plate to create either left- or right-handed circular polarized light. It is known that right-handed circularly polarized light induces positive helicity[25] in the photoelectrons ejected from the gold substrate and the reverse for the left-handed polarized light. The photoelectrons are known to be polarized by about 15%[26]. After passing through the organic layers, the electrons energy distribution is analyzed using a time-of-flight spectrometer.[27]

4 RESULTS

Figure 3 presents the grazing angle IR spectra obtained for the amide vibrations.[28] The amide I vibration is parallel to the molecular axis (at about 1665 cm^{-1}) while the amide II vibration at about 1550 cm^{-1} is perpendicular to the axis. The spectra are normalized at the peak of the 1665 cm^{-1} line. If the molecules are oriented normal to the surface, the intensity of the amide II component vanishes because of the metal substrate canceling of the transition dipole moment. Hence, the ratio between the intensity of the two peaks provides a direct measure of the tilt angle of the molecules relative to the surface normal.[29]

From Fig. 3 it is evident that there is a magnetic field effect on the relative intensities for both LC and DC layers. For the two LC layers shown (Figs. 3A,3C) South fields increase the average tilt angles while North fields decrease them. The reverse is true for the DC layers (Figs. 3B, 3D); namely, South fields decrease the average tilt angles while North fields increase them. Larger fields show larger tilt deviations. It takes 2 to 6 hours for the effect of the magnetic field on the tilt angle to reach its equilibrium. This time depends on the strength of the magnetic field and on the quality of the monolayer, namely the packing density of the layer and the size of the average tilt angle. Better quality layers require more time for equilibrium in magnetic fields. The above magnetic field orientation effects within the layers provide clear evidence that the films have magnetic properties and that opposite magnetization occur for different handedness. After all, the classic magnetization of iron is interpreted as orientation of domains. Looking on either the LC layer or the DC layer, the magnetic orientation of these artificial monolayers mimic orientational effects already found in biological membranes.[15]

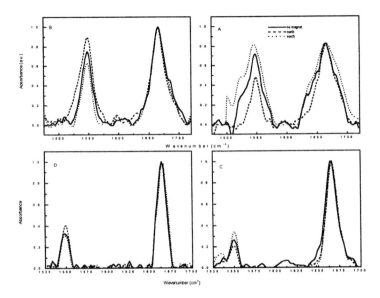

Figure 3: *The IR spectrum of the polyalanine layers of L and D molecules bound to the surface through the carbon terminal. The amide I and amid II bands are shown at about 1665 cm^{-1} and 1550 cm^{-1} respectively. The spectra are normalized at the peak of the 1665 cm^{-1} line. The spectra were taken either with a magnet having its North or South pole pointing towards the film (dashed and dotted lines respectively) or without magnetic field (solid line). The spectra shown in A and B were taken with a magnetic field of 4500 Gauss, while those at C and D were taken with 900 Gauss. Panels A and C correspond to the spectra of L layers while B and D correspond to D layers.*

Figure 4 presents the kinetic energy distributions for photoelectrons ejected with a left or right circular polarized laser (solid and dashed lines respectively). The spectra in panels A and C are obtained for the transmission of electrons through films of L- and D-polyalanine respectively, both bound to the surface through the C-terminal. Panel B corresponds to a film of D-polyalanine bound to the surface through the N-terminal. The results shown in Fig. 4A and 4C confirm earlier results that showed a large asymmetry for polarized electrons transmission through organized film of chiral molecules.[30] The sign of the asymmetry depends on the handedness of the molecules. Surprisingly, for a given handedness of the molecules the sign of the asymmetry switches upon reversing the way the molecules are adsorbed on the surface (from N to C terminated molecules). This is clearly seen in Figs. 4B and 4C. The observed asymmetry[31] in the transmission through the layer of photoelectrons produced by left and right circular polarization of photons changes from 0.09±0.02 to -0.10±0.02 (Fig. 4B and 4C respectively). Importantly, the observed 10% effect is induced by merely 15% polarization of the photoelectrons[26]. Thus, the selectivity to the incoming helicity of the electrons is as large as 70% and within experimental error could be even higher.

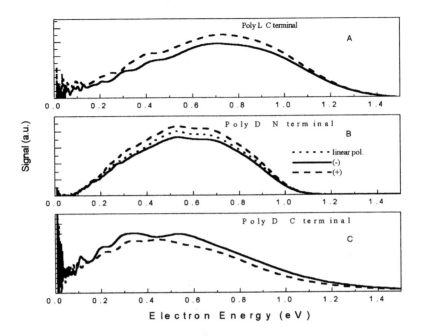

Figure 4: *The energy distribution for photoelectrons ejected with a left (negative spin polarization-red, solid) or right circular (positive spin polarization-blue, dashed) polarized laser. The electrons are transmitted through films of L and D-polyalanine both bound to the surface through the carbon terminal (A and C respectively), and through a film of D-polyalanine bound to the surface through the N-terminal (B).*

5 DISCUSSION

Several questions arise from the above experimental observations-
- Why are the layers magnetic?
- Why is there a relation between the direction of magnetism and the molecules' handedness?
- Why is there a large spin-selectivity for electron transmission, and
- Why does it change sign with both change of handedness and change of the direction of the molecular electric dipole?

Our model is based on the properties of closed packed monolayer of molecules having a dipole moment, as described above. The polyalanine molecules in solutions have a large electric dipole moment of about 50 Debye.[32,33] When the molecules are adsorbed as monolayers, by attaching either C or N terminals to the substrate, they must lose almost all their dipole moment. The simple electrostatic arguments given above indicate that the molecular dipole moment must be reduced by more than an order of magnitude. A direct indication on the value of the adsorbed molecular dipole moment can be obtained from the substrate's work function. From the high-energy cutoff of the photoelectron spectra (Fig. 4) it is evident that upon reversing the direction of the layer from being bound to the surface through the carbon versus being bound through the nitrogen (Figs. 4A, 4C versus 4B), the work function increases by only about 0.3 eV, while for molecules with a dipole moment of about 10 Debye or more the work function would have changed by many volts. This means that the dipole moment of the molecules was reduced by two orders of magnitude due to adsorption.[34] Consequently, the electronic structure of the molecules is substantially modified upon adsorption, probably by charge transfer between the metal substrate and the molecule. Apparently, the charge redistribution process results in unpaired electrons on the adsorbed molecules, namely the adsorbed molecules become paramagnetic.

Still, the above posed questions are not answered. Though a full description of these phenomena does not yet exist, the following simple model is compatible with all the observations and predicts correctly the direction of magnetization. Moreover, since the reported phenomena exhibit parallel effects to some biological membranes, the model may serve as a basis for some understanding of magnetic effect on biological systems.

Upon adsorption, there is a transient current through the helix, which discharges the electric dipole while inducing magnetism, like a classical electro-magnetic coil. The spins of simultaneously created unpaired electrons, adjust accordingly and stay polarized by support of exchange forces with the metal and the already polarized electrons of attached neighbor helices. The result of such an adsorption mechanism is a magnetic layer where the direction of the magnetic field is given by Ampere's law. For each adsorbed helix, the original electric dipole polarity and the handedness of the helix determine the direction of the magnetic field along the helix axis.

As can be seen in Figure 3, the helix axes are not well aligned normal to the surface. Therefore, when an external magnetic field is applied in the normal direction, it forces these axes to get either closer to the normal or away from it, depending on the direction of the external magnetic field. The observed magnetic effects in Figure 3 are consistent with the simple model described above.

The magnetic properties of the film are the reason for the large spin selectivity in the electron transmission. This effect is consistent with the effect in inorganic thin magnetic layers, known as the Giant Magneto Spin selectivity[35,36] and the related colossal magnetoresistance effect.[37,38] Hence, the magnetic properties of the organic films explain their high selectivity for transmission of spin polarized electrons, an effect that results from the cooperative nature of the film on the metal substrate. The seemingly surprising change in the direction of magnetization by the change of the molecular terminals on the gold surface is resolved by noticing the direction of the electric dipole discharge through the helices.

In the present work we find that the adsorption process of chiral species into layers converts electric dipoles into magnetic dipoles. Some biological membranes, which show similar magnetic orientation effect, are now suspected to have similar magnetic properties.

Acknowledgement
We are grateful to Prof. M. Fridkin and his group for helping us in the synthesis of the polyalanine. Partial support from the Israel Science Foundation and the US-Israel Binational Science Foundation is acknowledged.

REFERENCES

1. H. Ishii, K. Sugiyama, E. Ito, K. Seki, *Adv. Mater.* 1999, **11**,605.

2. S.N. Yaliraki, A.E. Roitberg, C. Gonzalez, V. Mujica and M.A. Ratner, *J. Chem. Phys.* 1999, **111**,6997.

3. Vilan, A. Shanzer and D. Cahen, *Nature,* 2000, **404**, 166.

4. I.H. Campbell, J.D. Kress, R.L. Martin, D.L. Smith, N.N. Barashkov, J.P. Ferraris, *Appl. Phys. Lett.* 1997, **71**,3528.

5. T. Verbiest, S. Van Elshocht, M. Kauranen, L. Hellemans, J. Snauwaert, C. Nuckolls, T.J. Katz, A. Persoon, *Science*, 1998, **282**,913.

6. C.P. Collier, E.W. Wong, M. Belohradsky, F.M. Raymo, J.F. Stoddart, P.J. Kuekes, R. S. Williams, J.R. Heath, *Science*, 1999, **285**,391.

7. J. Chen, M.A. Reed, A.M. Rawlett, J.M. Tour, *Science*, 1999, **286**,1550.

8. Vuillaume, B. Chen and R.M. Metzger, *Langmuir,* 1999, **15**,4011.

9. E. Punkka and R.F. Rubner, *J. Elect. Mat.* 1992, **21**,1057.

10. See for example: A. Yamamori, S. Hayashi, T. Koyama, and Y. Taniguchi, *App. Phys. Lett.* **78**, 3343 (2001); B. K. Crone, P. S. Davids, I. H. Campbell, D. L. Smith, *J. Appl. Phys.* 2000, **87**, 1974.

11. See for example: P. Peumans, V. Bulovic, S. R. Forrest, *Appl. Phys. Lett.* 2000, **76**, 2650; U. Bach, D. Lupo, P. Comte, J.E. Moser, F. Weissörtel, J. Salbeck, H. Spreitzer, and M. Grätzel, *Nature*, 1998, **395**,583.

12. D. G. Wu, D. Cahen, P. Graf, R. Naaman, A. Nitzan, D. Shvarts, *Chem. Eur. J.* 2001, **7**,1743.

13. A. Nitzan and I. Benjamin, *Acc. Chem. Res.,* 1999, **32**, 854.

14. G.R. Hutchison, M.A. Ratner, T.J. Marks, R. Naaman, *J. Phys. Chem. B.* 2001, **105**,2881.

15 D.-Ch Neugebauer, A.E. Blaurock and D.L. Worcester, *FEBS Lett.* 1977, **78**,31.

16. See for example: W. Zhao, G. Kerner, and M. Asscher, X. M. Wilde, K. Al-Shamery, and H.-J. Freund, V. Staemmler and M. Wieszbowska, *Phys. Rev. B*, 2000, **62**,7527.

17. M. Bruening, E. Moons, D. Cahen, A. Shanzer, *J. Phys. Chem.* 1995, **99**,8368.

18. S. Liao, Y. Shnidman, and A. Ulman, *J. Am. Chem. Soc.* 2000, **122**,3688.

19. D.M. Taylor and G.F. Bays, *Phys. Rev. E* 1994, **49**,1439.

20. M. Iwamoto, Y. Mitzutani, and A. Sugimura, *Phys. Rev. B*, 1996, **54**,8186.

21. C-X. Wu and M. Iwamoto, *Phys. Rev. B* 1997, **55**,10922.

22. D. Liufu, X. S. Wang, D. M. Tu, K. C. Kao, *J. Appl. Phys.* 1998, **83**, 2209.

23. E. Miranda and J. Sune, *Appl. Phys. Lett.* 2001, **78**,255.

24. In helix peptides, the transition moment of amide I band lies nearly parallel to the helix axis and that of amide II perpendicular. Transition moments, which lie parallel to the gold surface, cannot be detected in grazing angle FTIR. The ratio between amide I band (1665 cm^{-1}) and amide II band (1550 cm^{-1}) indicates to what extend the molecules in the monolayer are oriented perpendicular to the gold surface.

25. The helicity is defined for particles with momentum p and spin s, as the expectation value of $\dfrac{s \cdot p}{|s \cdot p|}$.

26. J. Kirschner, *Polarized Electrons at Surfaces*, (Springer-Verlag, 1985); F. Meier and D. Pescia, *Phys. Rev. Lett.* 1981, **47**, 374-377; F. Meier, G. L. Bona , S. Hufner, *Phys. Rev. Lett.* 1984, **52**, 1152; G. Borstel, M. Wohlecke, *Phys. Rev. B* 1982, **26**, 1148.

27. R. Naaman, A. Haran, A. Nitzan, D. Evans, and M. Galperin, *J. Phys. Chem.* 1998, *B* **102**,3658.

28. *"Peptides Polypeptides and Proteins"* E.R. Blout et. al. eds, John Wiley&Sons 1974, p. 379.

29. Since polyalanine is hydrophobic, it is difficult to clean the sample using chromatography and to obtain a single size polypeptide. Hence, the samples tend to vary in the distribution of the peptides length from batch to batch. This is expressed by the spread in tilt angles of the different layers, which was found to be larger for less uniform batches. However, the results obtained were consistent for all selected samples. Samples were rejected if the average tilt angle exceeded 50^0.

30. K. Ray, S.P. Ananthavel, D.H. Waldeck, R. Naaman, *Science*, 1999, **283**, 814.

31. The asymmetry parameter is defined as $A \equiv \dfrac{I(+P) - I(-P)}{I(+P) + I(-P)}$ where I(+P) and I(-P) are the transmission of the electron beam with spin angular momentum oriented parallel (+) and antiparallel (-) to its velocity vector.

32. W.G.J. Hol, P.T. van Duijnen and H.J.C. Berendsen, *Nature*, 1978, **273**, 443.

33. C. Park and W.A. Goddard III, *J. Phys. Chem. B* 2000, **104**,7784.

34. The change in the workfunction ($\Delta e\Phi$) is related to the dipole of the layer by the equation $\Delta e\Phi = 4\pi D$ where D is the dipole density. We assumed that the layer density was 5×10^{14} molecules/cm^2.

35. E. Vélu et. al. *Phys. Rev. B* 1988, **37**,668.

36. A. Filipe et. al. *Phys. Rev. Lett.* 1998, **80**,2425.

37. R. von Helmholt, et. al. *Phys. Rev. Lett.* 1993, **71**,2331.

38. S. Jin et. al. *Science*, 1994, **264**,413.

BARRIER CROSSING BY A FLEXIBLE LONG CHAIN MOLECULE - THE KINK MECHANISM

K.L. Sebastian

Department of Inorganic and Physical Chemistry, Indian Institute of Science, Bangalore 560012, India

1. INTRODUCTION

The calculation of the rate at which a particle trapped in a metastable state escapes is a very important problem, having applications in several areas of physics and chemistry and biology. Kramers[1] found solutions in the limit of weak friction and also in the limit of moderate to strong damping. A classic reference to the problem is the article by Chandrasekhar.[2] The recent progress on this topic has been surveyed in a detailed review by Hanggi et. al.[3] The reason for this extensive activity is that this forms a model for a chemical reaction occuring in a condensed medium. The Kramers problem for few degrees of freedom has also been the topic of study.[3] Here we consider a natural extension of the problem to a case where the number of degrees of freedom, N is very large ($N \to \infty$). Further, the way these are connected (to form a long chain molecule) leads to interesting new aspects to the problem that are not present in the case where there are only finite number of degrees of freedom (see Fig. 1).

The problem is of importance in biology as quite a few biological processes involve the translocation of a chain molecule from one side of a membrane to the other, through a pore in the membrane. Some examples are:

1. The translocation of proteins from the cytosol into the endoplasmic reticulum, or into mitochondria, or chloroplasts. Often, the proteins are hydrophilic and the pore in the membrane forms a hydrophobic region, through which it has to pass through,[4-7] resulting in an increase in the free energy for the portion of the chain inside the pore

2. In infection by bacteriophages, conjugative DNA transfer etc, long chain DNA molecules snake through pores in membranes.[8,9]

In the above processes, the translocation of the chain molecule seems to occur with ease, contrary to the expectation that one gets from the theoretical analysis available in the literature. In an interesting experiment Kasianowicz et. al.[10] forced long polynucleotides to move through a pore in a membrane and studied the time that it takes the molecule to cross the pore, as a function of the length of the molecule.

They found the time to be proportional to the length of the chain. More recently, Han and Craighead[11] studied the forced motion of very long DNA molecules through microfabricated channels and found that the activation energy for crossing shallow regions in the channel was independent of the chain length.

1.1 The problem and related earlier work

All the above problems involve the passage of a long chain molecule, through a region in space, where the free energy per segment is higher, thus effectively presenting a barrier for the motion of the molecule. This is what we refer to as the Kramers problem for a chain molecule. Muthukumar and Baumgartner[12] studied the movement of self avoiding polymer molecules between periodic cubic cavities separated by bottlenecks, the passage through which presents an entropic barrier to the motion. They find that there is an exponetial slowing down of diffusion with the number of segments N in the chain. Park and Sung[5] have studied the translocation through a pore. They analyze the passage through a pore on a flat membrane, with the effects of entropy included. They refer to the side that the polymer is initially in as the cis side and the other as the trans side. The free energy barrier has the functional form $F \sim -k_B T \ln (n(N - n))/2 + n\Delta\mu$, where n is the number of segments that have crossed from the cis side to the trans side and $\Delta\mu$ is the free energy change per segment in the process. As is obvious, this barrier is rather broad, and its width is proportional to N. Consequently, they

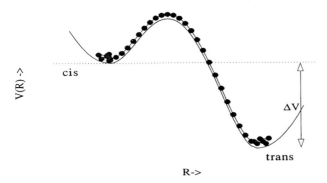

Figure 1:   *The potential energy per segment of the chain, plotted as a function of position*

consider the translocation process as being equivalent to the motion of the center of mass of the molecule. Using the result of the Rouse model that the diffusion coefficient of the center of mass is proportional to $1/N$, they effectively reduce the problem to the barrier crossing of single particle having a diffusion coefficient proportional to $1/N$. As the translocation involves motion of N segments across the pore, the time taken to cross in the case where $\Delta\mu = 0$, t_{cross} scales as N^3. They also show that in cases where there is adsorption on the trans side ($\Delta\mu < 0$), translocation is favored and then t_{cross} scales as N^2. Kumar and Sebastian[13] have analysed the free energy profile for translocation from the outer surface of a vesicle to the inner surface, for the case where the molecule is adsorbed on both the surfaces, pointing out that adsorption on the outer surface can lead to the non-existence of the entropic barrier, thereby facilitating the process. Park and Sung[14] have given a detailed investigation of the Rouse dynamics going over a broad potential barrier. They use multidimensional barrier crossing theory to study the motion of a chain molecule over a barrier, in the limit where the width of the barrier is much larger than the lateral dimension of the molecule. Lubensky and Nelson[15] study a case where the interaction of the segments of the polymer with the pore is strong. In this case, it is the dynamics of the portion that is inside the pore that is important and they show that this can lead to t_{cross} proportional to N. In recent papers, we have suggested[16] a kink mechanism for the motion of the chain, and we give details of this mechanism in this paper. We consider a polymer undergoing activated crossing over a barrier whose width w is larger than the Kuhn length l of the polymer, but small in comparison with the length Nl of the polymer. Thus, we assume $l << w << Nl$. This is the case in the experiments referred to above. For example, in the experiments of Kasianowicz et. al.,[10] the length of the pore is about 100 Å, while the Kuhn length for a single stranded DNA is perhaps around 15 Å.[15] Therefore, one is justified in using a continuum approach to the dynamics of the long chain. (It is possible to retain the discrete approach, and develop the ideas based on them, but this is more involved mathematically).

Our approach is the following: We describe the dynamics of the chain using the continuum version of the Rouse model, discussed in detail in the book by Doi and Edwards.[19] The barrier exerts forces on the segments and inclusion of this in to the Rouse model makes the equation a non-linear one. We refer to this equation as the non-linear Rouse model. The portion of the chain inside the barrier would be distorted in comparison with the portions that are outside and we refer to this distortion as the kink. In the non-linear Rouse model, the distortion is a special solution and as is usual in non-linear physics, we refer to it as a kink. Movement of the chain across the barrier is equivalent to the motion of the kink in the reverse direction. In the presence of a driving force (i.e. $\Delta\mu < 0$) , the kink moves with a finite velocity and hence the polymer would cross the barrier with t_{cross} proportional to N.

1.2 Kinks pinned in space

Traditionally, the usual non-linear models (for example, the ϕ^4 or the sine-Gordon model[20–22]) have potentials that are translationally invariant, and hence the kink can migrate freely in space. For example, the sine-Gordon model has the equation of motion

$-\frac{\partial^2 \phi}{\partial t^2} + \frac{\partial^2 \phi}{\partial x^2} = \sin(\phi)$. Note that the position co-ordinate enters the equation only through the derivative terms, and all points in space are completely equivalent. Kink solutions to this equation may be found by putting $\phi = \phi(x - vt)$, and these represents kinks that are free to move anywhere in space. In comparison, in the barrier crossing problem, the position of the barrier is fixed in space. This breaks the translational symmetry. As a result, the kink is pinned in space. However, the long chain molecule (modelled as a string[19]) can move in space and hence the kink migrates, not in space, but on the chain. This is the obvious and simple mechanism of translocation. Such a suggestion is new for barrier crossing (somewhat similar ideas are used in reptation) and we believe that this is a very useful idea in understanding polymer translocation. In the following, we develop this idea based upon the one dimensional version of the Rouse model.

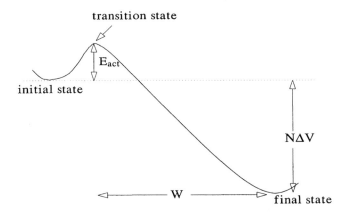

Figure 2: *The potential energy along the reaction co-ordinate. The E_{act} is independent of the length of the chain. After the barrier is crossed, there is a region of width W, with W proportional to N, which is to be crossed. The time required to cross this regions is t_{cross}*

2. THE FREE ENERGY LANDSCAPE

The considerations in this subsection are quite general and do not depend on the model that one uses to describe the polymer dynamics. We assume only that the polymer is flexible over a length scale comparable to the width of the barrier. We start by considering the free energy landscape for the crossing of the barrier. The barrier and the polymer stretched across it are shown in Fig. 1. The polymer has initially all its units on the cis side, where its free energy per segment is taken to be zero. So the initial state has a free energy zero in the free energy hypersurface shown in Fig. 2. In crossing over to the trans side, it has to go over a barrier, as in the Fig. 1. The transition state for the crossing can be easily found, from physical considerations, by remembering that the transition state is a saddle point - i.e. it is a maximum on the free energy surface in one direction, while in all the other directions it is a minimum. The transition state is shown in the Fig. 3. In the transition state, the configuration of the polymer is such that the free energy of the chain is an extremum, subject to the two constraints: (a) the end of the polymer on the trans side is located exactly at the point at which its free energy per segment is zero (b) the other end is on the cis side. This is the transition

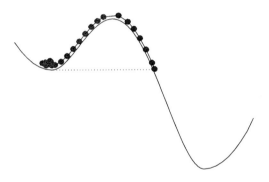

Figure 3 *The transition state*

state, because if one moves the end at the trans side either in the forward or in the backward direction (and the rest of the chain adjusted so that the free energy of the chain as a whole is a minimum), then the total free energy of the chain would decrease. Thus the transition state shown in the free energy hypersurface in the Fig. 2 has the configuration shown in the Fig. 3. Once the system has crossed the transition state, the chain is stretched across the barrier. The path of steepest descent then corresponds to moving segments from the cis side to the trans side, without changing the configuration of the polymer in the barrier region. As there is a free energy difference ΔV between the two sides, this would lead to a lowering of the free energy by ΔV per segment, and this leads to a path on the free energy surface with a constant slope, and of width W proportional to N (see the Fig. 2). Such a landscape implies that the barrier-crossing would involve two steps. The first step is going through the transition state by overcoming the activation barrier. Once this is done, there is a rather wide region of width proportional to the length of the chain. Traversing this is the second step. As this region has a constant slope, the motion is driven and it is similar to that of a Brownian particle subject to a constant force. Such a particle would take a time t_{cross}, proportional to N to cross this region. Till now, we considered the case where one end of the molecule overcomes the barrier first, which we refer to as end-crossing. It is also possible for a portion not at the end, to overcome the barrier, by forming a hairpin. For this, the molecule has to be flexible and the pore/channel wide enough to accommodate the hairpin. The scenario for hairpin crossing is similar, though the activation energy is higher for hairpin crossing. Within the Rouse model discussed below, for hairpin crossing, the transition state is equivalent to the one end crossing, repeated two times.[17] Hence the activation energy for the process is two times larger.[17] Once a hairpin crossing occurs, a kink-antikink pair is formed and the kink and the anti-kink separate on the chain, due to the driving force of the free energy gain. Then further crossing occurs by the movement of these two on the chain, and this too leads to a time of crossing proportional to N. In the following we make all these considerations quantitative, using the Rouse model to describe the dynamics of the chain.

3. THE NON-LINER ROUSE MODEL

We consider the continuum limit of the Rouse model, discussed in detail in the book by Doi and Edwards.[19] The discrete nature of the chain is ignored and it is approximated by a string. The position of a segment (bead) along the string is measured by the variable n, which is taken to be a continuous variable, having values ranging from 0 to N. The chain moves only in one dimension and the position of the nth bead at the time t is denoted by $R(n,t)$. In the Rouse model, this position undergoes overdamped Brownian motion. Its change with time is given by

$$\zeta \frac{\partial R(n,t)}{\partial t} = m \frac{\partial^2 R(n,t)}{\partial n^2} - V'(R(n,t)) + f(n,t). \tag{1}$$

In the above, ζ is a friction coefficient for the nth segment. The term $m \frac{\partial^2 R(n,t)}{\partial n^2}$ comes from the fact that stretching the chain would lower its entropy and hence increase its free energy. Consequently, the parameter $m = 3k_BT/l^2$ (see Doi and Edwards,[19] equation (4.5). They use the symbol k for our m) . As the ends of the string are free, the boundary conditions to be satisfied are $\left\{ \frac{\partial R(n,t)}{\partial n} \right\}_{n=0} = \left\{ \frac{\partial R(n,t)}{\partial n} \right\}_{n=N} = 0$. $V(R)$ is the free energy per segment of the chain for a segment at the position R. In the following, we shall take $V(R)$ to have a double well shape, as in Fig. 1. $f(n,t)$ are random forces acting on the nth segment. These have the correlation function $\langle f(n,t)f(n_1,t_1) \rangle = 2\zeta k_BT\delta(n-n_1)\delta(t-t_1)$(see Doi and Edwards,[19] equation (4.12)), so that the forces acting on different segments are uncorrelated. Further, the random force on each segment is white noise. The deterministic part of the equation (1), which will play a key role in our analysis, is obtained by neglecting the random noise term in (1). It is:

$$\zeta \frac{\partial R(n,t)}{\partial t} = m \frac{\partial^2 R(n,t)}{\partial n^2} - V'(R(n,t)). \tag{2}$$

This may also be written as:

$$\zeta \frac{\partial R(n,t)}{\partial t} = -\frac{\delta E[R(n,t)]}{\delta R(n,t)}, \tag{3}$$

where $E[R(n,t)]$ is the free energy functional for the chain defined by:

$$E[R(n,t)] = \int_0^N dn \left[\frac{m}{2} \left\{ \frac{\partial R(n,t)}{\partial n} \right\}^2 + V(R(n,t)) \right]. \tag{4}$$

In the following, we take

$$V(R) = \frac{k}{6}(R+a_0)^2(3R^2 - 2Ra_0 - 4Ra_1 + a_0^2 + 2a_0a_1). \tag{5}$$

This potential has two minima at $R = -a_0$ and at $R = a_1$ and a maximum at $R = 0$. If $a_1 > a_0$, then the minimum at $-a_0$ is less stable than the one at a_1. Hence, if the polymer is initially in the metastable minimum, it will eventually cross over to the more stable minimum.

3.1 The Activation Free Energy

The crossing of the polymer across the barrier involves two steps. The first is the process by which one end of the polymer goes over the barrier as a result of which the polymer gets stretched across the barrier. This process has a free energy of activation. The second step involves the motion of the polymer across the barrier, once it has been stretched across. The second step does not involve overcoming any barrier. In this subsection, we consider the first step and calculate the activation free energy. The activation free energy can be obtained from the free energy functional of Eq. (4). This free energy functional implies that at equilibrium, the probability distribution functional is $\exp\left[-\frac{1}{k_B T} \int dn \left\{ \frac{1}{2} m \left(\frac{dR}{dn} \right)^2 + V(R(n)) \right\} \right]$. If the polymer is on the left side of the barrier, then minimum free energy is attained by having $R(n) = -a_0$. This has a free energy that is equal to zero. If one end of the polymer is stretched across the barrier, then $V(R(n))$ is non-zero and the extremum free energy configuration is determined by $\frac{\delta E[R(n)]}{\delta R(n)} = 0$, which leads to the equation

$$m \frac{d^2 R}{dn^2} = V'(R) \tag{6}$$

Notice that this is just a Newton's equation for a ficticious particle of mass m moving in a potential $-V(R)$. This equation has to be solved, subject to the condition that one end of the polymer is on cis side of the barrier and the other end on the trans side. As we are interested in the case where the polymer is very long, we can find the extremum free energy configuration by finding a solution satisfying $R(-\infty) = -a_0$ and the other end of the polymer to be at a point with $R > R_{max}$, where R_{max} is the point where $V(R)$ has its maximum value. For the specific form of the potential given above $R_{max} = 0$. For the Newton's Eq. (6) the conserved energy is $E_c = \frac{1}{2}m \left(\frac{dR}{dn}\right)^2 - V(R(n))$. For the extremum path, $E_c = 0$. Thus, the particle starts at $R(-\infty) = -a_0$ and ends up at R_f where $R_f (> R_{max})$, is the point such that $V(R_f) = 0$. The free energy of this configuration is the activation free energy. As for this configuration, $\frac{1}{2}m \left(\frac{dR}{dn}\right)^2 = V(R(n))$, we find the activation free energy to be given by

$$E_{act} = \int_{-a_0}^{R_f} \sqrt{2mV(R)}dR. \tag{7}$$

As the parameter m is proportional to the temperature ($= 3k_BT/l^2$), we arrive at the general conclusion that the activation energy $E_{act} \propto \sqrt{T}$. The Boltzmann factor $e^{-\frac{E_{act}}{k_BT}}$ for the crossing of one end of the polymer over the barrier thus has the form $e^{-constant/\sqrt{T}}$. Further, we find that it is independent of N for large N.

For the barrier potential given above, for the forward crossing, there is a barrier with height $V_f = V(0) - V(-a_0) = \frac{1}{6}ka_0^3 (a_0 + 2a_1)$ and for the reverse process, the barrier height is $V_b = V(0) - V(a_1) = \frac{1}{6}ka_1^3 (2a_0 + a_1)$. The forward crossing, would lead to a net change in free energy per unit of the polymer, which is equal to $\Delta V = V(a_1) - V(-a_0) = \frac{1}{6}k (a_0 - a_1) (a_0 + a_1)^3$. Though our results are for this particular form of the potential, the ideas and the conclusions are quite general and of wide applicability.

For the potential of equation (5), we find $R_f = a_0(\gamma - \sqrt{\gamma^2 - \gamma})$ where $\gamma = (1 + 2\frac{a_1}{a_0})\frac{1}{3}$. Further,

$$E_{act} = \frac{\sqrt{m}ka_0^3}{6} \left[(3\gamma^2 + 1)\sqrt{1 + 3\gamma} - 3\gamma(\gamma^2 - 1)\ln\left(\sqrt{\gamma(\gamma - 1)}/\left(1 + \gamma - \sqrt{1 + 3\gamma}\right)\right)\right]. \tag{8}$$

4. THE KINK AND ITS MOTION

4.1 The kink solution and its velocity

Having overcome the activation barrier, how much time would the polymer take to cross it? We denote this time by t_{trans}. To calculate this, we first look at the mathematical solutions of the deterministic equation (2). The simplest solutions of this equation are: $R(n,t) = -a_0$ or with $R(n,t) = a_1$. These correspond to the polymer being on either side of the barrier. Thermal noise makes $R(n,t)$ fluctuate about the mean positions and this may be analyzed using the normal co-ordinates for fluctuations about this mean position. Each normal mode obeys a Langevin equation similar to that for a harmonic oscillator, executing Brownian motion. Further, the mean position (center of mass) itself executes Brownian motion.[19] In addition to these two time independent solutions, the above equation has a time dependent solution (a kink) too, which corresponds to the polymer crossing the barrier.

As is usual in the theory of non-linear wave equations, a kink solution moving with a velocity v may be found using the ansatz $R(n, t) = R_s(\tau)$ where $\tau = n - vt$.[23] Then the equation (2) reduces to

$$m \frac{d^2 R_s}{d\tau^2} + v\zeta \frac{dR_s}{d\tau} = V'(R_s). \tag{9}$$

If one imagines τ as time, then this too is a simple Newtonian equation for the motion of particle of mass m, moving in the upside down potential $-V(R)$. However, in this case, there is a frictional term too, and $v\zeta/m$ is the coefficient of friction. This term makes it possible for us to find a solution for quite general forms of potential. If there was no friction, then total energy would be conserved and such a solution would not exist for the potential given above.

We give results for the potential given in the Eq. (5). The solution of the equation (9), obeying the conditions $R_s(\tau) = -a_0$ for $\tau \longrightarrow -\infty$ and $R_s(\tau) = a_1$ for $\tau \longrightarrow \infty$ is

$$R_s(\tau) = \left(-a_0 + e^{\tau\omega(a_0+a_1)} a_1\right) \left(1 + e^{\tau\omega(a_0+a_1)}\right)^{-1}, \tag{10}$$

with $\omega = \sqrt{k/m}$. The solution exists only if the velocity $v = \frac{\sqrt{mk}}{\zeta}(a_0 - a_1)$. This solution is a kink, occurring in the portion of the chain inside the barrier. We shall refer to the point with $\tau = 0$ as the center of the kink. (Actually one has a one-parameter family of solutions of the form $R_s(\tau + \tau_0)$, where τ_0 is any arbitrary contant). As $\tau = n - vt$, the center of the kink moves with a constant velocity v. Note that this velocity depends on the shape of the barrier. Thus for our model potential, if $a_0 < a_1$, then $V_f < V_b$, and this velocity is negative. This implies that the kink is moving in the negative direction, which corresponds to the chain moving in the positive direction. That is, the chain moves to the lower free energy region, with this velocity. If the barrier is symmetric, then $a_0 = a_1$ (i.e., $V_f = V_b$) the velocity of the kink is zero.

4.2 The crossing time t_{cross}

For the polymer to cross the barrier, the kink has to go in the reverse direction, by a distance proportional to the length of the chain. As the kink moves with a velocity v, $t_{cross} \sim N/v$. Further, as v is proportional $\sqrt{mk}$, assuming $V(R)$ to be temperature independent we find $t_{cross} \sim N$. This too is a general conclusion, independent of the model that we assume for the potential.

If the barrier is symmetric, the kink velocity $v = 0$ - it does not move. This is not surprising as there is no free energy gain, by the movement of the kink. The kink can be anywhere on the chain, with the same total free energy. Thermal fluctuations, ofcourse, would drive the kink randomly- which means that the kink would execute a random walk. We do not give here the detailed analysis,[16] but only the final result: the result is that it executes a diffusive motion, with diffusion coefficient $D = \frac{3k_B T}{4\zeta a_0^3}\sqrt{\frac{m}{k}}$. In this case, the time required for the polymer to cross over the barrier is $t_{trans} \sim N^2$.

Park and Sung[5] consider the passage of a polymer through a pore for which the barrier is entropic in origin. Consequently it is very broad, the width being of the order of N. In comparison, we take the barrier to be extrinsic in origin and assume its width to be small in comparison with the length of the chain. The crossing occurs by the motion of the kink, which is a localized non-linear object in the chain whose width is of the same order as that of the barrier. As the kink is a localized object, its diffusion coefficient has no N dependence and hence our results are different from those of Park and Sung.[5] In the case where there is no free energy difference, our crossing time is proportional to N^2 (in contrast to N^3 of Park and Sung), while if there is a free energy difference, our crossing time is proportional to N (in contrast to N^2 of Park and Sung). In a very recent paper,[14] Park and Sung consider the Rouse dynamics of a short polymer surmounting a barrier. The size of the polymer is assumed to be small in comparison with the width of the potential barrier. Consequently, the transition state has almost all the beads at the top of the barrier, leading to the prediction that the activation energy is proportional to N. This leads to a crossing probability that decreases exponentially with N. In comparison, as found earlier, the free energy of activation does not depend on the length of the chain. Hence, the kink mechanism must be the favoured one for long chains.

4.3 The net rate

As the actual crossing is a two step process, with activation as the first step and kink motion as the second step, the net rate of the two has to be a harmonic mean of the two rates. However, though we have estimated the free energy of activation, it seems rather difficult to estimate a pre-factor for this process. Hence one can only make some general observations: For a very long chain, the motion of the kink has to become rate determining. In the experiments of Kasianowicz,[10] one directly observes t_{trans} and hence our considerations on kink motion must be directly applicable. Also, in the case of translocation of biological macromolecules (see the next section) there does not seem to be any free energy of activation and then the rate is determined by t_{trans} alone.

To verify the analysis given above, we performed computer simulations, the details of which are given elsewhere.[24] The simulations were done for a discrete model, for a variety of barrier heights and widths. In all the cases, the average time at which a given bead crosses, when plotted against the number of the bead, lead to a straight line, thus verifying the operation of the kink mechanism for a flexible chain.

Further, even though, the analysis given above is for a one dimensional version of the Rouse model, it is fairly straightforward to extend the analysis to a three dimensional situation, and argue that the kink exists. The operation of the kink mechanism, even in this three dimensional case too has been verified by simulations.[24]

5. TRANSLOCATION IN BIOLOGICAL SYSTEMS

If there was a high activation energy ($>> k_B T$) for the translocation, the process would be unlikely. However, translocation seems to be very efficient in biological systems, and hence it is probably a barrierless process. What is the mechanism by which this reduction is achieved? One possibility is for the segments to have a lower free energy when they are inside the membrane. But then, the chain would not leave the membrane and perhaps would get implanted there. The final destination where a biological long chain molecule ends up is determined by a sequence of units at the begining of the chain, referred to as the signal sequence. The way the sequence works is simple. If the pore is hydrophobic and the chain hydrophilic, then the signal sequence is hydrophobic, so that the signal sequence has a low free energy inside the pore. Analysis of this type of problem, using the Rouse model[17] is straightforward. One assumes that the segments at the end of the chain has a lower potential energy if they are inside the pore.

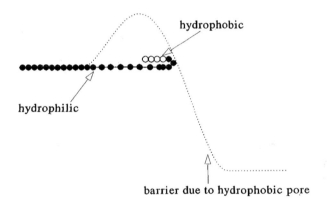

Figure 4: *The transition state for a hydrophilic chain with a hydrophobic signal sequence, passing through a hydrophobic pore. Compare with figure 14-14 of the book by Alberts et. al.*

Analysis of this model leads to a transition state that is shaped like a hook.[17] The hydrophobic part of the chain is completely in the short arm of the hook (see figure 3). The activation energy would depend on the length of the hydrophobic part, and if it is sufficiently long, the activation energy may become zero, so that the crossing can become a barrierless process. In such a scenario, the rate is determined by t_{cross}. The transition state, though it seems likely to occur in crossing between liquid-liquid interfaces, seems rather difficult to form in the case of passage through a pore as there are two difficulties: (1) the chain has to bend to form the hook (2) the pore has to be wide enough to accommodate the two strands of the hook simultaneously. In spite of these, nature does seem to use this as an inspection of the figure 14-14 of the book by Alberts *et. al.*[6] shows.

6. CONCLUSIONS

We have considered the generalization of the Kramers escape over a barrier problem to the case of a long chain molecule. It involves the motion of a chain molecule of N segments across a region where the free energy per segment is higher, so that it has to cross a barrier. We consider the limit where the width of the barrier w is large in comparison with the Kuhn length l, but small in comparison with the total length Nl of the molecule. The limit where $Nl << w$ has been considered in a recent paper by Park and Sung.[14] We use the Rouse model and find that the free energy of activation has a square root dependence on the temperature T, leading to a non-Arrhenius form for the rate. While in the short chain limit Park and Sung find the activation energy to be linearly dependent on N, we find that for long chains, the activation energy is independent of N. We also show that there is a special time dependent solution of the model, which corresponds to a kink in the chain, confined to the region of the barrier. In usual non-linear problems with a kink solution, the problem has translational invariance and the kink can therefore migrate. In our problem, the translational invariance is not there, due to the presence of the barrier and the kink solution is not free to move in space. However, the polymer on which the kink exists can move, though the kink is fixed in space. Thus, the polymer goes from one side to the other by the motion of the kink in the reverse direction on the chain. If there is no free energy difference between the two sides of the barrier, then the kink moves by diffusion and the time of crossing $t_{cross} \sim N^2$. If there is a free energy difference, then the kink moves with a non-zero velocity from the lower free energy side to the other, leading to $t_{cross} \sim N$. Our result that $t_{cross} \sim N$ is in agreement with the recent experiments of Kaisanowicze *et. al*[10] where DNA molecules were drawn through a nanopore by the application of a potential difference. We also consider the translocation of hydrophilic polypeptides across hydrophobic pores. Biological systems accompolish this by having a hydrophobic signal sequence at the end that goes in first. Our analysis leads to the conclusion that for such a molecule, the configuration of the molecule in the transition state is similar to a hook, and this is in agreement with presently accepted view in cell biology.[6]

7. ACKNOWLEDGEMENTS

I thank Professor S.Vasudevan for interesting discussions.

References

1 H.A. Kramers, *Physica (Utrecht)* **7**, 284 (1940).

2 S. Chandrasekhar, *Rev. Mod. Phys.*, **15**, 1 (1943).

3 P. Hanggi, P. Talkner and M. Borkovec, *Rev. Mod. Phys.*, 62, 251 (1990).

4 H. Riezman, *Science*, **278**, 1728 (1997).

5 W.Sung and P.J. Park, *Phys. Rev. Lett.*, **77**, 783 (1996), see also: P.J. Park and W. Sung, *J. Chem. Phys.*, **108**, 3013 (1998) and P.J. Park and W. Sung, *Phys. Rev.* **E 57**, 730 (1998).

6 B. Alberts, D. Bray, A. Johnson, J. Lewis, M. Raff, K. Roberts and P. Walker, *Essential Cell Biology*, Garland Publishing Inc, New York (1998).

7 S.M. Simon and G. Blobel, *Cell*, **65**, 371 (1991).

8 B. Dreiseikelmann, *Microbiological Reviews*, **58**, 293 (1994).

9 V. Citovsky and P. Zambryski, *Ann. Rev. Microbiol.* **47**, 167 (1993).

10 J.J. Kasianowicz, E. Brandin, D. Branton and D.W. Deamer, *Proc. Natl. Acad. Sci. USA,* **93**, 13770 (1996).

11 J. Han, S.W. Turner and H.G. Craighead, *Phys. Rev. Lett.*, **83**, 1688 (1999).

12 M. Muthukumar and A. Baumgartner, *Macromolecules*, **22**, 1937 (1989).

13 K.Kiran Kumar and K.L. Sebastian, *Phys. Rev.* **E62**, 7536 (2000).

14 P. Park and W. Sung, *J. Chem. Phys.*, **111**, 5259 (1999).

15 D.K. Lubensky and D.R. Nelson, *Biophysical Journal*, **77**, 99005 (1999).

16 K.L. Sebastian, *Phys. Rev.* **E 61**, 3245 (2000). K.L. Sebastian, *J. Am. Chem. Soc.* **122**, 2972 (2000)

17 K.L. Sebastian and A.K.R. Paul, *Phys. Rev.* **E 62**, 927 (2000).

18 K.L. Sebastian, *Phys. Rev.* **E 62**, 1128 (2000).

19 M.Doi and S.F. Edwards, *The theory of polymer dynamics*, Clarendon Press, Oxford, (1986). The model has the defect of not taking excluded volume interactions in to account.

20 R. Rajaraman, *Instantons and Solitons*, North Holland, Amsterdam (1982)

21 A. Scott, *Nonlinear Science*, Oxford University Press, Oxford (1999).

22 P.M. Chaikin and T.C. Lubensky, *Principles of Condensed Matter Physics*, Cambridge University Press, Cambridge, (1998).

23 There is an extensive literature on this topic. The most relevant to our discussion are: M. Buttiker and R. Landauer, *Phys. Rev.* **A 37**, 235 (1988), P. Hanggi, F. Marchesoni and P. Sodano, *Phys. Rev. Lett.*, **60**, 2563 (1988). F. Marchesoni, *Phys. Rev. Lett.*, **73**, 2394 (1994).

24 Alok K.R. Paul and K.L. Sebastian, to be published.

II Proteins, Lipids and Their Interactions

LIPID INTERACTION WITH CYTIDYLYLTRANSFERASE REGULATES MEMBRANE SYNTHESIS

S. Jackowski and I. Baburina[*]

Protein Science Division, Department of Infectious Diseases, St. Jude Children's Research Hospital, Memphis, TN 38105, USA
[*]Present address: Roche Diagnostics Corporation, Building D, 9115 Hague Road, P.O. Box 50457, Indianapolis, IN 46250, USA

1 INTRODUCTION

CTP:phosphocholine cytidylyltransferase (CCT) is a major regulator of phosphatidylcholine biosynthesis. Phosphatidylcholine is the major membrane phospholipid in eukaryotic cells and also is a precursor to the two other major membrane phospholipids, sphingomyelin and phosphatidylethanolamine. The dominant metabolic route for phosphatidylcholine biosynthesis in mammals is the cytidine diphosphocholine (CDP-choline) pathway (Figure 1)[1]. The concentrations of the biochemical intermediates in the CDP-choline pathway are not equal, with the phosphocholine pool being the largest under normal culture conditions and the CDP-choline pool being the smallest, thus illustrating that the conversion of phosphocholine to CDP-choline by CCT is the slow step and limits the overall rate of phosphatidylcholine headgroup formation. Changes in the cellular CCT activity have a direct impact on the rate of phosphatidylcholine synthesis and the rate of membrane formation.

2 DISTRIBUTION OF CCT ISOFORMS

2.1 Tissue Expression

Three isoforms have been identified: CCTα encoded by the *PCYT1A* gene in human[2] (*Pcyt1a* in mouse), and CCTβ1 or CCTβ2, splice variants arising from the human *PCYT1B* gene[3,4] (*Pcyt1b* in mouse). The mouse *Pcyt1a* gene was originally named *Ctpct*[5,6] but was later changed to agree with the human CCT gene nomenclature. The CCTα cDNA was first cloned from rat liver using degenerate PCR primers complementary to the yeast CCT gene[7] and an active CCTβ cDNA clone was identified in the public expressed sequence tagged database[3]. All three isoforms can complement a Chinese hamster ovary cell line with defective CCT activity which limits phosphatidylcholine production at elevated temperatures[8], illustrating that CCTα, CCTβ1 and CCTβ2 catalyze the same biochemical activity *in vivo*[4,9]. CCTα is ubiquitous and expressed in all tissue types examined thus far[4]. The CCTβ isoforms, on the other hand, are expressed at their highest levels in fetal tissue, placenta, ovary, testis and brain. The shorter β1 isoform is the major CCTβ expressed in placenta, liver and fetal lung, whereas the β2 isoform is present in brain. Adult lung tissue appears to have lost CCTβ expression[4,10] and normal wild-type macrophages do not express significant levels of CCTβ[11]. However, genetic deletion of the CCTα locus causes CCTβ2 gene expression to increase dramatically[11], thereby compensating for the loss of CCTα.

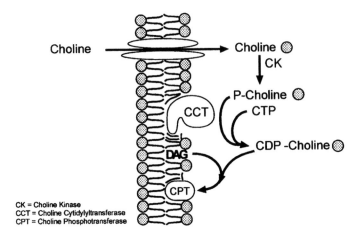

Figure 1   *Pathway of Phosphatidylcholine Biosynthesis. Choline (•) is incorporated into cells and phosphorylated by choline kinase, forming phosphocholine. Phosphocholine and cytidine triphosphate (CTP) are substrates for the next enzyme, the CTP:phosphocholine cytidylyltransferase (CCT) which produces cytidine diphoshocholine (CDP-choline). The choline phosphotransferase then combines CDP-choline together with diacylglycerol (DAG) to form phosphatidylcholine.*

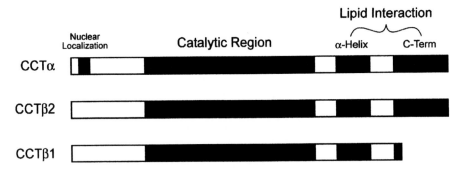

Figure 2   *Comparison of CCT Isoforms. The functional domains found in the CCTα, CCTβ1 and CCTβ2 proteins are indicated.*

CCTβ1 and CCTβ2 are identical proteins from the amino-terminus through residue 323, but differ from each other at their carboxyl termini (Figure 2)[4]. The predicted CCTβ2 protein has 369 amino acids whereas the CCTβ1 protein is smaller, with 333 amino acids[3]. The CCTβ1 represents a truncated version of CCTβ2 and lacks most of the phosphorylation sites that characterize the carboxyl terminal end of CCTβ2. CCTα, which has 367 amino acids, has a highly phosphorylated carboxyl terminus but the sequence varies considerably from that of CCTβ2. The CCTα and the CCTβ proteins also differ at their respective amino termini, but are strikingly similar in the central catalytic domain and the adjacent α-helical domain. Like CCTα, CCTβ1 and CCTβ2 require the presence of lipid regulators for maximum catalytic activity[3,4] and the helical domains of each protein mediate an interaction with lipids[3,12-15].

2.2 Cellular Location

A functional nuclear localization signal characterized by a cluster of positive charges distinguishes the amino terminus of CCTα[16] from CCTβ and, in fact, the majority of CCTα protein can be found in the cell nucleus in many cell types[4,16-18]. A smaller amount of CCTα associates with the endoplasmic reticulum[4,19,20] together with the CCTβ proteins[3,4]. The role of nuclear CCTα is currently being investigated by several groups and two hypotheses are current. One hypothesis favors the nucleus as a reservoir for inactive CCTα, whereby the protein translocates to the endoplasmic reticulum when immediately needed for membrane biosynthesis, as in the IIC9 cellular response to serum stimulation[19]. However, translocation between the nucleus and the cytoplasm may not be a general function of further cell cycle progression[17] as originally postulated[19]. Overexpression of CCTα in a variety of cell lines results in CCTα localization in the nucleus, without a detectable immunofluorescent signal outside of this organelle despite perturbation of the cell cycle[17]. The alternative hypothesis about the role of nuclear CCTα is based on the observation that CCTα associates with chromatin[21]. CCTα is active in the production of highly saturated phospholipids which are unique among the cellular phosphatidylcholine molecular species[22]. These phospholipids are actively synthesized in nuclei isolated during the transition from cell cycle arrest to growth in response to serum stimulation of IMR-32 neuroblastoma cells, indicating that at least a portion of the CCTα remains nuclear and catalytically active during the G_0 to G_1 transition.

We investigated the distribution of CCTα protein as a function of cell cycle progression through two cell cycles in the BAC1.2F5 mouse macrophage cell line. The cells were cultured on slides[4] and synchronized by removal of growth factor, i.e., colony stimulating factor-1 (CSF-1)[23]. Cell cycle progression was monitored using flow cytofluorometry of permeabilized cells stained with propidium iodide for quantitation of DNA content[23]. CCTα-specific antibodies were modified with Oregon Green™-fluorescent label and CCTβ-specific antibodies were labeled with Texas Red™ using a Molecular Probes kit. Cells were fixed, permeabilized and stained with the antibodies as described[4]. At least 100 cells per condition were examined using direct immunofluorescent confocal microscopy and the intensity gain for the fluorescent signal was maintained at the same setting during examination of cells. The use of direct immunofluorescence, rather than the more common indirect method, together with preparation of labeled antibodies with a 1:1 molar ratio, increased the sensitivity of the fluorescent signal and allowed us to visualize smaller quantities of CCT protein.

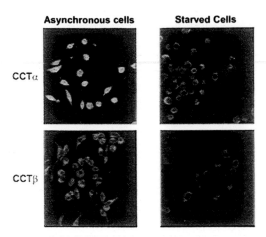

Figure 3   *CCT Localization by Direct Immunofluorescent Microscopy. Isoform-specific
antibodies were labelled directly with fluorescent probes and used to detect the
distribution of CCTα and CCTβ proteins in situ in either normally growing
asynchronous macrophage cells or after 18 hours of CSF-1 growth factor
withdrawal.*

As expected, the CCTα protein was distributed both in the nucleus and outside the
nucleus in asynchronous, proliferating cells (Figure 3).   CCTβ was exclusively located
outside the nucleus in these same cells.   In contrast, the CCTα protein was located
exclusively outside the nuclei of cells arrested in G_0/G_1 due to removal of CSF-1 from the
culture medium[23](Figure 3).   Also, the amount of CCTα protein per cell was significantly
reduced when cells were arrested by starvation for CSF-1, consistent with the observation
that CCTα gene expression is dependent on CSF-1 in the medium[24].

CCTα protein accumulated and translocated to the cell nuclei with time following re-
addition of CSF-1 (Figure 4).   Accumulation in the nucleus was not significant until 36
hours after CSF-1 re-addition, however, which was well after the synchronized population
progressed through two complete cycles of growth.   As a follow-up experiment, normally
proliferating asynchronous cells were examined for their cell cycle distribution which
indicated that 66.3% of the cells were in G_1, 28.6% in S and 5.1% in G_2 plus M phases.   In
all cases, CCTα was located both within and outside the nucleus, supporting the view that
CCTα translocation is not associated with a cell cycle event.   Rather, the data suggest that
the accumulation of CCTα protein in the nucleus may be a function of the quantity of
CCTα protein.   The amount of CCTα protein is reduced during CSF-1 starvation, and then
new CCTα protein is synthesized following return of CSF-1 to the growth medium and
stimulation of CCTα gene expression[24].   The endoplasmic reticulum is the site of new
protein synthesis and thus, new CCTα is likely to first accumulate at this site and then
move to the nucleus as the protein quantity increases.   Interestingly, the extended absence
of CSF-1 causes the cells to undergo a program of cell death called apoptosis.   In turn, one
might postulate that the absence of CCTα from the nucleus for an extended period may
contribute to the characteristic chromatin condensation and fragmentation associated with
apoptosis.

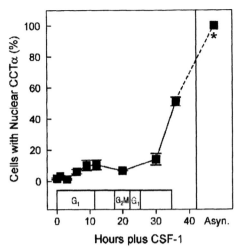

Figure 4 *Quantitation of CCT Positive Nuclei. Macrophage cells were synchronized in G_0/G_1 by CSF-1 withdrawal and then were examined for CCTα distribution following re-addition of CSF-1. Cell cycle progression was determined by flow cytofluorometry.*

2 BIOCHEMICAL REGULATION BY LIPIDS

Lipid interaction with CCT regulates enzymatic activity. Lipids bind tightly to CCT, even when the protein is purified in soluble form, and a detergent wash is necessary to reliably assay the lipid-activation or lipid-inhibition of the enzyme[14,23,25-29]. The catalytic fragment is not activated by lipid nor inhibited by delipidation, and CCT catalytic activity is compromised when the lipid regulatory domains are truncated[14,30,31], in contrast to the full-length protein. While the activity of the truncated protein is greater compared with the delipidated full-length protein, it is less than the full-length protein in the presence of lipid[30], particularly when the CCT constructs are assayed under identical conditions[31]. The first model for lipid regulation of CCT proposed that the helical domain was inhibitory to activity[13], perhaps due to steric interference of critical reaction sites[30], and suggested that it was the removal of this inhibition that was responsible for elevated activity. A second early model proposed that the re-alignment of the regulatory domain by lipid-binding caused a conformational change in the CTP substrate site, thereby increasing its affinity[14]. The key to the development of the second model was not only a comparison between the full-length, lipid-activated CCT and the truncated catalytic fragment, but also comparison with the delipidated, full-length protein. Close examination of the data in subsequent studies from the two laboratory groups reveals that both models are correct and are not mutually exclusive, and that the regulatory domain(s) perform both functions (Figure 5)[31]. Removal of the lipid regulatory domain(s) results in a dysregulated protein that lacks both positive and negative control over CCT catalytic activity based on the data from kinetic analyses[30] and careful specific activity measurements[31].

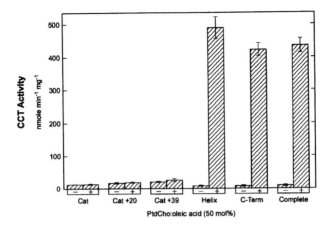

Figure 5   *Lipid Responsiveness of CCT molecular constructs. The catalytic region alone (Cat) was assayed using high CTP (10 mM) and was not stimulated by lipid. Two longer constructs with 20 additional residues (Cat +20) or 39 additional residues (Cat +39) at the carboxyl terminus also have the same activity and do not respond to lipid. Expression of either the helical domain (Hel) or the authentic carboxyl terminal domain (C-Term) confer slight inhibition in the absence of lipid or lipid activation, similar to the full-length protein (Complete). (Drawn from tabular data in reference[31].)*

3.1   Regulation by the Helical Domain

Two lipid interaction domains have been identified in mammalian CCTα and are contained within the helical region spanning residues 257 through 290, and within the carboxyl terminal region from residues 310 through 367[31]. The helical region has long been recognized as a site for membrane lipid binding and has been studied and reviewed extensively[32,33]. As a result, the limits of the membrane interaction site within the helical domain are delineated more precisely. The helical region is characterized by three 11-mer repeats, each of which is sufficient to form an α-helix and one individual 11-mer unit is capable of binding to phosphatidylcholine vesicles[34]. However, the tandem repeats actually form one contiguous 33-residue helix when associated with an optimizing lipid mixture[35] and the entire 30+ helical domain is required to activate[31] or inhibit the enzyme[25] in the absence of the carboxyl terminus. A peptide corresponding to the helical region lacks structure in solution and formation of the helical conformation requires addition of phosphatidylcholine vesicles[35], suggesting that the regulatory domain may be a random coil in the delipidated form of CCT and then convert to an α-helical conformation when lipid mixtures are added. One can envision the bulk of a random coil interfering with the reactive sites of the catalytic domain, accounting for the inhibitory nature of the delipidated helical region (Figure 5). The structuring of the helical region in the presence of lipid, then, likely influences the structure of the adjacent catalytic domain, slightly rearranging the CTP site to enhance binding of the substrate[14] or to promote the chemical events associated with the reaction[30], for example, CTP dephosphorylation, phosphocholine esterification of CMP, or exit of inorganic phosphate or CDP-choline. The CTP reactive site appears to be the relevant target for regulation since the activity of the delipidated enzyme can be revealed by very high concentrations of this nucleotide[14,30,31].

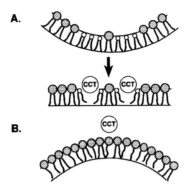

Figure 7 *CCT response to membrane curvature elastic stress. Panel A: The tendency of a monolayer to bend toward the polar aspect of the interface is influenced by type 2 lipids. The CCT amphipathic helix inserts into the monolayer half of the bilayer and the stress is relieved. Panel B: The insertion of type 1 lipids increases the tendency of a monolayer to bend away from the polar aspect of the interface. CCT does not insert into the type 1 monolayer.*

The helix is amphipathic and features hydrophobic amino acids clustered along one side of its long axis, with charged amino acids distributed along the opposite side. The hydrophobic portion of the helix is stabilized by interaction with the acyl chains of a phosphatidylcholine monolayer, positioning the charged residues of the helix among the charged phosphate and choline moieties of the phospholipid headgroups. The insertion of the CCT helix into one leaflet of a phosphatidylcholine bilayer is predicted from the alignment of its amino acids, although phosphatidylcholine alone will not activate the enzyme[31] and other phospholipids cannot substitute. Mixtures of type 2 lipids with phosphatidylcholine maximally activate CCT. Type 2 lipids include the zwitterionic phosphatidylethanolamines, fatty acids which in the presence of phosphatidylcholine are only partially deprotonated and neutral lipids, such as the diacylglycerols[36]. These lipids increase the curvature elastic stress of the lipid monolayer which, in turn enhances CCT binding to the membrane (Figure 6). The enhanced binding arises out of the increased amount of curvature elastic stress that is released by the splay of the lipid chains upon insertion of the hydrophobic strip of the binding helix.[36]. Type 1 lipids thereby disallow the CCT helix from inserting into the lipid monolayer due to their tendency to bend the phospholipid structure away from the polar interface[36]. Type 1 lipids thereby oppose the activation of CCT and compete kinetically with the type 2 lipid when enzyme activity is assayed *in vitro*[25]. Type 1 lipids include lysophosphatidylcholine and the alkylphospholipid antineoplastic agents[36].

The membrane curvature elastic stress hypothesis incorporates many of the features of previous hypotheses for CCT activation by lipids and, in doing so, resolves many of the associated concerns. CCT binds to phosphatidylcholine vesicles that contain negatively charged lipids, such as phosphatidylserine or fatty acid, and electrostatic interaction between the anionic lipids and the positively charged amino acid residues in the amphipathic helix has been suggested previously as an underlying mechanism[37,38]. However, electrostatic interactions cannot account for enzyme activation by uncharged lipids, such as diacylglycerol. Similarly, CCT inhibitors such as lysophosphatidylcholine and the synthetic alkyl analogs ET-18-OCH$_3$[25] or HexPC[26], are characterized by phosphocholine moieties and thus argue against inhibition due to alteration of the phospholipid headgroup packing of the phosphatidylcholine vesicle[27]. Rather, activators or inhibitors can be grouped into two distinct classes which either promote negative or positive elastic stress, respectively, of a phosphatidylcholine bilayer and the magnitude of the effect is a function of the mole percent of the incorporated amphiphile[36]. Lateral stress of a phospholipid monolayer is determined by the repulsion of headgroups, the interfacial tension and the outward chain pressure of the acyl groups. The curvature elastic stress of a bilayer, in turn, is determined by the relative degree of lateral stress of the back to back monolayers. A subsequent study confirms the basic tenets of the membrane curvature elastic stress hypothesis whereby CCT binding and activation correlate very well with relaxation of the curvature strain energy of several chemically distinct type 2 lipid systems[39]. In this study, however, cholesterol does not activate CCT and the authors attribute this to its inability to alter the surface hydrophobicity despite its tendency to increase the negative strain on a phosphatidylcholine monolayer[39]. Alternatively, if one considers cholesterol as a rigid lath shaped amphiphilic molecule, then it cannot splay and fill the regions beneath the CCT. This aspect would explain cholesterol's poor activation of CCT, despite cholesterol's ability to change the rigidity of the membrane so that any curvature of the membrane would be sensitive to its mechanical properties.

3.2   Regulation by the Carboxyl Terminal Domain

The carboxyl-terminal region is also known as the phosphorylation domain and only recently has it been demonstrated to interact with lipids[31]. The helical domain and the carboxyl-terminal region interact with lipids in different ways (Figure 5), as indicated by lipid protection of the full-length protein from proteolysis[12] and using CCTα molecular constructs[31]. The helical domain is fully protected from chymotrypsin proteolysis in the presence of phosphatidylcholine vesicles, while the carboxyl terminus is exposed and susceptible to degradation[12]. Comparative analysis of CCTα proteins with only one domain tethered to the catalytic core reveals that the helical domain is responsive to both neutral and anionic lipids embedded in phosphatidylcholine vesicles, and the carboxyl domain is responsive to anionic lipids only, preferably mixtures of phosphatidic acid and Triton X-100 detergent[31]. Either domain can fully activate CCTα when the other is absent from the protein, and either domain is inhibitory when delipidated[31]. The fact that either domain confers the same range and degree of regulation on CCT activity (Figure 5) argues against a specific inhibitory interaction between either regulatory domain and the catalytic core in the absence of lipid, and supports a general steric hindrance by an unstructured peptide. One would postulate that the carboxyl terminal domain, then, takes on organized structure in the presence of lipid although there is no experimental evidence as yet. The two domains can regulate CCTα independently[31] or coordinately[38], depending on the lipids that are present. The carboxyl domain can also interfere with lipid activation via the helical domain[31], by conferring negative cooperativity on the kinetics of the lipid activation[29]. Either interaction between the two regulatory domains, or competition for binding to a common anionic lipid may account for the apparent dampening of the enzyme response by the carboxyl terminal region. The high degree of similarity of the carboxyl termini of CCTβ2 and CCTα suggests that CCTβ2 would be regulated in an analogous

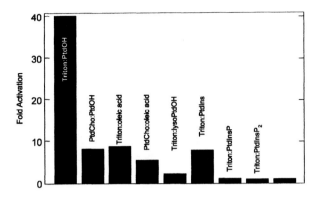

Figure 8 *The CCT Carboxyl Terminus Prefers Phosphatidic Acid. CCTα protein lacking the helical domain but retaining the authentic carboxyl terminus was assayed for the response to various anionic lipids (20 mol %) mixed with Triton X-100 or phosphatidylcholine. (Redrawn from tabular data [31].)*

manner, and that CCTβ1 would not be responsive to low concentrations of phosphatidic acid due to the absence of this unique domain.

The rodent CCTα carboxyl terminus is phosphorylated reversibly at 19 serine residues[40] whereas the human CCTα domain has 13 serines, all of which are potential phosphorylation sites and several kinases are capable of utilizing CCTα as substrate[15,41]. CCTβ2 is also highly phosphorylated at the carboxyl-terminus and has 20 potential phosphorylation sites, whereas CCTβ1 has only 5 phosphorylation sites[4]. CCTα is phosphorylated in a cell cycle-dependent manner[23]: the population of CCTα molecules is phosphorylated least in early G_1, and attains a higher phosphorylation state as cells progress further through G_1, S and G_2M phases. Increased phosphorylation coincides with a protein shift on denaturing polyacrylamide gels to a slower-migrating molecular species and there are at least three phosphorylated species that have been identified by gel-shift[23]. When cells are arrested in M phase by a cell-cycle blocker, only hyperphosphorylated CCTα is evident as characterized by the slowest migrating gel forms. Phosphorylation is inversely proportional to CCTα activity in immune precipitates[23] and in *in vitro* assays[15,42]. The enzyme gains activity after G_1 initiation and peaks at mid-G_1 phase, then becomes progressively less active as phosphorylation increases through the remainder of the cell cycle. Cellular CCT activity is at its lowest level in M phase, corresponding to the hyperphosphorylated state. In cells that are proliferating asynchronously, decreased CCTα phosphorylation often correlates with a higher rate of phosphatidylcholine biosynthesis and a tighter association between CCT and the particulate fraction in permeabilized cells[43]. Dephosphorylation does not drive the apparent membrane translocation, however, but rather, it has been suggested that the CCT protein may be a substrate for phosphatases following membrane association[44]. Dephosphorylated CCT may, in turn, be a substrate for protease degradation[10,45].

4 CELLULAR REGULATION OF CCT ACTIVITY

4.1 Feedback Regulation by Cell-Associated Lipids

Type 2 and type 1 lipids regulate CCT *in vivo* (Figure 10). Exogenous fatty acids immediately stimulate CCT and phosphatidylcholine biosynthesis in a variety of cells and tissues[43,46-49]. Stimulation of phosphatidylcholine biosynthesis also follows treatment of cells with phospholipases C or D, generating diacylglycerol or phosphatidic acid, respectively, *in situ* at the cell membrane[43,48-50]. It is difficult to prove, however, that enhanced membrane phosphatidylcholine synthesis is a direct result of type 2 lipid invasion of the CCT membrane environment rather than a cellular signaling event, such as activation of a CCT phosphatase. Since membrane phosphatidylcholine synthesis is essential for cell survival, a basal rate of CCT activity is always operating and further activation may be limited by the availability of an inactive CCT pool, thus dampening the maximal CCT activity achievable while probing the system experimentally. Overproduction of phosphatidylcholine due to increased CCT activity also does not yield an overt accumulation of membrane[51-53], except in a few specific cell types under unique circumstances. For example, the loading of macrophages with free cholesterol stimulates formation of lamellar membrane bodies[54]. Rather, the cellular phospholipid content is usually controlled by phospholipase degradation which offsets any biosynthetic increase to maintain homeostatic control over the absolute amount of cellular membrane[52,53]. Lastly, the type 2 activators identified thus far are readily metabolized and converted to inactive molecular species. The intracellular free fatty acid pool is small and endogenous fatty acids are either modified to acyl-coenzyme A molecules, bound to fatty acid binding proteins, or incorporated into phosphatidylcholine, sphingomyelin or triacylglycerol, all of which are inactive in regulating CCT activity. However, exogenous unsaturated fatty acids can remodel pre-existing phosphatidylethanolamine, which also can, in turn, stimulate CCT. Similarly, phosphatidic acid or diacylglycerol, either from *de novo* synthesis or generated *in situ* in the membrane by phospholipase treatment of cells, can become part of the phospholipid pool or the triacylglycerol pool[55] and thus provide only transient, localized stimulation of CCT activity. The plasticity of the system is ultimately beneficial for cells, allowing the population of CCT proteins to respond immediately near sites of membrane modification or damage.

Negative regulation by treatment of cells with type 1 lipid, such as lysophosphatidylcholine, clearly causes inhibition of CDP-choline and *de novo* phosphatidylcholine production[56]. Inhibition by this lysolipid does not interrupt cell growth, however, because of its conversion to phosphatidylcholine. On the other hand, the non-metabolizable type 1 lipids, namely the alkylphospholipid antineoplastic agents, reveal persistent CCT inhibition and foster the development of a quantifiable cellular phenotype. The synthetic alkylphospholipids, such as ET-18-OCH$_3$ (edelfosine), are structural analogs of lysophosphatidylcholine, a natural CCT inhibitor[25]. Lysophosphatidylcholine does not cause apoptosis because it bypasses the CCT inhibition of the CDP-choline pathway and provides an alternate source for the membrane phospholipid necessary for survival[57]. Both a genetic model and a molecular model empower the hypothesis. A mutant Chinese hamster ovary cell line with a conditional defect in CCT[8] undergoes apoptosis under the non-permissive condition[58]. The apoptosis caused by genetic inactivation of CCT can be prevented by supplementing cells with a source of phosphatidylcholine or lysophosphatidylcholine to maintain membrane integrity[57]. Induced overexpression of CCTα in an engineered human HeLa cell line prevents the apoptosis caused by exogenous alkylphospholipids, proving that the target is indeed CCT. Addition of supplementing lysophospholipid to the medium does not increase survival further[59], indicating that both methods rescue the same intracellular process. Altogether, the data strongly support the concept that inhibition of CCT, or the CDP-choline pathway, is a primary effector mechanism in apoptosis. Thus future investigation of CCT as a cellular target in ligand-initiated apoptosis, for example in response to FAS ligand or TNFα, is warranted. Inactivation of CCT by proteolytic degradation[10,45], reduced expression[11], or inhibition of the CCT-lipid interaction, all would decreased membrane phosphatidylcholine synthesis.

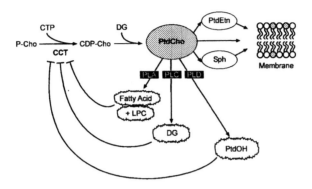

Figure 10 *Feedback Regulation of CCT. The CDP-choline pathway of phosphatidylcholine biosynthesis is regulated by CCT. Phosphatidylcholine is degraded by phospholipase A (PLA), phospholipase C (PLC) or phospholipase D (PLD), releasing lipid products (☆) which either activate or inhibit CCT.*

4.2 Regulation of CCT Expression

While it is clear that CCT is strictly dependent on lipid for maximum activity, high-level overexpression of the truncated, dysregulated and compromised catalytic fragment can complement cells with defective endogenous CCT[13]. These data demonstrate that despite a need for higher cellular CTP[14,30], despite reduced catalytic efficiency[15,30], and without the ability to associated with cellular lipids[14,30], CCTα can still be active in cells[13,31]. Expression levels of CCT, then, become another factor in regulating the rate of phosphatidylcholine production, in addition to the nature of CCT association with lipid. When CCTα expression is knocked-out in mouse macrophages by genetic inactivation of the *Pcyt1a* locus, CCTβ2 expression increases dramatically to compensate for the loss of CCTα, allowing for normal macrophage development and apparently normal overall mouse development[11]. However, the macrophage CCTβ2 total activity is significantly lower compared to that otherwise attributed to CCTα, and the macrophages are restricted in their response to the challenge of free cholesterol loading. Downregulation of CCTα expression by proteolytic degradation has been observed in pancreatic acini treated with cholecystokinin[45] and in alveolar type II cells treated with TNFα[10]. Cholecystokinin stimulates the release of digestive enzymes and alveolar cells specialize in secreting disaturated phosphatidylcholine, a major component of lung surfactant. In both cases, cellular phosphatidylcholine synthesis is reduced significantly when CCT protein levels are reduced, supporting a role for CCT expression in the regulation of cellular phospholipid content and function.

5 CONCLUSION

Alteration of CCT activity changes the flux through the phosphatidylcholine biosynthetic pathway and CCT is regulated biochemically by different lipid mixtures. CCT is activated by anionic lipids or neutral lipids when presented to the enzyme in the context of a phosphatidylcholine vesicle. The CCT helical domain mediates interaction with this lipid structure. CCT activity is inhibited by inclusion of lysolipids in the phosphatidylcholine vesicles by subverting enzyme insertion into the vesicle. CCT is also activated by phosphatidic acid when presented to the enzyme in a mixture with a non-ionic detergent. The carboxyl terminus mediates the second type of lipid interaction and suggests that CCTα and CCTβ2 regulation *in vivo* is not strictly dependent on association with membrane bilayers. The distribution of the CCTα, CCTβ1 and CCTβ2 isoforms in tissues likely contributes to the membrane characteristics and function of cellular compartments, particularly CCTα in the nucleus. Evidence argues against cell-cycle regulation of CCTα translocation outside of the nucleus. Changes in the cellular lipid environment govern the response of CCT through physical association between the protein and the membrane, allowing the cells to produce more or less phosphatidylcholine as needed.

Abbreviations: CCT, CTP:phosphocholine cytidylyltransferase; CTP, cytidine diphosphate; CDP-choline, cytidine diphosphocholine; ET-18-OCH$_3$, octaethyleneglycolmonohexadecxyl ether; HexPC, hexadecylphosphocholine; CK, choline kinase; CPT, choline phosphotransferase; DAG or DG, diacylglycerol; P-Cho, phosphocholine; PtdCho, phosphatidylcholine; PtdEtn, phosphatidylethanolamine; Sph, sphingomyelin; LPC, lysophosphatidylcholine; PtdOH, phosphatidic acid

Acknowledgements

We thank George Attard (University of Southampton) and Richard Templer (Imperial College) for valuable discussion and keen interest. The research is supported by the National Institutes of Health Grant GM45737, Cancer Center (CORE) Support Grant CA21765, the American Lebanese and Syrian Associated Charities, the Royal Society, and the Engineering and Physical Sciences Research Council.

References

1   C. Kent, *Annu. Rev. Biochem.*, 1995, **64**, 315.
2   G.B. Kalmar, R. J. Kay, A.C. LaChance and R.B. Cornell, *Biochim. Biophys. Acta*, 1994, **1219**, 328.
3   A. Lykidis, K.G. Murti and S. Jackowski, *J. Biol. Chem.*, 1998, **273**, 14022.

4   A. Lykidis, I. Baburina and S. Jackowski, *J. Biol. Chem.*, 1999, **274**, 26992.
5   M.S. Rutherford, C.O. Rock, N.A. Jenkins, D.J. Gilbert, T.G. Tessner, N.G. Copeland and S. Jackowski, *Genomics*, 1993, **18**, 698.
6   W. Tang, G.A. Keesler and I. Tabas, *J. Biol. Chem.*, 1997, **272**, 13146.
7   G.B. Kalmar, R.J. Kay, A. LaChance, R. Aebersold and R.B. Cornell, *Proc. Natl. Acad. Sci. U. S. A.*, 1990, **87**, 6029.
8   J.D. Esko, M.M. Wermuth and C.R.H. Raetz, *J. Biol. Chem.*, 1981, **256**, 7388.
9   T.D. Sweitzer and C. Kent, *Arch, Biochem. Biophys.*, 1994, **311**, 107.
10   R.K. Mallampalli, A.J. Ryan, R.G. Salome and S. Jackowski, *J. Biol. Chem.*, 2000, **275**, 9699.
11   D. Zhang, W. Tang, P.M. Yao, C. Yang, B. Xie, S. Jackowski and I. Tabas, *J. Biol. Chem.*, 2000, **275**, 35368.
12   L. Craig, J.E. Johnson and R.B. Cornell, *J. Biol. Chem.*, 1994, **269**, 3311.
13   Y. Wang and C. Kent. *J. Biol. Chem.*, 1995, **270**, 18948.
14   W. Yang, K.P. Boggs and S. Jackowski, *J. Biol. Chem.*, 1995, **270**, 23951.
15   R.B. Cornell, G.B. Kalmar, R.J. Kay, M.A. Johnson, J.S. Shanghera and S. L. Pelech, *Biochem. J.*, 1995, **310**, 699.
16   Y. Wang, J.I.S. MacDonald and C. Kent, *J. Biol. Chem.*, 1995, **270**, 354.
17   C.J. DeLong, L. Qin and Z. Cui, *J. Biol. Chem.*, 2000, **275**, 32325.
18   Y. Wang, T.D. Sweitzer, P.A. Weinhold and C. Kent, *J. Biol. Chem.*, 1993, **268**, 5899.
19   I.C. Northwood, A.H. Tong, B. Crawford, A.E. Drobnies and R.B. Cornell. *J. Biol. Chem.*, 1999, **274**, 26240.
20   M. Houweling, Z. Cui, C.D. Anfuso, M. Bussiere, M.H. Chen and D.E. Vance, *Eur. J. Cell Biol.*, 1996, **69**, 55.
21   A.N. Hunt and G.C. Burdge, *Biochem. Soc, Trans.*, 1998, **26**, S223.
22   A.N. Hunt, G.T. Clark, G.S. Attard and A.D. Postle, *J. Biol. Chem.*, 2001, **276**, 8492.
23   S. Jackowski, *J. Biol. Chem.*, 1994, **269**, 3858.
24   T.G. Tessner, C.O. Rock, G.B. Kalmar, R.B. Cornell and S. Jackowski, *J. Biol. Chem.*, 1991, **266**, 16261.
25   K.P. Boggs, C.O. Rock and S. Jackowski, *J. Biol. Chem.*, 1995, **270**, 7757.
26   K.P. Boggs, C.O. Rock and S. Jackowski, *Biochim. Biophys. Acta*, 1998, **1389**, 1.
27   R.B. Cornell, *Biochemistry*, 1991, **30**, 5881.
28   M.M. Luche, C.O. Rock and S. Jackowski, *Arch. Biochem. Biophys.*, 1993, **301**, 114.
29   W. Yang and S. Jackowski, *J. Biol. Chem.*, 1995, **270**, 16503.
30   J.A. Friesen, H.A. Campbell and C. Kent, *J. Biol. Chem.*, 1999, **274**, 13384.
31   A. Lykidis, P. Jackson and S. Jackowski, *Biochemistry*, 2001, **40**, 494.
32   J.E. Johnson and R.B. Cornell, *Mol. Membr. Biol.*, 1999, **16**, 217.
33   R.B. Cornell and I.C. Northwood, *Trends Biochem. Sci.*, 2000, **25**, 441.
34   J. Yang, J. Wang, I. Tseu, M. Kuliszewski, W. Lee and M. Post, *Biochem. J.*, 1997, **325**, 29.
35   J.E. Johnson and R.B. Cornell, *Biochem. J.*, 1994, **33**, 4327.
36   G.S. Attard, R.H. Templer, W.S. Smith, A.N. Hunt and S. Jackowski, *Proc. Natl. Acad. Sci. U.S.A.*, 2000, **97**, 9032.
37   R.B. Cornell, *Biochemistry*, 1991, **30**, 5873.
38   R.S. Arnold and R.B. Cornell, *Biochemistry*, 1996, **35**, 9917.
39   S.M. Davies, R.M. Epand, R. Kraayenhof and R.B. Cornell, *Biochemistry*, 2001, **40**, 10522.
40   J.I.S. MacDonald and C. Kent, *J. Biol. Chem.*, 1994, **269**, 10529.

41 M. Wieprecht, T. Wieder, C. Paul, C.C. Geilen and C.E. Orfanos, *J. Biol. Chem.*, 1996, **271**, 9955.
42 Y. Wang and C. Kent, *J. Biol. Chem.*, 1995, **270**, 17843.
43 H. Tronchere, M. Record, F. Terce and H. Chap, *Biochim. Biophys. Acta*, 1994, **1212**, 137.
44 M. Houweling, H. Jamil, G.M. Hatch and D.E. Vance, *J. Biol. Chem.*, 1994, **269**, 7544.
45 G.E. Groblewski, Y. Wang, S.A. Ernst, C. Kent and J.A. Williams, *J. Biol. Chem.*, 1995, **270**, 1437.
46 R.K. Mallampalli, R.G. Salome, C.H. Li, M. VanRollins and G.W. Hunninghake, *J. Cell Physiol.*, 1995, **162**, 410.
47 H. Tronchere, F. Terce, M. Record, G. Ribbes and H. Chap, *Biochem. Biophys. Res. Commun.*, 1991, **176**, 157.
48 C. Kent, *Biochim. Biophys. Acta*, 1997, **1348**, 79.
49 J.M. Clement and C. Kent, *Biochem. Biophys. Res. Commun.*, 1999, **257** 643.
50 F. Terce, M. Record, G. Ribbes, H. Chap and L. Douste-Blazy, *J. Biol. Chem.*, 1988, **263**, 3142.
51 F. Terce, M. Record, H. Tronchere, G. Ribbes and H. Chap, *Biochim. Biophys. Acta*, 1991, **1084**, 69.
52 C.J. Walkey, G.B. Kalmar and R.B. Cornell, *J. Biol. Chem.*, 1994, **269**, 5742.
53 I. Baburina and S. Jackowski, *J. Biol. Chem.*, 1999, **274**, 9400.
54 Y. Shiratori, M. Houweling, X. Zha and I. Tabas, *J. Biol. Chem.*, 1995, **270**, 29894.
55 S. Jackowski, J. Wang and I. Baburina, *Biochim. Biophys. Acta*, 2000, **1483**, 301.
56 K.P. Boggs, C.O. Rock and S. Jackowski, *J. Biol. Chem.*, 1995, **270**, 11612.
57 J.D. Esko, M. Nishijima and C.R.H. Raetz, *Proc. Natl. Acad. Sci. U.S.A.*, 1982, **79**, 1698.
58 Z. Cui, M. Houweling, M.H. Chen, M. Record, H. Chap, D.E. Vance and F. Terce, *J. Biol. Chem.*, 1996, **271**, 14668.
59 I. Baburina and S. Jackowski. *J. Biol. Chem.*, 1998, **273**, 2169.

MODELS AND MEASUREMENTS ON THE MONOLAYER BENDING ENERGY OF INVERSE LYOTROPIC MESOPHASES

A. M. Squires, J. M. Seddon and R. H. Templer

Department of Chemistry, Imperial College, London SW7 2AY, UK

1 INTRODUCTION

Aqueous dispersions of amphiphilic molecules such as lipids can form a rich array of mesophase structures, both lamellar and non-lamellar. A given lipid system can form different mesophase geometries, depending on the water content, as well as other factors such as temperature, pressure and pH. Here we present work carried out on a lipid system which is a mixture of lauric acid, LA, and dilauroylphosphatidylcholine, DLPC, in the molar ratio of 2:1 LA:DLPC. The mesophase structures formed by this system are shown in Figure 1. The lamellar phase simply consists of a stack of flat bilayers separated by layers of water. In the inverse hexagonal phase, cylinders of water surrounded by lipid monolayers stack to make a 2-dimensional hexagonal array. The three inverse bicontinuous cubic phases each consist of a single continuous lipid bilayer at the centre of which runs a mathematical surface known as a Triply Periodic Minimal Surface (TPMS).

If we change the geometry of a structure by adding or removing water, we have to bend the constituent lipid monolayers into different shapes, and this changes the energy of the system. Our research models and measures this as a function of monolayer curvature.

An understanding of the bending energy of lipid monolayers enables us to predict certain properties of a lipid system, such as the mesophase geometry and lattice parameter adopted in excess water. In addition, certain biological processes involving changes in membrane topology require intermediate structures of higher curvature, so a study of the energy associated with curvature will help us to understand the energetics of such transformations in biological systems. Furthermore, there is considerable evidence that bending energy itself is used in the regulation of the activity of a number of membrane-bound proteins.

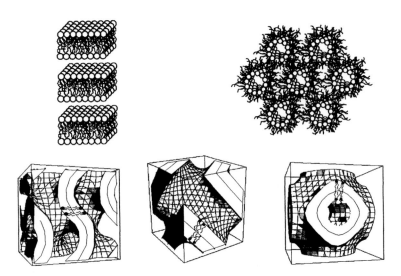

Figure 1. *Structures of the liquid crystalline mesophases formed by 2LA/DLPC/water. Top left: Lamellar (L_α). Top right: Inverse Hexagonal (H_{II}). Bottom: Inverse Bicontinuous Cubic (Q_{II}), from left to right: Gyroid ($Q_{II}{}^G$), Diamond ($Q_{II}{}^D$) and Primitive ($Q_{II}{}^P$).*

2 BACKGROUND

2.1 Theory

Our treatment assumes that lipid molecules have a preferred chain splay, and therefore that a given lipid would prefer to form monolayers of a certain constant curvature. If the monolayer is forced to adopt a structure of a different geometry, then its curvature would have to deviate away from this value, and in each lipid molecule the chains regions would be squashed or splayed into a different shape. This incurs an energetic cost known as the *curvature elastic energy*. We would expect the curvature elastic energy to increase in the sequence spheres < cylinders < saddles < flat planes[1].

At any point on a surface, the curvature can be described by two principle curvatures, c_1 and c_2. In this work we combine these two curvatures to give the mean curvature, $H = \frac{1}{2}(c_1+c_2)$, and the gaussian curvature, $K = c_1 \times c_2$. The dependence of curvature elastic energy on the curvature of the monolayer is assumed to take the following quadratic form, taken from Helfrich's Ansatz[2]

$$g_c = 2\kappa(H-H_0)^2 + \kappa_G K \tag{1}$$

where g_c is the curvature elastic energy per unit area of the monolayer. The curvature elastic behaviour of the lipid system is characterised by three constants, H_0, κ and κ_G. These are, respectively, the spontaneous mean curvature, and the bending stiffness moduli associated with mean and gaussian curvature. Here, we present work seeking to measure the values of these three quantities for the 2LA/DLPC system at 42°C.

The surface at which we make our measurements of monolayer curvature is known as the 'pivotal' (or 'area-neutral') surface. This surface corresponds to the position on the lipid molecules where the area per molecule does not change as the monolayer is bent at a constant temperature. [3, 4].

2.2 Experimental Techniques

In this work we present results obtained using three different ways of probing the curvature elastic behaviour of 2LA/DLPC. These are as follows.

2.2.1 Excess water point. Over a certain range of temperatures and water contents, the 2LA/DLPC system forms one of three inverse bicontinuous cubic (Q_{II}) phases. Below excess water conditions, a Q_{II} phase is geometrically constrained to adopt a certain lattice parameter, where the monolayers have a greater curvature than their preferred value. The addition of water allows the lattice to swell, increasing the lattice parameter and so lowering the monolayer curvatures and thus the curvature elastic energy of the system. As more water is added, the lattice parameter continues to increase until the curvature elastic energy reaches a minimum. Beyond this point, there is no longer an energetic incentive for the mesophase to swell any further. Thus at higher hydrations, the lattice parameter of the mesophase remains constant, and the mesophase co-exists with excess water. Measurement of the lattice parameter in excess water enables us to determine the geometry and curvatures of the system at which the curvature elastic energy is minimised[1].

2.2.2 Addition of a hydrophobic liquid. The use of a hydrophobic liquid to relieve packing stress follows an approach first used by Rand et al[5] to determine the bending energy of DOPE in the H_{II} phase.

2LA/DLPC forms Q_{II} phases in excess water only because of the packing frustration energy associated with the formation of the H_{II} phase; this is the energy associated with the variations in lipid chain extension in order to satisfy the requirement that the mesophase fills all space[6]. Purely from curvature elastic considerations, the H_{II} phase is more stable than the Q_{II} phases, a general result for systems where $0 > \kappa_G / \kappa > -1$[7]. (The value of κ_G / κ for 2LA/DLPC does indeed fall in this range, as we shall show later.) When a hydrophobic liquid such as an alkene is added to a type II lyotropic liquid crystal, it preferentially partitions into regions of higher packing stress. This can lower the packing frustration energy enough to induce the formation of the H_{II} phase instead of the inverse bicontinuous cubic. If enough alkene is added, the system reaches an energetic minimum, and no further change in lattice parameter is observed as still more alkene is added. This method allows us to calculate H_0, which is the value of the mean curvature at this point.

2.2.3 Osmotic stress measurements. The osmotic stress technique gives a direct measurement of the amount of energy required to remove water from a mesophase. The mesophase is placed in contact with water of a (known) lower chemical potential. This sucks water out of the mesophase until the energetic advantage in moving water to a lower chemical potential is balanced by the increase in energy of the mesophase as it is dehydrated. The technique has been used to measure curvature elastic parameters in the H_{II} phase in DOPE[5] and 1-MO[7], and in Q_{II} phases in 1-MO[8]. Osmotic stress measurements give information on curvature elastic parameters in these phases because the energetic cost of dehydrating the mesophase directly reflects the energy required to increase the monolayer curvature around progressively thinner water channels.

2.3   Previous Work on the 2LA/DLPC System

The phase behaviour of the 2LA/DLPC/water system has already been studied[9]. Its Temperature-Composition phase diagram is shown in Figure 2.

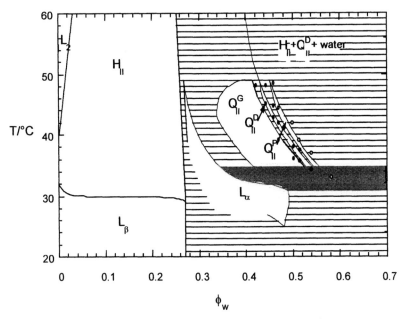

Figure 2. *Temperature-Composition phase diagram of 2LA/DLPC/water.*

The position of the pivotal surface for 2LA/DLPC has been determined from lattice parameter / water content data[9]. The area per molecule at the pivotal surface and the volume of the chain regions below it were found to take the values of $A_n=108\pm2\text{Å}^2$ and $v_n=1230\pm30\text{Å}^3$ respectively. We note that the 'molecule' in this stage refers to an 'average' unit consisting of one DLPC and two LA molecules. The total volume per molecule, v has also been determined by density measurements, and was found to be 1735 Å^3.

3   EXPERIMENTAL METHODS

3.1 Sample Preparation

DLPC and LA were obtained from Avanti Polar Lipids (Alabaster, AL) and Fluka respectively, each at >99% purity, according to the manufacturers. The purity of both lipids had been confirmed by thin-layer chromatography. The lipids were left overnight in a desiccator before use to remove residual moisture. 2LA/DLPC mixtures were prepared by co-dissolving the appropriate ratio of LA and DLPC in cyclohexane (BDH chemical supplies, Dorset, UK), and removing the solvent by freezing to liquid nitrogen temperature and then freeze-drying overnight. To each was added an appropriate amount of milipore-filtered distilled water. The sample was then incubated overnight at 42°C, which is approximately 10°C above the chain melting temperature. It was then allowed to cool to room temperature before being homogenised by centrifuging to force the components through a 1mm diameter hole in a Teflon 'hourglass'. The sample was centrifuged through this device approximately 10 times, until it appeared homogeneous.

3.1.1 Addition of a hydrophobic liquid. For the samples containing alkene, the lipids were co-dissolved in an exact weight of cyclohexane. To this was added a weighed amount of the alkene. The alkene used was 1-octadecene (Sigma chemical company, St Louis, MO) in a cyclohexane solution of known concentration (approximately 10% by weight). The mixture was frozen in liquid nitrogen and freeze-dried overnight. The following day, the sample was re-weighed to ensure that no lipids or alkene had been removed by the freeze-drying procedure. The sample was then hydrated to 67% water by weight, incubated overnight at 42°C, allowed to cool, then homogenised using the centrifugation procedure described above.

This experiment requires that there is no significant change in monolayer properties caused by penetration of the alkene into the lipid monolayer. It has been shown that using alkene molecules of chain length greater than those of the amphiphiles minimises the possibility of such penetration[7]. Since we use the 18-C alkene 1-octadecene while the LA and DLPC lipid chains are both 12-C, we assume that no significant chain penetration occurs.

3.1.2 Osmotic stress method. The osmotic stress method began with hydrated lipid samples of 60% water by weight. Lipid samples of around 0.1ml were osmotically stressed overnight at 42°C using a solution of polyethyleneglycol (PEG) 6000 (BDH, Dorset, UK), through a semi-permeable membrane made of Visking tubing (Medicell, London, UK) which had been pre-treated by boiling three times in micro-pure water. After osmotic stress, the sample was rapidly transferred to a sealed container. Some was then analysed using x-ray diffraction. The remainder was weighed, both before and after freeze-drying, to determine the water content.

3.2 Sample Analysis

3.2.1 Mass analysis. In order to determine the water content of a sample after osmotic stress, the sample was first weighed. It was then frozen at -20°C for >3hours, and freeze-dried in ice/water/salt overnight in order to remove all water, and the dry sample re-weighed.

3.2.2 X-ray diffraction. The mesophase and lattice parameter of a sample were determined by x-ray crystallography. The sample was analysed either in a glass capillary tube of diameter 1.5mm, wall thickness 0.01mm (Pantak ltd., Reading, UK) which was sealed with heatshrink tubing (RS supplies, UK), or else in a sealed cell made from a Teflon spacer of thickness 1mm, with Mylar windows attached using double-sided tape, clamped between two aluminium spacers in a brass sample cell holder. The sample was kept at the required temperature in the x-ray machine for 20 minutes before the start of the diffraction experiment, and several successive x-ray diffraction exposures were taken, in order to ensure equilibration. Exposure times of between 2 and 20 minutes were used.

X-rays were produced on a GX-20 rotating anode generator (Enraf-Nonius, Netherlands) run at 20kV and 25mA, monochromated and focused using a nickel filter and Franks double mirror optics to select for Cu K_α X-rays ($\lambda = 1.54$ Å). The beam was isolated and further attenuated with a set of adjustable tungsten slits in the sample chamber, to produce a beam of dimensions 0.15 mm x 0.1 mm, with a typical flux of 1 million photons per second.

The sample temperature was regulated using peltier heating. This was computer-controlled via a Scorpion K4 micro-controller, which also enabled automated shutter operation. The equipment provided temperature control up to 80 °C, to an accuracy of ±0.03 °C. The sample chamber and optics were kept under vacuum, to minimise both loss in beam intensity and scatter due to air. The sample chamber included a variable-length flight tube with a lead beam stop at the end. This set-up allowed the X-ray detector to be placed between 100 and 300 mm from the sample, keeping an evacuated X-ray path throughout.

The x-ray diffraction pattern was collected on a (Zn,Cd)S:Ag phosphorescent screen, intensified using a high-voltage image intensifier, and then focused onto a CCD camera (Wright instruments) which relayed the image back to the control computer. The diffraction pattern was analysed using the TV4 software program, originally written by Prof. S. M. Gruner (Cornell University) and Prof. E. F. Eikenberry (Paul Scherrer Institute, Switzerland), with more specific alterations made for our purposes by Neil Warrender and Prof R. Templer, at Imperial College.

The lipid samples were powder-like, and therefore gave diffraction patterns which appeared as a series of concentric rings. The radius of each ring can be converted to a scattering angle for a single reflection, and therefore into a d-spacing for a particular set of planes within the mesophase. A given mesophase gives a set of reflections where the d-spacing ratios are characteristic of the symmetry of the mesophase, and the actual values may be used to calculate its lattice parameter. This was carried out by the TV4 software package. The H_{II} phase was characterised by the $\sqrt{1}$, $\sqrt{3}$, $\sqrt{4}$ reflections, the Q_{II}^D phase by the $\sqrt{2}$, $\sqrt{3}$, $\sqrt{6}$, $\sqrt{8}$ and $\sqrt{9}$ reflections, the Q_{II}^G phase by $\sqrt{6}$, $\sqrt{8}$ and a broad peak from the $\sqrt{20}$, $\sqrt{22}$, $\sqrt{24}$ and $\sqrt{26}$ reflections, and the Q_{II}^P phase by the $\sqrt{2}$, $\sqrt{4}$, and $\sqrt{6}$ reflections. The beamline was calibrated with silver behenate, and could measure lattice parameters to an accuray of ±0.1Å.

4 RESULTS, CALCULATIONS AND DISCUSSIONS

4.1 Excess Water Measurements

4.1.1 Results. Previous measurements at fixed hydration on 2LA/DLPC [9] indicate that the excess water point lies close to a water volume fraction of $\phi_w = 0.55$ at 35°C, and that the water volume fraction at the excess water point decreases with increasing temperature (see Figure 2). A sample of 2LA/DLPC was therefore prepared with a water volume fraction of $\phi_w = 0.595$ (60% water by weight). A temperature-scan was carried out on this sample, with x-ray diffraction patterns taken at intervals of 2°C. The phase sequence and lattice parameters determined from the diffraction patterns are shown in Figure 3.

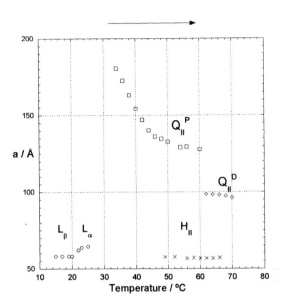

Figure 3. *Temperature-scan of 2LA/DLPC/60% water by weight. Diffraction patterns were obtained using 10-minute exposures, with an equilibrium time of 20 minutes at each 2-degree increment. The temperature-scan was carried out in order of increasing temperature. Between 28 and 34°C, no well-defined reflections were observed. This was probably because the system was not forming a sufficiently ordered structure.*

The strong temperature-dependence of the lattice parameter indicates that the sample is co-existing with excess water throughout.

4.1.2 Calculations In order to model the lattice parameters of the Q_{II} phases, we use a parallel-interface model where the pivotal surface for each monolayer lies a uniform distance ξ_n away from the underlying minimal surface. Throughout this section we locate the pivotal surface by using values for molecular dimensions quoted in Section 2. These are $A_n = 108 \pm 2 \text{Å}^2$, $v_n = 1230 \pm 30 \text{Å}^3$ and $v = 1735 \text{ Å}^3$.

We can calculate ξ_n for each lattice parameter by solving the following equation at each different lattice parameter[9]

$$\frac{\langle v_n \rangle}{\langle v \rangle}(1 - \phi_w) = 2\sigma\left(\frac{\xi_n}{a}\right) + \frac{4}{3}\pi\left(\frac{\xi_n}{a}\right)^3 \tag{2}$$

The water volume fraction within the mesophase, ϕ_w, is calculated using the following equation, whose derivation is outlined elsewhere[10]:

$$(3)$$

$$\phi_w = 1 - \frac{4v^3\sigma^3\left(a^3 A_n^3 v^6 \sigma^6 + \sqrt{v^{12}\sigma^9\left(a^6 A_n^6 \sigma^3 - (a^2 A_n^2 \sigma - 6\pi v_n^2 \chi)^3\right)}\right)^{\frac{1}{3}}}{a^2 A_n^2 v^4 \sigma^4 - 6\pi v^4 v_n^2 \sigma^3 \chi + \left(a^3 A_n^3 v^6 \sigma^6 + \sqrt{v^{12}\sigma^9\left(a^6 A_n^6 \sigma^3 - (a^2 A_n^2 \sigma - 6\pi v_n^2 \chi)^3\right)}\right)^{\frac{2}{3}}}$$

In these equations, σ is the dimensionless surface area of the TPMS in the unit cell, and has values of 3.0910, 1.9189 and 2.3451 for Q_{II}^G , Q_{II}^D and Q_{II}^P respectively.[9]. The corresponding values of χ, the Euler characteristics of the TPMS, are $-8,-2$ and -4.

Equation 1 can be re-written using values for the mean and gaussian curvatures for a Q_{II} phase, assuming a pivotal surface lying parallel to the TPMS[1]. This gives a first-order expression for the surface-averaged curvature elastic energy per unit area of the monolayer, $<g_c>$, in terms of ξ_n .

$$\langle g_c \rangle = 2\kappa H_0^2 + (\kappa_G - 4\kappa H_0 \xi_n)\left[\frac{A_n \xi_n - v_n}{\xi_n^2 (v_n - A_n \xi_n /3)}\right] \qquad (4)$$

The excess water point corresponds to an energetic minimum. Thus differentiating Equation 4 with respect to ξ_n and equating to zero gives us a relationship between κ_G/κ and H_0 at the excess water point, where $\xi_n = \xi_0$.

$$\frac{\kappa_G}{\kappa} = 2H_0 \xi_0 \left[\frac{3 - 2\langle A_n \rangle \xi_0 /\langle v_n \rangle + (\langle A_n \rangle \xi_0 /\langle v_n \rangle)^2}{3 - 3\langle A_n \rangle \xi_0 /\langle v_n \rangle + (\langle A_n \rangle \xi_0 /\langle v_n \rangle)^2}\right] \qquad (5)$$

For example, at 42°C, the system is in a Q_{II}^P phase, with lattice parameter $a(Q_{II}^P)=147.0\pm0.5\text{Å}$. From Equation 3 we can calculate the water volume fraction, ϕ_w $=0.518\pm0.001$. We note that this is the water volume fraction *within the mesophase*, and that the value is smaller than 0.595, the water volume fraction of the sample overall. This confirms that the mesophase is coexisting with excess water. From these values of ϕ_w and the lattice parameter a, we can calculate ξ_n in excess water (ξ_0), and thus from Equation 5 determine the quantity $H_0(\kappa/\kappa_G)$. This treatment yields the following values at 42°C.

$$\xi_0 = 10.93 \pm 0.01\text{Å}$$
$$H_0(\kappa/\kappa_G) = 0.0238 \pm 0.0002\text{Å}^{-1}.$$

We can perform the same calculations on the rest of the data shown in Figure 3, to see how the quantity $H_0(\kappa/\kappa_G)$ varies with temperature. The estimated values are shown in Figure 4.

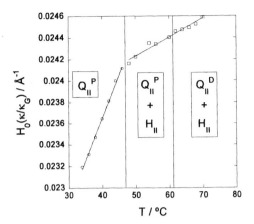

Figure 4. *Variation of $H_0(\kappa/\kappa_G)$ with temperature for 2LA/DLPC.*

The graph shown a linear increase in $H_0(\kappa/\kappa_G)$ with temperature up to 48°C, given by $H_0(\kappa/\kappa_G)$ (Å^{-1}) = 0.0204 + 8.0 x 10^{-5} T(°C). Above 48°C, the graph shown a change in behaviour; $H_0(\kappa/\kappa_G)$ still shows a linear increase with temperature, but with a different gradient, and the behaviour is described by $H_0(\kappa/\kappa_G)$(Å^{-1})=0.0234 + 1.7 x 10^{-5} T(°C). The sudden change in behaviour occurs at the point at which the H_{II} phase first appears. The most likely explanation for this is that a slight partitioning of LA and DLPC is occurring, which leads to a difference in lipid composition between the two phases.

4.2 Addition of Alkene

4.2.1 Results. Throughout this section we refer to the alkene content of a sample in terms of n_a, the number of molecules of 1-octadecene per 2LA/DLPC, calculated from the masses of the constituent components using values of m_w(2LA/DLPC) = 1022.5g mol^{-1} and m_w(1-octadecene) = 252.5g mol^{-1}. The lattice parameter of the H_{II} phase formed with samples of different alkene contents in excess water varies with the alkene content as shown in Figure 5.

As we can see, the lattice parameter, $a(H_{II})$, increases with alkene content, n_a, up to a certain point, the 'excess alkene point'. This corresponds to an energetic minimum where the mean curvature H is equal to the spontaneous mean curvature, H_0. Beyond this point, $a(H_{II})$ remains constant. From Figure 5, we calculate that the excess alkene point corresponds to values of n_a=0.42±0.06 and $a(H_{II})$ =79.6±0.6Å.

4.2.2 Calculations. We use the values quoted before, of A_n=108±2Å^2, v_n=1230±30Å^3 and v =1735 Å^3. We also define a new quantity, v_n', which represents an average value for the overall volume of the hydrophobic region below the pivotal surface, including both the chain regions and the alkene. Given that for 1-octadecene, m_w(1-octadecene) = 252.5g mol^{-1} and ρ(1-octadecene) = 0.788g ml^{-1}, and that n_a=0.42 at the excess alkene point, we can calculate that, at this point, v_n'=1450±40Å^3.

We can now calculate the radius of the pivotal surface at the excess alkene point, R_p, using Equation 6. The derivation of this equation is given elsewhere[10].

$$R_p = \frac{-2v'_n + \sqrt{\dfrac{2\sqrt{3}a^2 A_n^2}{\pi} + 4v'^2_n}}{2A_n} \qquad (6)$$

From our experimental results, this gives a value of R_p = 30±1Å. We assume that the spontaneous mean curvature for 2LA/DLPC, H_0, is equal to the mean curvature at this point, which is given by $H = 1/(2R_p)$. From this we estimate H_0 = -0.0167±0.0006Å^{-1}.

We can now combine this with our value of 0.0238 for the quantity $H_0(\kappa/\kappa_G)$ at this temperature, to give a value for the ratio (κ_G/κ) of κ_G/κ = -0.70±0.03. We note that this does indeed fall in the predicted range of $0 > \kappa_G/\kappa > -1$ mentioned earlier.

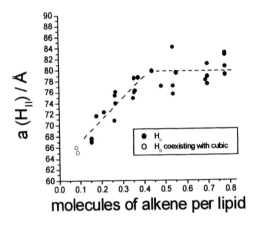

Figure5. *Effect of alkene content on the lattice parameter of the H_{II} phase in 2LA/DLPC at 42°C.*

4.3 Osmotic Stress Experiments

4.3.1 Results. The analysis of each sample following osmotic stress both by weight and by x-ray diffraction yields two sets of data; both the water content and the phase and lattice parameter can be plotted as a function of the osmotic pressure applied by the polymer solution. The polymer solution, PEG 6000, has been characterised at a range of concentrations[11]. Online datahttp://aqueous.labs.brocku.ca/data/peg6000 is available for the osmotic pressure applied, Π, as a function of the weight% of polymer, and it can be described by the empirical fit $log(\Pi /Pa) = a + b(wt\%)^c$, with the parameters a=5.12, b=0.28 and c=0.59.

This equation was used to calculate the osmotic pressure applied, and plots for the water content and lattice parameters of the 2LA/DLPC samples following this osmotic stress are shown in Figure 6.

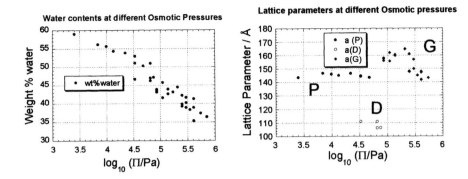

Figure 6. *Water contents (left) and lattice parameters (right) of osmotically stressed 2LA/DLPC at 42°C.*

4.3.2 Calculation of energy. The energy of the mesophase after osmotic stressing to different levels of dehydration may be calculated from the osmotic stress data as follows.

- The weight% of water in the sample is converted to a volume of water per 2LA/DLPC, v_w.
- The osmotic pressure Π is plotted against this quantity. This is shown in Figure 7.
- A smooth curve is fitted through the data in order to give us a function which we can integrate. The function chosen in this case has the form $\Pi/Pa = a \times (v_w/Å^3)^{-b}$. The fit is shown in Figure 7, with the parameters $a = 3.1215 \times 10^{20}$ and $b = 4.936$.

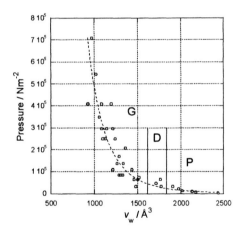

Figure 7. *Π vs water content for osmotic stress of 2LA/DLPC at 42°C.*

- The function is integrated to give the energy required to dehydrate to a given water content, per 2LA/DLPC. (In this case we assume that this is the increase in curvature elastic energy.)

$$E = -\int \Pi dv_w \tag{7}$$

Using the curve fit chosen, this gives

$$E = E_0 + 10^{-30} \frac{a}{b-1} v_w^{-(b-1)} = E_0 + (7.9296 \times 10^{-11}) v_w^{-3.936} \tag{8}$$

Here, E is the energy per 2LA/DLPC, in J, E_0 is the integration constant, and v_w the volume of water per 2LA/DLPC, in Å^3.

4.3.3 Calculation of κ and κ_G. While it is possible to extract values of κ and κ_G from the data for the integrated energy calculated as described in the previous section [13], there are statistical disadvantages to analysing the data in this way[10]. It is more appropriate to fit the data for Π directly, rather than E.

Before this can be done, we need a model for the actual geometry of the pivotal surface at which we will be measuring the mean and gaussian curvatures of the monolayers. There is a choice of two possible geometries[14]. The first is a 'parallel interface' model where the pivotal surface lies a uniform distance ξ_n from the underlying TPMS, as described in Section 4.1.2. The second is a 'constant mean curvature' interface, where the distance of the pivotal surface from the TPMS may vary over the mesophase, but where its mean curvature H is assumed to be uniform.

Parallel interface model. Using the parallel interface model, we can obtain an expression for Π in terms of ξ_n. The derivation for this is contained elsewhere[10], and it yields the following result.

$$\Pi = \frac{-2\sqrt{\frac{6}{\pi}}\ \sigma_0\ (v_n - A_n\xi_n)\ \sqrt{-(\sigma_0 v_n)0 + A_n\sigma_0\xi_n}\ \left(2H_0^{\,2}\kappa + \dfrac{(-v_n + A_n\xi_n)(\kappa_G - 4H_0\kappa\xi_n)}{\xi_n^{\,2}\left(v_n - \dfrac{A_n\xi_n}{3}\right)}\right)}{A_n\ \sqrt{-\left(\chi\xi_n^{\,2}(-3v_n + A_n\xi_n)\right)}} \tag{9}$$

We use the parameters for v, v_n and A_n given earlier, and a value of $H_0 = -0.0167 \pm 0.0006\text{Å}^{-1}$ measured in the experiments involving the addition of alkene, described in the previous section. The curve fit can thus provide estimes of the two parameters κ and κ_G. This has been carried out for the osmotic stress data from the Q_{II}^{G} phase. The data and curve fit are shown in Figure 8. The curve fit shown yielded values of $\kappa = 1.4 \pm 0.4 \times 10^{-19}$ J, and $\kappa_G = -1.0 \pm 0.3 \times 10^{-19}$ J.

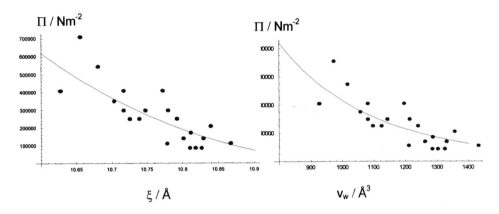

Figure 8. *Left: Osmotic pressure Π as a function of the distance ξ_n of a (parallel) pivotal surface from the underlying minimal surface, with curve fit. Right: Osmotic pressure Π as a function of the volume of water per lipid, v_w, and curve fit for a pivotal surface of constant mean curvature. Both data sets calculated from measurements taken on osmotically stressed 2LA/DLPC at 42°C.*

Constant mean curvature model. Assuming the constant mean curvature model, we can generate an equation for the energy per lipid, E, as a function of the volume of water per lipid, v_w. Again, the derivation of this is contained elsewhere[10]. The expression generated is as follows.

$$E = A_n \left[\frac{\pi \chi A_n^2 \left(1 - \frac{v_w}{v+v_w}\right)^2 \kappa_G}{2v^2 \left(\sum_{i=0} \sigma_i \frac{v_n^{2i}(1-(v_w/(v+v_w)))^{2i}}{v^{2i}}\right)^3} + 2\kappa \left\{-H_0 - \frac{A_n \left(\sum_{i=0} \zeta_i \frac{v_n^{2i+1}(1-(v_w/(v+v_w)))^{2i+1}}{v^{2i+1}}\right)}{2v \left(\sum_{i=0} \sigma_i \frac{v_n^{2i}(1-(v_w/(v+v_w)))^{2i}}{v^{2i}}\right)}\right\}^2 \right] \quad (10$$

Values for the coefficients ζ_i and σ_i for the different Q_{II} phases are given elsewhere[14]. This equation can be differentiated with respect to v_w to give an expression for the osmotic pressure Π. We can then use this to fit data of Π against v_w to generate κ and κ_G. In practice, the expression for Π is too cumbersome to copy down and then input into a separate data analysis package, so the expression is generated in a software package such as

Mathematica, which can use this output directly as a curve fit. We used this to fit the same data as in the previous section, with the same values for v, v_n and A_n, and still using a value of $H_0 = -0.0167\pm0.0006\text{Å}^{-1}$. The data and curve fit are shown in Figure 8. This yielded values of $\kappa=3\pm1\times10^{-20}$ J, and $\kappa_G=-9\pm2\times10^{-21}$ J.

We can compare the values we have obtained in this way with bending moduli measured on DLPC bilayer vesicles[15, 16], which provide a value of 0.91×10^{-19} J for bilayer bending, or 4.6×10^{-20} J for the curvature elastic modulus of the monolayer (which we can equate to κ). Our estimate for κ is three times as great as this if we calculate it assuming a parallel pivotal surface, but *smaller* by one-third if we assume a pivotal surface of constant mean curvature.

Our results therefore indicate that is is possible to measure approximate values for curvature elastic parameters using the techniques described, and that these agree with values obtained using other techniques to within an order of magnitude. However, our results clearly suggest that we should introduce a note of caution. Our interpretation of the osmotic stress data is heavily model-dependent; the estimated value of κ varies by a factor of 5, depending on whether we assume a parallel pivotal surface or one of constant mean curvature.

Acknowledgements. We wish to acknowledge funding from the EPSRC.

References

1   R. H. Templer, B. J. Khoo, and J. M. Seddon, *Langmuir*, 1998, **14**, 7427.

2   W. Helfrich, *Z. Naturforsch*, 1973, **28c**, 693.

3   M. M. Kozlov, S. Leikin, and R. P. Rand, *Biophysical Journal*, 1994, **67**, 1603.

4   R. H. Templer, *Langmuir*, 1995, **11**, 334.

5   R. P. Rand, N. L. Fuller, S. M. Gruner, and V. A. Parsegian, *Biochemistry*, 1990, **27**, 76.

6   G. L. Kirk, S. M. Gruner, and D. L. Stein, *Biochemistry*, 1984, **23**, 1093.

7   H. Vacklin, B. J. Khoo, K. H. Madan, J. M. Seddon, and R. H. Templer, *Langmuir*, 2000, **16**, 4741.

8   H. Chung and M. Caffrey, *Nature*, 1994, **368**, 224.

9   R. H. Templer, J. M. Seddon, N. A. Warrender, A. Syrykh, Z. Huang, R. Winter, and J. Erbes, *Journal Of Physical Chemistry B*, 1998, **102**, 7251.

10   A. Squires, 'PhD Thesis', Imperial College, 2001.

11   R. P. Parsegian, R. P. Rand, N. L. Fuller, and D. C. Rau, *Methods in Enzymology*, 1986, **127**, 400.

12   http://aqueous.labs.brocku.ca/data/peg6000

13   S. C. Roberts, PhD thesis, *Imperial College, London*, 1998.

14   R. H. Templer, J. M. Seddon, P. Duesing, R. Winter, and J. Erbes, *Journal of Physical Chemistry: B*, 1998, **102**, 7262.

15   L. Fernandez-Puente, I. Bivas, M. D. Mitov, and P. Meleard, *Europhysics Letters*, 1994, **28**, 181.

16   P. Meleard, C. Gerbeaud, P. Bardusco, N. Jeandaine, M. D. Mitov, and L. Fernandez-Puente, *Biochimie*, 1998, **80**, 401.

HEMOLYTIC AND ANTIBACTERIAL ACTIVITIES OF LK PEPTIDES OF VARIOUS TOPOLOGIES : A MONOLAYER AND PM-IRRAS APPROACH.

S. Castano[1]; B. Desbat[1]; H. Wróblewski[2] and J. Dufourcq[3]

[1] LPCM-CRCM, UMR CNRS 5803, University Bordeaux I, 33405 Talence, France
[2] UMR CNRS 6026, University Rennes I, 35042 Rennes, France
[3] CRPP-CNRS, 33600 Pessac, France

1 INTRODUCTION

Most living species secrete a wide range of cytotoxic peptides which play a critical role against external agression either with specific antimicrobial activity or aspecific cytotoxicity. Most of these natural peptides enhance the permeability of the biological membranes generally by a direct perturbation of the lipid matrix[1-4]. Intensive studies have demonstrated that i) peptides generally display a positive net charge and a high amphipathic character which allow them to adopt mainly amphipathic α-helical or β-sheet secondary structures in the membrane environment[1-6]; ii) the enhancement of these parameters increases their binding ability and hence their membrane activity; iii) binding is only the first step: for the same amount of peptide bound, very different activities are obsered, i.e. an efficiency for increasing permeability can be defined. Therefore numerous rationalized amphipathic peptides have been designed to mimic these natural active compounds. The very minimal requirement to get an amphipath being to associate properly

Peptide	Putative amphipathicity	Sequence 1 5 10 15
i.a. LK$_{15}$	α helix	K L L K L L L K L L L K L W K
Alterned (KL$_7$)K	β sheet	Dns-K L K L K L K L K L K L K L K$_{CONH2}$
Regular LK$_{16}$	3$_{10}$ helix	K L L K L L K L L K L L K L W K
Scrambled LK$_{15}$	no	L K L L L L K L L K L K L W K

Table 1: *Name, putative amphipathicity and sequence of the studied peptides*

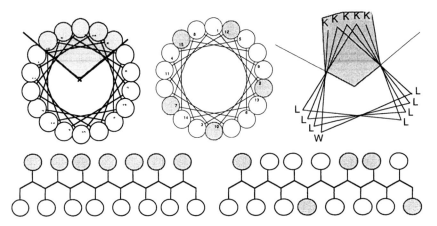

Figure 1 :*Representation of the peptide, from top to bottom and left to right: Schiffer-Edmundson's helical wheels of i.a.LK$_{15}$ and scrambled LK$_{15}$; vertical projection of the putative 3$_{10}$ helix of regular LK$_{16}$; β-sheet representations of alterned (KL)$_7$K and scrambled LK$_{15}$. White circles: L, leucine residue; grey circles: K, lysine residue.*

two residues of opposite properties, this led to the design of several families of peptides out of which those composed of only leucine (L), and lysine (K), representative of apolar residues or polar and charged ones respectively. Such LK peptides proved to be efficient, versatile and relevant to the problem of getting new active compounds which further improve our understanding of their mechanisms[7-10]. Ideally secondary amphipathic structures were obtained upon folding such peptides only playing with the composition, i.e. the L/K molar ratio, and the charged residue periodicity, either 2, 3 or 3.6[7-17].

Here we compare the role of the amphipathic topology of a series of LK peptides on their hemolytic and antibacterial activities. Four peptides of 15 or 16 residues were then designed with nearly identical composition (L/K ≈ 2 molar ratio) to generate an ideally amphipathic α-helix (i.a.LK$_{15}$), a 3$_{10}$ helix (regular LK$_{16}$) and a non amphipathic peptide (scrambled LK$_{15}$); the ideally β-sheeted analog (alterned (KL)$_7$K) differs since it requires a 1:1 L:K composition (table1, figure1). We chose *Spiroplasma melifferum* mollicutes as bacterial targets because they proved to be sensitive to many membrane active natural peptides and allowed monitoring of toxin activity at different levels[18]. To adress the problem of the mechanism of interaction and perturbation of lipids by these peptides, they were allowed to interact with lipid monolayers whose compositions mimic biological environments. The peptides secondary structures and orientation, bound to the lipids were determined *in situ* using the Polarization Modulation Infra-Red Reflection Absorption Spectroscopy (PM-IRRAS) technique[19-23].

2 METHODS AND RESULTS

2.1 Hemolytic and antibacterial activities of the peptides

The hemolytic activity of the peptides was measured on human erythrocytes as already described[24]. The dose-response curves show very similar sigmoïdal shapes, total lysis occuring for concentrations <20µM. The lethal doses for 50% of lysis, LD_{50} values, listed in table 2, vary in a ten fold range with the following hierarchy:

i.a.LK15 < alternated $(KL)_7K$ < regular LK_{16} < scrambled LK_{15}.

The antibacterial activity of the peptides, characterized by minimal inhibition concentrations (MICs) on cultured *S. melliferum* BC3 mollicutes[25], were determined by growing the bacteria in the presence of peptide. All assays were performed in triplicate and the results are summarized in table 2. All the peptides are active with MICs≤100µM. The hierarchy of MICs differs from that observed for hemolysis:

i.a.LK15 ≈ regular LK_{16} << alternated $(KL)_7K$ ≈ scrambled LK_{15}.

The concentration range of bacterial activity is 2 to 66 fold higher than for hemolysis. This could be due to the different lipid composition of the two cell membranes and to the growth medium which leads to peptide loss by aspecific interactions.

2.2 Peptide insertion into lipid monolayers

Since both biological activities vary differently according to the peptide topology, the interaction of the peptides with model membranes such as lipid monolayers at the air/water interface of a Langmuir trough was investigated. Monolayers of two different compositions were formed to mimic the two different biological systems: a zwitterionic monolayer of DMPC to mimic the erythrocyte membrane, since it was already found that PC lipids can account for a parallel variation of cytotoxic lytic activity[21,23], and a monolayer of natural

Peptide	Biological activity		Lipid affinity (10^6 M^{-1})	
	LD_{50} (µM)	MIC (µM)	DMPC	Sm lip
i.a. LK_{15}	0,5	6,25	20,5	506
Alternated $(KL_7)K$	1,5	100	1,2	n.d.
Regular LK_{16}	3,0	6,25	0,5	508
Scrambled LK_{15}	5,0	100	0,5	665

Table 2: *Hemolytic and antibacterial activities of the peptides on erythrocytes (LD_{50}) and on Spiroplasma melliferum BC3 (MIC) respectively, and peptide affinities for lipid monolayers (DMPC and Spiroplasma lipids (Sm lip)). n.d.: not determined.*

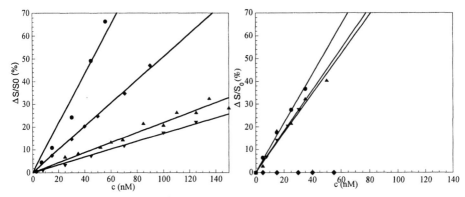

Figure 2: *Relative film surface increases due to peptide insertion at $\Pi=30mN/m$ in preformed DMPC (left) and Sm lipids (right) monolayers on increasing peptide bulk concentration in the subphase:(20mM Tris, 130mM NaCl, HCl, pH=7.5 , T=25°C :*
-●- i.a.LK$_{15}$,   -◆- alternated (KL)$_7$K,   -▼- regular LK$_{16}$, -▲- scrambled LK$_{15}$,

lipids extracted from the spiroplasma membrane (Sm lipids). The ability of the peptides to insert into the compressed lipid monolayers were determined by measuring the relative film surface increase ($\Delta S/S_0$) after peptide injection into the subphase. Data in figure 2 shows the different abilities of the peptides to insert into the lipid monolayers according to the lipid environment and the peptide topology. The hierarchy of insertion into the DMPC monolayer follows that of hemolytic activity. In the case of the Sm lipids, the hierarchy differs: i.a.LK$_{15}$ > regular LK$_{16}$ ≈ scrambled LK$_{15}$.

The most striking difference is for (KL)$_7$K which induces a very large surface increase on DMPC while there is no change on the Sm lipids. But it is important to notice that if the lack of surface increase observed for this peptide characterizes a lack of insertion, it does not rule out the potentiality of peptide <u>adsorption</u> underneath the Sm lipid monolayer.

2.3 PM-IRRAS structural study of the peptides in interaction with lipid monolayers

PM-IRRAS spectroscopy is a powerful technique to determine the secondary structure and the orientation of the peptides *in situ* in the lipid environment[19-23]. The 'subtracted spectra' presented in figure 3 result from the subtraction of the pure lipid spectrum from that of the mixed peptide/lipid one and then allow one to extract the characteristic contributions of the peptide bound. Clearly the peptide structure and orientation depends both on the lipid environment and on the peptide topology. In the case of i.a.LK$_{15}$ the main amide I absorption band, centered around 1655 cm^{-1}, whatever the lipid environment, is attributed to an α-helix structure[26-29] as expected from the peptide design. Simulation of orientations proves an orientation of the α-helix that is flat at the interface. These conclusions are in

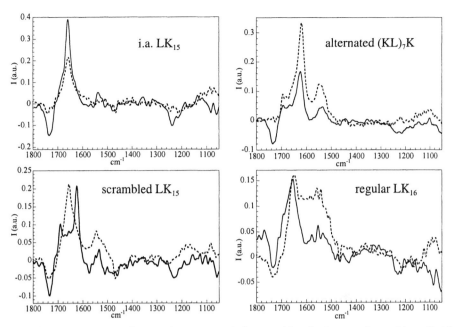

Figure 3: *PM-IRRAS subtracted spectra of the peptides in interaction with a lipid monolayer composed of either DMPC (—) or Sm (---) lipids.*

agreement with previous studies[21,23,24]. The amide I mode of the alternated $(KL)_7K$ is split into an intense band around 1622 cm[-1] and a weaker one around 1693 cm[-1], both characteristic of antiparallel β-sheets[26-29] oriented flat to the interface whatever the lipid environment[13,21,23,24]. Since PMIRRAS spectroscopy is only sensitive to molecules located at the interface, the spectrum obtained on Sm lipid environment proves the interaction of alternated $(KL)_7K$ with the monolayer even if no lateral surface increase was observed. The different spectra obtained for the scrambled LK_{15} indicate a versatile structure and orientation dependent upon the lipid environment. Inserted into the DMPC monolayer, the peptide folds into antiparallel β-sheets (amide 1 bands around 1625 cm[-1] and 1695 cm[-1])[27-30] flat oriented at the interface[13,21,23,24], while inserted into the Sm lipid environment, if folds into an α-helix (amide 1 band around 1655 cm[-1])[26-29] with a tilted orientation ($\approx 30°$)[21,23,24], both structures displaying no amphipathicity. Similarly for regular LK_{16} the spectra vary dependent upon the lipid environment. On the DMPC monolayer, the amide 1 band of the peptide is mainly characteristic of a tilted ($\approx 40°$) α-helix but not of the expected amphipathic 3_{10} helix[30-33]. On Sm lipids the broad amide 1 band may contain some 3_{10} helix contribution[30-33], but the structure is not well defined.

2.4 Peptide affinities for lipid monolayers

The determination of the peptide secondary structure and orientation allows an estimate of the molecular area for each peptide in lipid environment and then to calculate the affinity constants for lipid monolayers using a previously proposed equation and $\Delta S/S_0$ data obtained in § 2.2[13,17] (table 2). No affinity can be estimated for the alternated $(KL)_7K$ on Sm lipids, since it does not insert into the monolayer. In similar conditions, the affinities for Sm lipids are 20 to 1500 fold higher than those measured for DMPC. This well known effect differentiating interactions with erythrocytes and bacteria is first related to the strong electrostatic interaction due to the presence of negative charge in Sm lipids. It is also noteworthy that if a strong affinity for lipids is required to get the aspecific hemolytic and antibacterial activities, this is not the sole parameter of importance since there is no simple correlation between lipid affinities and biological activities.

3 CONCLUSION

We show that all the rationally designed LK peptides have cytotoxic activities which strongly differ according to the peptide topology without correlation between the hierarchies of hemolytic and antibacterial activities. As anticipated the estimated affinities for the negatively charged bacterial lipids are significantly higher than for DMPC, then it is not peptides affinity but peptides structures when bound which govern activities on Sm. Indeed such structures, orientations and the affinities vary according to the peptide topology and are modulated by the lipid environment which implies the need to define such structures in the most relevant conditions compared to those of the target cells. Nevertheless, new conclusions clearly appear:

i) The ideally amphipathic α-helix of i.a.LK_{15} is the most active structure both in hemolysis and in antibacterial assays, the membrane destabilization is then probably due to accumulation of flat oriented α helices on the outer leaflet whatever the lipid environment.

ii) The ideally amphipathic β-sheet of alternated $(KL)_7K$ is as efficient on erythrocytes as i.a.LK15, due to its flat insertion mainly governed by hydrophobic interactions[34]. However it is demontrated here that the same peptide is almost inactive on bacteria where it only adsorbs on the outer leaflet of the membrane due to the main contribution of the electrostatic effect. This striking difference is at odds with what might be expected since numerous natural peptides essentially structured with a β-hairpin are strongly antibacterial[35-38]. We propose that it is the ability to generate large antiparallel intermolecular aggregates on the membrane which is unfavorable for antibacterial activity due to the inability of these large aggregates to penetrate into the membrane.

iii) Both the scrambled LK_{15} and regular LK_{16} display very versatile structures, varying according to the lipids, they penetrate into the membranes and better in charged bacterial lipids. Their very weak hemolytic activity could be related to their very low amphipathicity (the hydrophibic moment μ_H= 0.04 and 0.15 for their α-helical and β-sheeted actual structures respectively).

iiii) The regular LK_{16} was designed to fold into an ideally amphipathic 3_{10} helix but it does not behave as expected. However some 3_{10} helical content did occur but only when bound to bacterial lipids. This could be due to electrostatic interactions which force K residues to be in register with charged polar head groups, it results in μ_H= 0.58 , a value higher than that of the i.aLK_{15}. peptide in Hα (μ_H= 0.46). It then fits with the previously defined criteria for being active and, indeed, it is strongly antibacterial. This is at odd with the conclusions on the first designed $(LRL)_n$ peptides, which were perhaps too long[39,] but it reinforces the more recent findings[40].

Finally in generating rationally designed *de novo* active compounds, more structural data are required to fully ensure the main conclusions that it is the intermolecular packing which should be controlled to get highly antimicrobial β-sheeted peptides and better characterization is needed to understand the 3_{10} helix antibacterial mechanisms.

Aknowledgements

This work was supported in part by the GDR 790 from the CNRS.

References

1   I. Cornut, E. Thiaudière and J.Dufourcq, *in*: *The Amphipathic Helix*. Epand RM (ed), CRC Press, Boca Raton, FL, 1993, pp 173.

2   G. Saberwal and R. Nagaraj, *Biochim. Biophys. Acta*, 1994, **1197**,109.

3   W.L. Maloy, U.P. Kari, *Biopolymers*, 1995, **37**, 105.

4   P. Nicolas and A. Mor, *Annu. Rev. Microbol.*,1995, **49**, 277.

5   R.M. Kini and H.J. Evans, *Intern. J. Pept. Prot. Res.*, 1989, **34**, 277.

6   K. Matsuzaki, *Biochim. Biophys. Acta*, 1999, **1462**, 1.

7   6   W. F.De Grado and J. D. Lear, *J. Am. Chem. Soc.*, 1985, **107**, 7684.

8   S. E.Blondelle and R. A. Houghten, *Biochemistry*, 1992, **31**, 12688.

9   J. A. Reynaud, J. P. Grivet, D. Sy, and Y. Trudelle, *Biochemistry*, 1993,**32**, 4997.

10   I. Cornut, K.Büttner, J. L. Dasseux and J. Dufourcq, *FEBS Lett.*, 1994, **349**, 29.

11   A. Brack and G.Spach, *J. Am. Chem. Soc.*, 1981, **103** , 6319.

12   S. E. Blondelle, J. M. Ostrech, R. A.Houghten and E. Pérez-Paya, *Biophys. J.*,1995, **68**, 351.

13   S. Castano, B. Desbat, and J. Dufourcq, *Biochim.Biophys.Acta*, 2000, **1463**, 65.

14   T. Kiyota, S. Lee and G. V. Sugihara, *Biochemistry*, 1996, **35**, 13196.

15   A. Kitamura, T. Kiyota, M. Tomohiro, A. Umeda, S. Lee, T. Inoue, G. Sugihara, *Biophys. J.*, 1999, **76**, 1457.

16 M. Dathe, T. Wieprecht, H. Nikolenko, L. Handel, M. L. Maloy, D. L. Mc Donald, M. Beyermann and M. Bienert, *FEBS Lett.*, 1997, **403**, 208.

17 S. Castano, B. Desbat, M. Laguerre, and J. Dufourcq, *Biochim. Biophys. Acta*, 1999, **1416**, 176.

18 L. Beven and H. Wróblewski, *Res.Microbiol*, 1997,**148**, 163.

19 D. Blaudez, T. Buffeteau, J.C. Cornut, B. Desbat, N. Escafre, M. Pezolet and J.M. Turlet, *Thin Solid Films* ,1994, **242**,146.

20 D. Blaudez, J.M. Turlet, J. Dufourcq, D. Bard, T. Buffeteau and B. Desbat, *J. Chem. Soc. Faraday Trans.*, 1996, **92**, 525.

21 I. Cornut, B. Desbat, J.M. Turlet and J. Dufourcq, *Biophys. J.*, 1996, **70**, 305.

22 M. Boncheva and H. Vogel, *Biophys. J.*, 1997, **73**, 1056.

23 S. Castano, B. Desbat, I. Cornut, P. Méléard and J. Dufourcq, *Lett. Pept. Sci.*, 1997, **4**, 195.

24 S. Castano, I. Cornut, K. Büttner, J.L. Dasseux and J. Dufourcq, *Biochim. Biophys. Acta*, 1999, **1416**, 161.

25 L. Béven and H. Wróblewski, *Res. Microbiol.*, 1997, **148**, 163.

26 T. Miyazawa, *in: Poly-α-amino acids*, Fasman GD (ed), vol 1, Marcel Dekker Inc New York, 1967, p 69.

27 W.K. Surewicz, H.H. Mantsch and D. Chapman, *Biochemistry*, 1993, **32**, 389.

28 E. Goormaghtigh, V. Cabiaux and J.M. Ruysschaert, *in: Subcellular Biochemistry: Physicochemical Methods in the Study of Biomembranes*, Hilderson HJ, Ralston GB (eds), vol 23, Plenum Press, New York, 1994, pp 363.

29 S. Krimm and W.C. Reisdorf, *Faraday Discuss.*, 1994, **99**, 181.

30 S.M Miick., G.V. Martinez, W.R. Fiori, A.P. Todd and G.L. Milhauser, *Nature*, 1992, **359**, 653.

31 W.R. Fiori, S.M. Miick and G.L. Milhauser, *Biochemistry*, 1993, **32**, 1957.

32 G.V. Martinez and G.L. Milhauser, *J.Struct.Biol.*, 1995, **114**, 23.

33 L.K. Tamm and S.A. Tatulian, *Quarterly Rev. Biophys.*, 1997, **30**, 365.

34 S. Castano, B. Desbat and J. Dufourcq, *Biochim. Biophys. Acta*, 2000, **1463**, 65.

35 L.H Kondejewski., S.W. Farmer, D.S. Wishart, C.M. Kay, R.E. Hancock, R.S Hodges. *J Biol Chem*,1996, **271**, 25261.

36 L.H Kondejews

37 ki., S.W. Farmer, D.S. Wishart, R.E. Hancock, R.S Hodges, *Int .J .Pept Protein Res*, 1996, **47**, 460.

38 S. Thennarasu, R. Nagaraj, *Biochem. Biophys. Res. Commun.*, 1999, **254**, 281.

39 Z. Oren, J. Hong, Y. Shai, *Eur. J. Biochem.*, 1999, **259**, 360.

40 T. Iwata, S. Lee, O. Oishi, H. Aoyagi, M. Ohno, K. Anzai, Y. Kirino, G. Sugihara, *J.Biol.Chem.*, 1994, **269**, 4928.

41 K. H. Mayo, J. Haseman, H. C. Young and J. W. Mayo, *Biochem J.*, 2000, **349**, 717.

A NOVEL APPROACH FOR PROBING PROTEIN-LIPID INTERACTIONS OF MscL, A MEMBRANE-TENSION-GATED CHANNEL

P. C. Moe and P. Blount

Department of Physiology, University of Texas-Southwestern Medical Center
Dallas, TX 75390-9040
U.S.A.

1 ABSTRACT

The bacterial mechanosensitive channel MscL, a homo-pentameric protein that resides in the cytoplasmic membrane, protects the cell from acute hypo-osmotic stress by opening in response to increased membrane tension. Early work demonstrated that only the structural protein and a lipid membrane are required for the activity of this channel. Subsequent characterization, much of it directed by the crystallization of the *M. tuberculosis* protein, focused on a genetic dissection of internal contributions to the channel activity and yielded an intramolecular gating model based on hydrophobic interactions. However, because the membrane matrix provides the medium through which the 'agonist', tension, is delivered to the channel we have inaugurated work to investigate the role of the lipid environment on the activation of this channel. Here we show that the MscL channel activity can be reconstituted in lipid membranes of discrete composition, and that the gating-tension threshold can be determined for these systems. This technique now enables the dissection of the various lipid effects such as lateral pressure, bilayer thickness, and head-group chemistry, on the function of a membrane protein. Here we describe the methodology required to address this question and present preliminary data demonstrating a specific lipid effect on the function of this mechanosensitive channel.

2 INTRODUCTION

The bacterial mechanosensitive (MS) channel, MscL, was originally isolated from *E. coli* membranes. Subsequent investigation found orthologues of MscL scattered across a wide evolutionary range of the eubacteria, and even into the Archaea.[1, 2] Numerous lines of evidence suggest that this channel serves as a bacterial "emergency valve" protecting the organism from acute hypotonic stress.[3-5] The channel itself is exquisitely simple: consisting solely of a homo-pentamer comprised of two-transmembrane subunits.[6] In the *E. coli* channel these subunits are only 136 amino acids long providing a facile model for a dissection of the mechanosensation phenomenon. In fact, at the time of this work, MscL remains the only mechanosensitive channel to be cloned, reconstituted and available for molecular characterization. Early work toward such characterization concentrated on intramolecular manipulation through random mutagenesis. Ablation of the MscL channel evoked no obvious phenotype because of redundant functions within the cell; therefore channel-mutants were screened for a gain-of -function (GOF) phenotype resulting from channels that gated inappropriately. These experiments quickly yielded a number of mutants, nearly all of which clustered to the proximal portion of the first transmembrane element, and comprised changes to charged and/or more hydrophilic residues.[7] In fact, when these residues are projected on a helical wheel model, the most severe are found to line one face.[8]

After a structure was solved for an orthologue from the bacterium, *Mycobacterium tuberculosis* these genetic results, and other evidence from biochemical experiments, coalesced to yield a gating model for the channel.[9-12] When GOF mutations were projected onto the *M. tuberculosis* structure they were noted to line the channel lumen along the first transmembrane element. In fact, among the most severe mutant recovered, valine 21 in *E. coli*, mapped to a residue resting at the closest stricture of the lumen. It was hypothesized that this valine, and several proximal residues, constituted a key element in the MscL gating mechanism: a hydrophobic lock. If this lock was disrupted by the substitution of a charged or more hydrophilic residue, the consistent result was the creation of a GOF mutant. Although this part of the puzzle appears incontrovertible, there remains the question of the ultimate conformational state of the open channel, and equally prickly; how does the membrane transfer the tension to the channel? For this answer we have turned to the membrane itself.

For many years the lipid bilayer that delimits a cell from its environ was referred to as a "sea of lipids" in which the "important players" *i.e.* proteins, were suspended and carried out their various vital tasks. This invokes an image of the bilayer as a vaguely homogeneous structure that, at best, serves to define the boundaries of the cell, and provides a milieu for hydrophobic proteins whose wont it is to reside in such an environment. We are now beginning to find evidence that this outlook was entirely too facile. The sheer complexity of the lipid bilayer, even as found in bacteria belies this supposed simplicity. Additionally, it is also known that most organisms closely regulate the constituency, and architecture of their lipid membranes.

Why would organisms, at great energetic cost, manipulate their membrane? Perhaps the answer lies in the complex diversity of lipid species available to most organisms. Biophysical studies have clearly shown the impact on bilayer mechanics of a head group substitution of glycerol for ethanolamine. The former readily adopts a bilayer conformation under physiological conditions, while the latter adopts an inverse hexagonal (non-bilayer) conformation. Surprisingly, phosphatidylethanolamine, at ~60%, constitutes the single most prevalent lipid in *E. coli* membranes. What is the role of such an apparent *non-sequitor*? Also, how is the tension or stress within a lipid membrane transduced to a channel opening as in the case of MscL? What effects do the lipid headgroups and pressures internal to the membrane have on the activity of such channels?

Here, using the bacterial mechanosensitive channel - MscL- as a reporter of membrane energetics we explore a possible modulatory role for regulation of membrane lipid composition by means of a refined technical approach. This technique includes reconstitution of the channel into defined lipid systems coupled with the biophysical analysis of the membrane under tension. Not only is this approach ideally suited for probing the lipid milieu itself but also for systematically dissecting the critical aspects of the protein-lipid interface that govern gating of MscL. At present, a handful of seemingly disparate theories compete to define the nature of these interactions: headgroup chemistry, bilayer lateral pressure, and hydrophobic mismatch induced by bilayer thinning have all been invoked. As we are now poised to winnow these theories, we present preliminary data showing a marked effect of different lipid compositions on the activity of the MS channel MscL

3 METHODS AND RESULTS

3.1 Molecular Biology and Protein Purification

To facilitate purification of the *E. coli* MscL protein the structural gene was modified by PCR to encode a string of six histidine residues appended to the carboxyl-terminus (MscLH$_6$). This construct was expressed from the moderate copy-number vector pB10b, under the control of an inducible promoter, in the *mscL*-null *E. coli* strain PB104.[8] Previous work has demonstrated that the moderate expression level from this vector does not perturb the ultimate activity of the protein or induce the formation of inclusion bodies, both indicative of protein misfolding.

Host cells were grown at 37°C in Luria-Bertani medium containing ampicillin (100 μg/ml) to maintain the plasmid and IPTG (1 mM) to induce expression of the channel protein (Figure 1). Cells were harvested while in logarithmic growth (~0.6 OD$_{600}$), and washed with KMg buffer (50 mM Kpi pH 7.2, 5 mM MgSO$_4$, and 1 mM DTT). Cells, resuspended to 20% w/v in KMg buffer plus 1 mM PMSF, 1 mg/ml lysozyme and 0.5 μg/ml DnaseI, were incubated at room temperature with gentle mixing. The resultant lysate was then subjected to 2 passages through a French pressure cell at 16,000 psi to complete the lysis. This and most further steps were performed at 4°C.

The crude lysate was cleared by centrifugation for 10 min at 6,000 x g. The total cell membrane fraction was collected by subsequent centrifugation at 200,000 x g for 30 min. This membrane fraction was solubilized on ice with the detergent octyl-β-glucopyranoside (OG) to facilitate the extraction of MscLH$_6$. OG was selected because it is a fairly mild detergent, and, at least for MscL, allows for the recovery of a functional protein (Figure 1). It should be noted however, that some membrane proteins are not compatible with OG, fairing much better with detergents such as *n*-dodecyl-β-D-maltoside (DDM); the only requirement is that the detergent be easily removed by dialysis (see below). The extraction buffer consisted of 300 mM NaCl, 50 mM KPi pH 8.0, 20 mM imidazole pH 8.0, plus 3% OG. The membrane suspension was thoroughly, but gently, homogenized in a glass dounse, and cleared by centrifugation for 20 min at 230,000 x g.

MscLH$_6$ was purified from the membrane homogenate by metal-chelation chromatography with Ni-NTA resin (Qiagen) using the batch column method. Briefly, 500 μl of the washed resin, equilibrated with extraction buffer plus 1% OG, was incubated, at room temperature, for 30 min with the homogenate. The suspension was then transferred to a gravity column and washed first with extraction buffer plus 1% OG and finally with elution buffer (50 mM NaPi pH 7.2, 300 mM NaCl, 1 mM PMSF and 1% OG) and a step-gradient of histidine. Elution of the MscLH$_6$ was effected at 200 mM histidine and found to be essentially homogeneous by SDS-PAGE. However, the specific activity of the MscLH$_6$ protein was not of concern because earlier works had demonstrated unequivocally that MscL, alone, is necessary and sufficient for the observed channel activity.[6] Total protein in the eluted fraction was quantified by modified Bradford assay (Pierce).

3.2 Reconstitution

Various lipids, and lipid mixtures, were assayed for their ability to support the reconstitution of the MscLH$_6$ as evidenced by patch-clamp. In this inaugural study, to establish a baseline for future experimentation, we chose two common, biologically relevant phospholipids: dioleoyl phosphatidylcholine (DOPC), and dioleoyl phosphatidylethanolamine (DOPE). These lipids are found in membranes, have the biologically favored 18-carbon unsaturated acyl chains, and are well-characterized biochemically. DOPE is found as a predominant lipid in the *E. coli* plasma membrane; ironic given its proclivity for assuming a non-bilayer configuration in solution. DOPC, although not represented in the *E. coli* envelope, was selected for these experiments because of its relatively neutral nature; it contributes little to the lateral pressure within the membrane, or charge to the surface of the bilayer. A stepwise increase in the ratio of DOPC: DOPE was performed to determine the effect of DOPE on MscLH$_6$ activity. The maximum ratio achieved was 1:3, ostensibly due to disruption of the bilayer configuration at DOPE concentrations above this ratio.

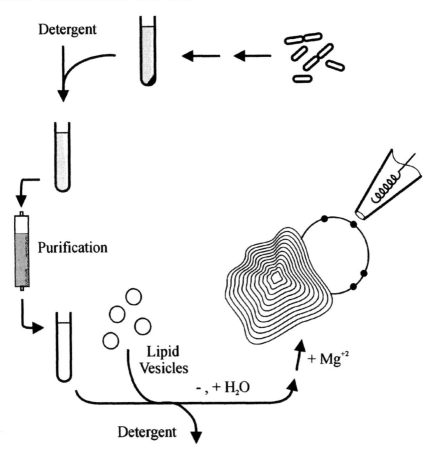

Figure 1 *Functional reconstitution of hexa-His tagged proteins into a defined phospholipid membrane.*

Lipid vesicles suitable for reconstitution (Figure 1) were prepared as described previously with modifications.[13-15] Lipids, dissolved in chloroform and stored under argon at -20°C, were aliquoted to a 13 mm glass tube. When lipids were to be mixed, they were mixed as a chloroform solution at this step. The chloroform was evaporated under a stream of argon for 30 min at room temperature. In the course of evaporation, the tube was rotated to yield a thin film of lipid that was deposited as the chloroform passed off. This lipid film was resuspended to 20 mg/ml in buffer (10 mM TrisCl pH 7.2, 1 mM EDTA, 1 mM EGTA) by vortex until it presented as a milky suspension. Lipid vesicles were formed by bath sonication to near-clarity. The time required to accomplish this varied from 10 min to greater than 45 min, but could be accelerated by addition of a few grains of OG.

Aliquots of the $MscLH_6$ fraction were combined with lipid vesicles at a 1:500 protein-to-lipid mass ratio. Removal of detergent and reconstitution of MscL into the lipid was effected by dialysis against 2x 1000 vols of buffer (100 mM NaCl, 0.2 mM NaEDTA, Tris Cl 5 mM pH 7.2, NaN_3 0.02%) containing a few microliters of Calbiosorb beads (Calbiochem) to sequester the detergent. The resultant proteoliposomes were collected by a 20 minute centrifugation at 30 psi in an Airfuge (Beckman-Coulter Instruments). The pellet was resuspended to 1.3 mg/μl in 5% ethylene glycol, 10 mM MOPS pH 7.4, separated to 5 μl aliquots on a plastic cover-slip, and desiccated overnight at 4°C.

Desiccated proteoliposomes were rehydrated for at least 2 h in Buffer A (150 mM KCl, 0.1 mM EDTA, 10 μM $CaCl_2$, 5 mM HEPES pH 7.2) at a lipid concentration of 90 mg/ml. Unilamellar blisters, suitable for access by patch clamp electrode were induced in the proteoliposomes by incubation in Buffer A plus 30 mM $MgCl_2$ (Figure 1).

Although this technique has proven quite repeatable and trouble-free with the lipids used to date, some difficulties would be anticipated as the range of lipid species widens. Fundamental to this technique is the formation of accessible blisters for patch-clamp. Some lipids, or mixtures of lipids, may prove disinclined to form these unilamellar structures. Varying the protein:lipid ratios, or the Mg^{2+} concentration in the buffer, are two approaches that have been employed with success. Phase transition behavior may also present a problem, especially as one canvasses lipids with transition temperatures well above the 0 - 4°C at which these reconstitutions normally take place. While an obvious approach might be to raise the temperature the welfare of the MscL protein must be kept in mind. Finally, although the lipid may blister readily, formation of a gigaohm resistance seal with the glass electrode (an imperative for the patch-clamp technique, described below) may not be possible. This seal is thought to involve the interaction of the lipids with the glass so such a problem might be addressed by selecting electrodes of different glass composition, pretreatment of the electrode with acid, or other chemical modifications. These potential difficulties should not adversely affect the efficacy of this technique.

3.3 Biophysical Characterization

Excised, air-cleared, patches were examined at room temperature under symmetrical conditions in Buffer A plus 30 mM $MgCl_2$. Recordings were performed at +20 mV (electrode). Data were acquired at a sampling rate of 20 kHz with a 5 kHz filtration using an AxoPatch 200B amplifier in conjunction with Axoscope software (Axon Instruments). A piezoelectric pressure transducer (World Precision Instruments) was used to measure the pressure response of the channels.[15]

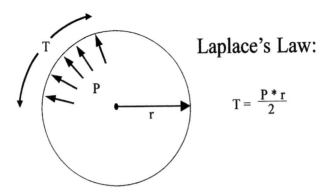

Figure 2 *Laplace's Law defines the tension (T) developed within a bilayer as a function of the radius of curvature (r), and pressure (P) across the membrane.*

Because the MscL channel is thought to gate in response to *tension* in the bilayer, rather than *pressure* across the membrane, it was necessary to find a way to derive the actual tension of the membrane patch. As depicted in Figure 2, Laplace's law shows that there is an intimate relationship between pressure and tension that is dictated by the curvature of a surface. Therefore, for any given pressure indicated by the manometer, there could be a very different curvature to the membrane patch depending on its nature and its position in the electrode, thus generating differing tension loads in the patch. Fortunately Laplace's law provides the solution: directly measure the patch radius and the pressure and solve for tension!

To facilitate visualization of the membrane patch the tip of the electrode was fused to an angle that placed it parallel to the focal plane while in the recording chamber. A 60x DIC oil-immersion objective, coupled with a Nikon DXM CCD camera and 5x relay lens, was used to capture images of the membrane patch (Figure 3). The patch radius was determined from the captured images using a commercial computer assisted design package IntelliCAD 2000 (CADopia).

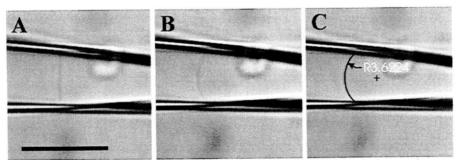

Figure 3 *Morphometric analysis of the membrane patch permits the calculation of tension by Laplace's Law. The scale bar represents 10 microns.*

Boltzmann data sets were constructed by increasing the pressure in a stepwise fashion until all MscL channels present in the patch were in the open state, or saturated (Figure 4). At this saturating pressure the open probability (P_O) is unity. As the pressure was held at each step, the average current was noted (an indicator of the number of channels active at that pressure), and an image of the patch was captured for further analysis.

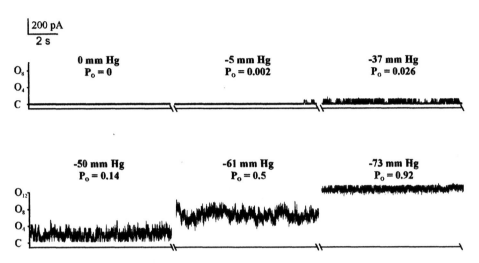

Figure 4 *Electrophysiological record of E.coli MscL reconstituted in DOPC membranes highlighting the increase in open probability (P_O) as a function of pressure.*

The two lipids used in this study (*see above*) were selected with an eye to some current theories explaining MscL gating. For instance, one approach holds that membrane thinning, due to lipid bilayer strain, and resultant hydrophobic mismatch between lipid and protein is the salient stimulus.[1] If true, then changing the ratio of DOPE in the bilayer might mimic this condition thus resulting in a lower gating threshold for the reporter – MscL. Alternatively, manipulating possible headgroup interactions may provide the key. It has been suggested that some DOPE might be required by MscL to satisfy hydrogen bonds with extra-membranous loop regions thus stabilizing the channel in the bilayer, and perhaps even facilitating the flow of membrane tension into the channel[16]. Finally, by virtue of its inherent geometry, addition of DOPE will promote a negative curvature strain in the bilayer, thus increasing the lateral pressure within the membrane. This increased energy within the bilayer is expected to affect the dynamics of lipid-protein interaction as well as impinge on the function of the proteins themselves.

Figure 5 depicts two data sets that are each representative of the lipid conditions presented. In parallel experiments (n=6 each lipid ratio) it was noted empirically, and consistently, that at a DOPC: DOPE ratio of 3:1 (●) far less tension was required to gate MscL than at a 1:3 ratio (■). Hence, higher complements of DOPE apparently result in more global effects on the membrane that are manifested in the higher gating tension. One interpretation of these data is that MscL responds to the decrease in lateral pressure in a stressed membrane as approximated in the 3:1 ratio. Alternatively there may be direct interactions between DOPC and MscL that increase the sensitivity of the protein to membrane tension or reduce the energy of the gating transition.

One should bear in mind that these experiments are yet in their infancy and future experimentation is necessary to unravel these clues. However, the observation that variations in lipid composition so notably affect mechanosensitive channel gating reflects the importance of the membrane and its influence on membrane-protein interactions; even suggesting the possibility that membrane composition, known to change during cell growth and in response to different environmental conditions, could modulate the function of this and other proteins.

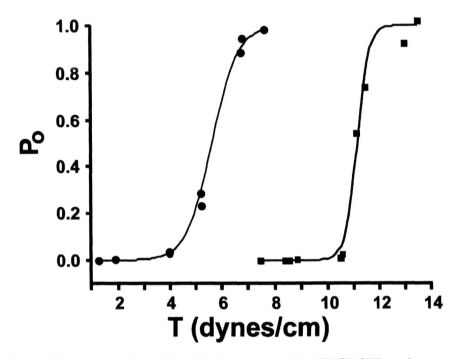

Figure 5 *Boltzmann analysis of E.coli MscL reconstituted in DOPC/DOPE membranes at two ratios; 3:1 (●) and 1:3 (■).*

4 CONCLUSION

What may we take away from these preliminary experiments? Firstly, The bacterial mechanosensitive channel MscL provides an exquisitely simple, yet robust, reporter for studying the biophysical properties of lipid bilayers. Here we describe a system that enables a rigorous, step-wise examination of the various lipid constituents of biological membranes. Any molecule that will support the functional reconstitution of the MscL channel, and subsequent visualization, can be studied with this technique. Further, as the biophysical characterization of lipid bilayer systems becomes more defined, the technique can be reversed to effectively study the intramolecular gating of MscL and other membrane-bound channels and transporter systems.

Secondly, these preliminary results show the dramatic effect wrought by merely adjusting the ratio of two common lipids in a binary system. It is well known that biological systems manipulate the lipid constituency of the membrane in response to homeostatic demands. These early results also show that DOPE, a major lipid fraction of bacterial membranes, though not a bilayer forming lipid, may serve a modulatory role in the bacterial envelope.

Finally, this technique lends itself to elucidation of the contributions made to the total energetic landscape of a biological membrane by lipid head-group moieties, acyl chain length, and even the degree of acyl chain saturation. Which, or what combinations of these influence the flow of tension-energy from lipid to protein? Put another way, what do mechanosensitive channels like MscL actually sense? Future experiments are being designed with these questions in mind.

Acknowledgements

This research was supported by grants from Robert A. Welch Foundation Grant I-1420, U.S. Air Force Grant F49620-01-1-0503, and NIH Grants GM61028 and DK60818.

References

1.   O. P. Hamill and B. Martinac, *Physiol. Rev.* **81**, 685 (2001).
2.   A. Kloda and B. Martinac, *Biophys. J.* **80**, 229 (2001).
3.   I. R. Booth and P. Louis, *Curr. Opin. Micro.* **2**, 166 (1999).
4.   P. Blount, S. I. Sukharev, P. C. Moe, S. K. Nagle and C. Kung, *Biol. Cell* **87**, 1 (1996).
5.   P. Blount and P. C. Moe, *Trends Microbiol.* **7**, 420 (1999).
6.   S. I. Sukharev, P. Blount, B. Martinac, F. R. Blattner and C. Kung, *Nature* **368**, 265 (1994).
7.   X. Ou, P. Blount, R. J. Hoffman and C. Kung, *Proc. Natl. Acad. Sci. U.S.A.* **95**, 11471 (1998).
8.   P. Blount, S. I. Sukharev, M. J. Schroeder, S. K. Nagle and C. Kung, *Proc. Natl. Acad. Sci. U.S.A.* **93**, 11652 (1996).
9.   P. C. Moe, G. Levin and P. Blount, *J. Biol. Chem.* **275**, 31121 (2000).
10.   B. Martinac, *Cell. Physiol. Biochem.* **11**, 61 (2001).
11.   S. Sukharev, M. Betanzos, C. S. Chiang and H. R. Guy, *Nature* **409**, 720 (2001).
12.   G. Chang, R. H. Spencer, A. T. Lee, M. T. Barclay and D. C. Rees, *Science* **282**, 2220 (1998).
13.   S. I. Sukharev, B. Martinac, P. Blount and C. Kung, Methods: *A Companion to Methods in Enzymology* **6**, 51 (1994).
14.   P. C. Moe, P. Blount and C. Kung, *Mol. Microbiol.* **28**, 583 (1998).
15.   P. Blount, S. I. Sukharev, P. C. Moe, B. Martinac and C. Kung, *Methods in Enzymology* **294**, 458-482 (1999).
16.   D. E. Elmore and D. A. Dougherty, *Biophys. J.* **81**, 1345 (2001).

FOLDING OF THE α-HELICAL MEMBRANE PROTEINS DsbB AND NhaA

Daniel E. Otzen

Department of Life Sciences, Aalborg University, Sohngaardsholmsvej 49, DK – 9000 Aalborg C

1 INTRODUCTION

The biophysical properties of integral membrane proteins are not as straightforward to study as those of their soluble counterparts, due to the difficulties in obtaining sufficient quantities for characterization. This is linked to their absolute need for a membrane-mimicking environment to stabilize and solubilize the native structure. A closer study of the mechanisms by which membrane proteins are inserted into the membrane may shed more light on the unique structural and functional facets underlying this class of proteins and allow us to manipulate conditions to improve yields of native protein. Pioneering studies on the α-helical integral membrane protein bacteriorhodopsin by Booth and co-workers [1-3] have demonstrated that *in vitro* folding of membrane proteins can be carried out and followed spectroscopically. Bacteriorhodopsin is a robust protein, resistant to 8M guanidinium chloride [4], but it remains denatured in SDS when transferred from an acid-denatured state [4]. The protein is then refolded by mixing with an excess of lipid vesicles. Both folding yields and the kinetics of folding are modulated by the properties of the membrane environment, such as bending rigidity [5] and lateral pressure [6].

To assess the generality of these observations, it is important to extend such studies to other proteins. Here I describe initial folding studies on two integral membrane proteins from the inner membrane of *E. coli*, namely the 176-residue disulfide bond reducing protein B (DsbB) and the 388-residue sodium-hydrogen antiporter A (NhaA). DsbB oxidizes the periplasmic protein DsbA [7], allowing transfer of oxidative potential to other reduced proteins in the periplasm, while NhaA exploits the proton gradient across the inner membrane to pump sodium ions out of the cell [8]. Unlike bacteriorhodopsin, the structures of these proteins are not known in atomic detail, but DsbB is predicted to contain 4 transmembrane helices, while 2-D crystals of NhaA have permitted electron cryo-microscopy analysis, revealing 12 transmembrane helices[9]. None of the proteins contains an exogenous chromophore, precluding absorption spectroscopy, but their aromatic residues provide convenient structural probes for fluorescence spectroscopy. Both proteins can be at least partially reconstituted from the SDS-denatured state into vesicles and the insertion reaction followed in real time. The kinetics are clearly modulated by environmental parameters such as ionic strength and pH as well as vesicle size.

2 METHODS

DsbB and NhaA were expressed and purified as described [7,10]. The genes for both proteins were cloned into expression vectors which inserted a His$_6$ tail onto the C-terminus of the protein, allowing a one-step purification on a Ni-NTA column (Amersham-Pharmacia). Both proteins were eluted using a 30-300 mM imidazol gradient and concentrated with Centriprep tubes (Millipore). The concentrated protein solution was twice diluted with buffer (50 mM sodium phosphate pH 8.0, 300 mM NaCl and 0.02% (w/v) dodecyl-maltoside) and concentrated again, in order to remove imidazol which interferes with far-UV CD spectroscopy. Unless otherwise stated, all experiments were carried out in 50 mM sodium phosphate pH 8.0 and 300 mM NaCl. Emission spectra were recorded on a PTI QM-2000-7 fluorometer with 1 µM protein and circular dichroism spectra on a Jasco J-715 spectropolarimeter with 10 µM protein in a 1 mm cuvette. Activity measurements of DsbB were carried out with 3.5 µM DsbA [11] and 30 µM decyl-ubiquinone in 50 mM sodium phosphate pH 6.0 and 300 mM NaCl at 25°C. Excitation was at 298 nm and emission at 350 nm.

Stopped-flow experiments were carried out on an SX-18 MV stopped-flow spectrometer (Applied Photophysics) using an excitation wavelength of 280 nm and a 320-nm cut-off glass filter. 1 volume of protein in detergent (typically 5 mM SDS) was mixed with 10 volumes of lipids to give a final concentration of 5 mg/ml lipid (ca. 8 mM), 0.45 mM SDS and 1 µM protein.

All lipids were from Avanti Polar Lipids. Vesicles of 50-200 nm size were prepared using an extruder (Northern Lipids) according to the manufacturer's instructions. All extrusions were carried out at least 10°C above the lipid's T_m. Unless otherwise stated, all experiments were done with 50 nm vesicles.

3 RESULTS AND DISCUSSION

3.1 Unfolding DsbB and NhaA in SDS and reconstitution in vesicles

Since membrane proteins are difficult to refold once denatured, the first issue to address is whether the denatured protein can be reconstituted into vesicles. Denaturation is best carried out in SDS, since low concentrations are required, which do not significantly affect vesicle structure upon mixing. In addition, membrane proteins generally retain significant α-helical structure in SDS, and this may crudely mimic *in vivo* folding, in which α-helices appear to be formed prior to membrane insertion, when released through the translocon [12].

DsbB and NhaA are purified in 0.4 mM dodecyl maltoside (DM). This detergent is preferable to *e.g.* decyl maltoside and *n*-octyl-glucoside, which lead to substantial loss of activity of DsbB (M. Bader and J. Bardwell, personal communication). DsbB and NhaA are both easily refolded from 5 mM SDS when mixed with 10-fold molar excess DM. Even boiling in SDS for 10 minutes followed by refolding into DM at room temperature does not destroy DsbB activity (data not shown).

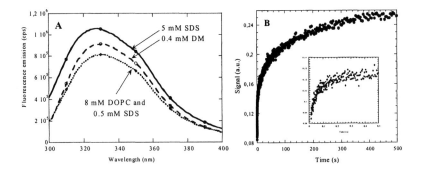

Figure 1   *(A) Fluorescence spectra of NhaA. The spectrum in DOPC and SDS was recorded after the protein was unfolded in 5 mM SDS and then diluted 10-fold into 8 mM DOPC. (B) Time-profile of NhaA folding when mixed with excess DOPC vesicles at 25 C. The inset depicts the first 0.5 seconds in more detail.*

There are clear structural differences between the SDS- and DM-state. When NhaA is transferred from 0.4 mM DM to 5 mM SDS, the fluorescence spectrum blue-shifts from 330 to 327 nm and increases in intensity (Figure 1A), suggesting alterations in tertiary structure[1]. CD spectra also indicate an increase in secondary structure (data not shown). Both sets of data are consistent with increased solvation of the aromatic residues by the apolar interior of SDS micelles, accompanied by increased α-helix formation. SDS is known to induce helical structure to facilitate micellar solvation of the protein backbone [14].

When the SDS-denatured NhaA is transferred to the lipid DOPC (giving 0.45 mM SDS and 8 mM DOPC), the spectral maximum shifts back to 329.5 nm and the intensity drops to slightly below that in DM, suggesting that at least partial renaturation has taken place. The slight drop in intensity may reflect the environmental differences between DM and DOPC. Fluorescence studies with DsbB yield similar results. In addition, activity assays on DsbB (the oxidation of DsbA, leading to a drop in Trp-fluorescence [11]) show a complete loss of activity in SDS and a 60% regain after mixing with vesicles under the above conditions. The 40% gap may reflect a misfolded fraction, similar to the observations for diacyl glycerate kinase [15]. It could also represent a fraction of the protein population with its active site (the two periplasmic loops) oriented towards the interior of the vesicles, which would isolate it from the water-soluble DsbA.

3.2   Following the folding kinetics of DsbB and NhaA

Given that some renaturation of DsbB and NhaA takes place under equilibrium conditions, we would expect to be able to follow the renaturation kinetically. Indeed, there is a substantial increase in the fluorescence signal when SDS-denatured DsbB

---

[1] The critical micelle concentration of SDS is 7 mM in water, but this falls to below 1 mM in 300 mM NaCl due to electrostatic screening of the head groups [13].

and NhaA are mixed with excess vesicles in a stopped-flow apparatus (Figure 1B). For both proteins, three well-separated relaxation phases are observed over a period of several hundred seconds, with rate constants k_1 around 2-20 s^{-1}, k_2 around 0.1-2 s^{-1} and k_3 around 0.02-0.1 s^{-1}. These signals represent some interaction between protein and vesicles, since the control experiment (SDS mixed with vesicles) shows no significant relaxation signal. Increasing the SDS concentration from 5 mM to 10 mM prior to mixing with vesicles does not change the kinetics significantly; however, higher concentrations of SDS (giving a final mole fraction of more than ca. 20% in the SDS-lipid mixture) leads to vesicle disrupture and a loss of protein folding signal (data not shown). Therefore we use 5 mM SDS as a standard denaturing condition. It does not appear possible to denature the proteins further in SDS; heating the SDS-protein solution to 90°C prior to carrying out the refolding experiment did not change the refolding kinetics (data not shown).

If the proteins are not denatured, but retain their native structure in a buffer containing DM prior to mixing with vesicles, only 1-2 phases are seen with much smaller amplitudes (data not shown). This suggests that some structural rearrangement has to take place for the protein to insert into the membrane even from the native state, but the conformational changes are largest when the protein starts from a more disordered state. We therefore initiate all folding reactions from the SDS-denatured state.

The existence of several phases in the refolding time profile indicates that the proteins have to pass through several structural rearrangements to reach the final state. Multi-state folding is not without precedence for α-helical membrane proteins. Bacteriorhodopsin passes through three intermediate states *en route* to the native retinal-bound state [1]. Native-like α-helical content is only attained in the second bacteriorhodopsin intermediate [3], whereas essentially all helical structure of DsbB and NhaA appears to be gained within the deadtime of mixing (data not shown). Thus, for the latter two proteins, the rate limiting folding steps seem to involve alterations in tertiary interactions within the membrane (probably through helical docking steps), rather than insertion of helices into the membrane.

3.3 Modulation of the folding kinetics

The kinetics of folding of both DsbB and NhaA can be modulated by various parameters. Vesicle size, controlled by the pore size of the filters used in the extrusion process, plays a substantial role. As the vesicle size increases from 50 nm to 200 nm, the folding rates of DsbB decline substantially while those of NhaA increase (Figure 2 A). Solvent conditions also influence folding. Increasing the ionic strength slows down DsbB's refolding rates (both the intermediate and slow phases) but has little effect on those of NhaA (Figure 2B). However, all rates decline uniformly with pH (Figure 2C). In contrast, folding rates for bacteriorhodopsin peak at pH 6 when the protein is refolded into mixed DMPC/DHPC micelles [5]; in this case, the recovery yield also decreases at extreme pH, unlike DsbB and NhaA (judging from the amplitudes of the refolding reaction).

Why does vesicle size affect the folding kinetics? Diminishing the vesicle size will increase surface curvature, exposing chinks to the hydrocarbon region and increasing the hydration of the interfacial region. This may affect folding reactions adversely or not, depending on the properties of the membrane protein in question. For example, those membrane proteins, which contain a significant number of polar groups in the parts residing in the interfacial region, may undergo rearrangements associated with folding more easily if the groups are well hydrated. Based on this interpretation, DsbB requires extensive hydration whereas NhaA does not.

More clues may be obtained from the two proteins' response to solvent changes, namely increasing ionic strength and pH. The effect of increasing ionic strength is complex. Obviously, there will be increased screening of charges, which will attenuate electrostatic interactions on the protein surface as well as between the protein and the vesicle. Screening phenomena will give rise to a linear dependence between the logarithm of the folding rate and the square root of the ionic strength, and such linearity appears to exist for DsbB (Figure 2B). In addition, small ions lead to a certain amount of dehydration, because their own solvation by a large shell of water molecules effectively lower the concentration of water molecules available for binding to lipid molecules. The ions can also interact directly with the lipid headgroups, and this in turn stabilizes more ordered phases [16]. As a consequence, melting transition temperatures are elevated, though this will not affect the order of the liquid phase significantly.

The results from the ionic strength dependence suggest that DsbB relies on extensive hydration and/or electrostatic interactions internally and/or with the vesicle to facilitate folding, whereas NhaA does not. This is consistent with the data on vesicle size dependence.

Altering the pH will not affect screening significantly. Protons mainly affect lipid order in the gel phase like other ions, so that lowering the pH increases T_m [17]. All our experiments with DOPC and DLPC are well above T_m (-18°C for DOPC and -1°C for DLPC), so vesicle phase order is probably not affected much. Furthermore, only negligible changes in lipid head group hydration are seen over the pH range 3-9 [18]. Therefore a direct pH effect on the membrane protein is more likely. The linear decline of log k with pH suggests a simple protonation model, in which titratable groups are repelled by the vesicle headgroups in the deprotonated form. The most likely candidates for this scenario are Asp and Glu; if this is indeed the case, the ionized side chains will be repelled by the negatively charged phosphate groups. Ionization of Asp and Glu probably also underlies the pH-dependence of SDS denaturation, in which unfolding kinetics are much more rapid at low pH (D.E.O., unpublished observations), where repulsion by the sulfate groups does not occur.

4   CONCLUSIONS

It appears possible to follow the renaturation of the α-helical proteins DsbB and NhaA from the SDS-denatured state by conventional stopped-flow analysis. The kinetics probably involve several consecutive structural transitions, which for both proteins are sensitive to changes in the solvent and vesicle, but to different degrees. Thus, the response of membrane proteins to manipulations in the folding environment is dictated

Figure 2   *The folding kinetics of NhaA (empty and filled circles) and DsbB (empty and filled squares) in DLPC are modulated by (A) extrusion size, (B) ionic strength and (C) pH. For clarity, only the rate constants of the two slowest phases are shown and data are fitted to exponential functions.*

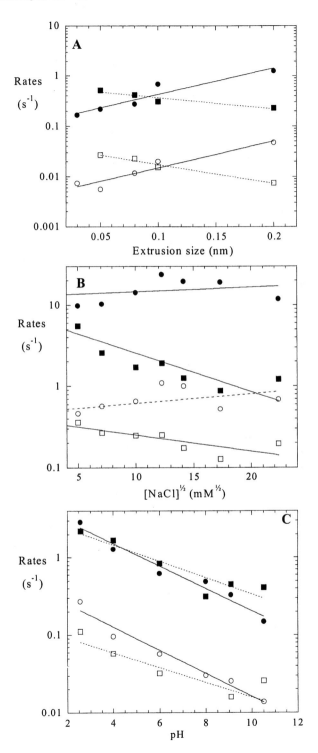

by the properties of the individual protein. These properties may include the amino acid composition in the interfacial region and the hydration and electrostatic properties in this area.

Acknowledgements

I am most grateful to Drs. J. Bardwell (DsbB) and W. Kühlbrandt and C. Ziegler (NhaA) for providing expression plasmids and providing very helpful advice on protein expression and purification. This work is supported by a grant from the Danish Technical Science Research Council.

Abbreviations

DLPC, L-α-dilauroylphosphatidylcholine, DM, dodecyl maltoside, DOPC, L-α-dioleoylphosphatidylcholine, SDS, sodium dodecyl sulfate.

References

1. P.J. Booth, *Biochim. Biophys. Acta,* 2000, **1460**, 4.
2. S.J. Allen, J.-M. Kim, H.G. Khorana, H. Lu and P.J. Booth, *J. Mol. Biol.,* 2001, **308**, 423.
3. M.L. Riley, B.A. Wallance, S.L. Flitsch and P.J. Booth, *Biochemistry,* 1997, **36**, 192.
4. K.-S. Huang, H. Bayley, M.-J. Liao, E. London and H.G. Khorana, *J. Biol. Chem.,* 1981, **256**, 3802.
5. P.J. Booth *et al.*, *Biochemistry,* 1997, **36**, 197.
6. A.R. Curran, R.H. Templer and P.J. Booth, *Biochemistry,* 1999, **38**, 9328.
7. M. Bader, W. Muse, D.P. Ballou, C. Gassner and J.C.A. Bardwell, *Cell,* 1999, **98**, 217.
8. E. Padan and S. Schuldiner, *J. Exp. Biol.,* 1994, **196**, 443.
9. K.A. Williams, *Nature,* 2000, **403**, 112.
10. Y. Gerchman, A. Rimon and E. Padan, *J. Biol. Chem.,* 1999, **274**, 24617.
11. M. Wunderlich and R. Glockshuber, *Prot. Sci.,* 1993, **2**, 717.
12. A.E. Johnson and M.A. van Waes, *Ann. Rev. Cell Dev. Biol.,* 1999, **15**, 799.
13. B. Jönsson, B. Lindman, K. Holmberg and B. Kronberg. *Surfactants and polymers in aqueous solutions,* (Wiley & Sons, New York, 1998).
14. W.L. Mattice, J.M. Riser and D.S. Clark, *Biochemistry,* 1976, **15**, 4264.
15. J.K. Nagy, W.L. Lonzer and C.R. Sander, *Biochemistry,* 2001, **40**, 8971.
16. G. Cevc and D. Marsh, *J. Phys. Chem.,* 1983, **87**, 376.
17. G. Cevc, *Biochemistry,* 1987, **26**, 6305.
18. J.M. Seddon, G. Cevc and D. Marsh, *Biochemistry,* 1983, **22**, 1280.

FhuA, AN ESCHERICHIA COLI IRON TRANSPORTER AND PHAGE RECEPTOR

P. Boulanger, L. Plançon, M. Bonhivers and L. Letellier

Institut de Biochimie et Biophysique Moléculaire et Cellulaire, UMR CNRS 8619 Université Paris Sud, Bât 430, 91 405 Orsay cedex, France

1 INTRODUCTION

Gram-negative bacteria are surrounded by two membranes, which confine the periplasm and the peptidoglycan, a structure responsible for the cell shape and rigidity. The inner (cytoplasmic) membrane is constituted of a phospholipid bilayer in which proteins are embedded. Ions and solutes are actively transported through the inner membrane by means of channels, pumps and transporters. ATP and the electrochemical gradient of protons generated by the electron transfer chain are used as the driving force for these transports.[1] The outer membrane constitutes a permeability barrier which protects the cell against noxious agents and allows exchanges of solutes with the external medium. Its lipid composition is atypical since the inner and outer leaflets contain phospholipids and lipopolysaccharides respectively. Non-selective porins are the most abundant class of outer membrane proteins. These pore-forming proteins allow the passive diffusion of small (< 600 Da) hydrophilic molecules. Selective porins, also permit the diffusion of small hydrophilic solutes but they contain specific binding sites for the ligands. The high affinity transporters (also called ligand-gated porins or TonB-receptors) are involved in the uptake of molecules that are present at very low concentration in the growth medium and which are too large to diffuse through the porins.[2]

Among the essential nutrients that are transported by these transporters is ferric iron. Gram-negative bacteria seeking to colonize aerobic environments at physiological pH are faced with the practical insolubility of ferric iron. To cope with this shortage they synthesize and secrete iron chelators (siderophores) that bind ferric iron with high affinity or use siderophores present in the growth medium. Iron-siderophores are transported across the outer membrane by high affinity transporters, each of them recognizing a specific siderophore. Outer membrane transport requires energy provided by the electrochemical gradient of protons generated by the electron transfer chain in the cytoplasmic membrane. This energy is transduced to the outer membrane transporters by a protein complex (TonB-ExbB-ExbD) anchored in the cytoplasmic membrane.[3] FhuA, is one of the seven ferric iron outer membrane transporters that are expressed in *Escherichia coli*. We present recent data obtained by our group on its *in vitro* functional and structural properties. We discuss these properties in light of its three dimensional structure.

2 FhuA: FROM STRUCTURE TO *IN VITRO* FUNCTION

1.1 *in vitro* Functionality of FhuA

FhuA is a 78.9 kDa protein that transports iron coupled to the hydroxamate siderophore ferrichrome (MM 688 Da). Besides its physiological function, FhuA is also the receptor for different phages (T1, T5, UC-1, Φ80) and for the antibiotic albomycin and the toxin colicin M.[4] FhuA is the first outer membrane transporter which was purified in a functional form and characterized *in vitro*.[5] Using a fluorescent DNA intercalator YO-PRO-1, we demonstrated that the interaction of phage T5 with purified FhuA, solubilized in non ionic detergent, resulted in the release of its double-stranded DNA (121 kbp) into the surrounding medium within a few seconds.[5] Furthermore, FhuA which shows no channel activity when incorporated into planar lipid bilayers was converted into an open channel upon phage binding.[6] Ferrichrome prevented channel opening and DNA ejection, indicating that it interacted with FhuA *in vitro*. From equilibrium dialysis experiments we calculated a K_d of ferrichrome binding to FhuA of 2.1 +/- 0.5 nM.[7]

1.2 Three-Dimensional Structure of FhuA

Our view of how iron-siderophore transporters function has considerably improved since the recent determination of the 3D structure (2.5A° resolution) by X ray diffraction of FhuA by two different groups.[8,9] FhuA, as well as FepA,[10] another member of the iron transporters family, share the same unexpected and unique structural organization. FhuA is composed of a barrel domain consisting of 22 antiparallel β-strands (residues 161 to 714) lodged in the membrane and of an amino-terminal globular domain (residues 1 to 160) that folds inside the barrel and occludes it. This "plug" or "cork" domain, which spans most of the interior of the β-barrel, consists of a four-stranded β-sheet and four short helices. It is connected to the β-barrel and to hydrophilic loops facing the external medium and involved in ligand binding. Ferrichrome binds to the top of the cork domain and phage T5 to one of the largest external loop. Few structural changes are observed within the cork domain upon ferrichrome binding except for a short helix (residues 24-29) located in the periplasmic pocket in the ligand-free conformation of FhuA which completely unwinds upon ferrichrome binding.[8,9]

1.3 Role of the N-terminal "Cork" Domain of FhuA in Structure and Function

The complex and very unusual structure of FhuA raises the question of the connectivities of the β-barrel and cork domains and of their respective role in the protein function. To address these questions we expressed and purified a recombinant FhuA Δ021-128 (FhuA Δ) missing almost all the cork domain and compared its properties to those of wild type FhuA (FhuA WT).[7]

1.1.1 The Cork Domain is not Essential for the Function of FhuA. Previous *in vivo* studies had shown that FhuA Δ could bind phage T5 and transport ferrichrome indicating that the mutated protein was targeted to the *E. coli* outer membrane and folded so as to retain functionality.[11] *In vitro* studies indicated that the purified protein was also functional. However phage T5 binding and DNA release required concentrations of FhuA Δ two orders of magnitude higher than those required for FhuA WT. Phage T5 binds to one of the hydrophilic loop of FhuA facing the external medium. The connectivities between the loop and the cork are therefore important for the phage receptor activity of FhuA. Ferrichrome could also bind to FhuA Δ since its addition prior the phage prevented DNA release. However, the affinity of the protein for ferrichrome was strongly decreased since its binding to FhuA Δ was not detectable in equilibrium dialysis experiments even when concentrations of ferrichrome up to 30 μM were used. These results are in accordance with the fact that residues belonging to the binding site of ferrichrome are deleted in FhuA Δ.[8,9]

1.1.2 The Cork Domain Contributes to the Stability of the Protein. Studies of the effects of proteases, denaturants and temperature gave strong indications of the reduced stability of FhuA Δ compared to FhuA WT. Indeed, the part of the cork remaining in FhuA Δ was fully cleaved by trypsin within 5 min whereas FhuA WT remained protected from cleavage for at least 1 h. Furthermore, and in contrast to FhuA WT, FhuA Δ was not protected from proteolysis by ferrichrome. The two proteins also showed a different pattern of migration on PAGE and of accessibility to monoclonal antibodies. FhuA WT remained folded up to 60°C whereas FhuA Δ was denatured at all temperatures studied between 4° and 100°C. Differential scanning calorimetry experiments corroborated these results. The pattern of thermal denaturation of FhuA WT was indicative of the presence of two well-resolved structural domains which corresponded to the unfolding of the cork and loops ($T_1 = 65$ °C) and of the barrel ($T_2 = 75$°C). Ferrichrome had a strong stabilizing effect on the loops and cork since it shifted the first transition to 71.4°C. Removal of the cork destabilized the protein since a unique transition at 61.6°C was observed even in the presence of ferrichrome. The cork is connected to the β-barrel and to the hydrophilic loops by nine salt bridges and about 60 hydrogen bonds. Given the data obtained on FhuA Δ it is likely that they contribute significantly to the stability of the β-barrel.

These results are corroborated by recent data showing that the N-terminal domains of FhuA and FepA can be genetically exchanged or deleted without loss of the siderophores and phage binding capacity of the proteins.[12,13] Altogether they indicate that the cork and the β-barrel of high affinity transporters behave as autonomous domains. What is then the function of the cork? It could enhance the binding capacity of the siderophores. In addition it may constitute a physical barrier preserving the cell from entry of noxious compounds which are excluded from porins.

3 INVOLVEMENT OF FhuA IN THE FIRST STEPS OF INFECTION OF *E.COLI* BY PHAGE T5

Infection of Gram-negative bacteria by phage starts by recognition of the host and binding to a specific outer membrane receptor. Almost all outer membrane proteins are potential receptors for phages.[14] In the case of tailed phage, this binding triggers conformational changes that are transmitted along the tail to the capsid, allowing its opening and the release of the viral genome, which is then transferred, base pair after base pair *via* the tail through the host envelope. The tailed phage T5 has proved to be a well-suited model to study phage-host interactions and DNA transfer since almost all these steps can be reconstituted *in vitro*.

3.1 FhuA forms a Functional Complex with Phage T5 Receptor-Binding Protein pb5

Binding of phage T5 to *E.coli* cells is mediated by specific interactions between FhuA and pb5, a 67.8 kDa protein located upstream of the central straight tail fiber at the distal part of the phage tail. A histidine-tagged form of pb5 was overproduced, purified and characterized.[15] The purified protein is monomeric and mostly organized as β-sheets (51 %). pb5 functionality was attested *in vivo* by its ability to impair infection of *E.coli* cells by phage T5 and Φ 80 when added before the phages. pb5 also prevented growth of bacteria on iron-ferrichrome suggesting that it alters either the binding of iron-ferrichrome to FhuA or its transport.

pb5 was also functional *in vitro* since addition of equimolar concentration of pb5   to purified FhuA prevented DNA release from phage T5. Direct interaction of pb5 with FhuA was demonstrated by isolating a pb5/FhuA complex using size exclusion chromatography. Its stoichiometry, 1   mol pb5/ 1 mol FhuA, was deduced from its molecular mass determined by analytical ultracentrifugation and confirmed by SDS-PAGE analysis. SDS-PAGE and differential scanning calorimetry   experiments highlighted the strong stability of the complex. i) the thermal denaturation of the complex occurred at 85 °C while pb5 and FhuA were denatured at 45°C and 74 °C respectively. ii) the complex was not dissociated by 2% SDS even when the temperature was raised up to 70 °C. The strength of the association between pb5 and FhuA is reminiscent of the irreversible nature of the phage adsorption step in the infectious process. Such stabilization is probably correlated to conformational changes occurring in both proteins. The availability of a stable and well defined complex opens the way to crystallographic studies and to the investigations of conformational changes occurring into siderophore receptors and phage receptor-binding proteins upon binding of a phage to its host.

3.2 Delivering Phage T5 Genome into Liposomes

In an attempt to decipher the mechanisms of phage DNA transport FhuA was reconstituted into unilamellar liposomes containing the fluorescent DNA intercalator YO PRO 1. Addition of phage T5 resulted in an increase of fluorescence consistent with the binding of the phage to reconstituted FhuA and transport of part of the DNA into the vesicles.[16] The high resolution of cryo-electron microscopy allowed the visualization of the phage-proteoliposome interactions and demonstrated unequivocally that the phage genome was delivered into the liposome.[17] Figure 1 (left) shows a typical cryo-electron micrograph. T5 phages are bound to the proteoliposomes by the tip of their tail. The capsid of one phage is filled with a dark gray striated material that corresponds to densely packed DNA. A second capsid is partially filled with DNA. Another is empty as evidenced by its lower electron density. The proteoliposome contains a dark gray material similar to that found in the capsids, an indication that the DNA has been transferred into the vesicle. The morphology of the liposome is not disturbed although the concentration of DNA within the vesicles reached values as high as 130 mg/ml.

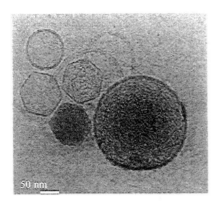 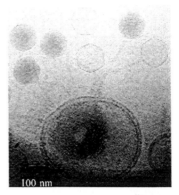

Figure 1 *Phage T5 DNA Transfer into FhuA-containing Liposomes (from reference 18 and 19)*

Furthermore, many DNA strands could be transferred from several phages and condensed into a unique compact toroidal structure when the proteoliposomes contained the DNA-condensing agent spermine (Figure 1, right).[18,19]

The route used by phage DNA to cross membranes is still a matter of debate. A model was initially proposed which supposed that the naked DNA would diffuse directly through the FhuA channel.[6] The 3D structure of FhuA makes now this model less realistic. *In vivo* studies argued in favor of a central role of pb2, the protein forming phage T5 straight tail fiber, in channelling DNA through the envelope.[20] Furthermore, cryo-electron microscopy and recent electron tomography images unambiguously show that the straight tail fiber crosses the lipid bilayer.[17] A reasonable hypothesis would be that the interaction of pb5 with FhuA triggers conformational changes in pb2, allowing its insertion in the membranes and the formation of a DNA channel. *In vitro* studies are now in progress taking advantage of the recent expression of pb2 in *E.coli* and of its purification.

4 CONCLUSIONS

Although major advances have been made in deciphering the mechanism by which high affinity transporters bind ligands, we are far from understanding how iron-siderophores are transported across the outer membrane. This transport is an energy-dependent process requiring a cytoplasmic membrane-anchored complex consisting of three proteins TonB, ExbB and ExbD. (Ton complex). Ton-dependent transport systems are widely spread. To date more than 20 outer membrane proteins whose function depends upon TonB have been identified in Gram-negative bacteria. However we are missing structural as well as functional data on these proteins. The next challenge will certainly be the expression, purification and functional reconstitution of these proteins together with the transporters in an energized *in vitro* system.

References

1   W.Cramer and D.Knaff. (1990). In *Energy transduction in biological membranes* (Springer-Verlag, ed.), pp. 246, New York.
2   H. Nikaido, *Science*, 1994, **264**, 382.
3   G.S. Moeck and J.W. Coulton, *Mol Microbiol*, 1998, **28**, 675.
4   V. Braun, *Biol Chem*, 1997, **378**, 779.
5   P. Boulanger, M. Le Maire, M. Bonhivers, S. Dubois, M. Desmadril and L. Letellier, *Biochem.*, 1996, **35**, 14216.
6   M. Bonhivers, A. Ghazi, P. Boulanger and L. Letellier, *Embo J*, 1996, **15**, 1850.
7   M. Bonhivers, M. Desmadril, G.S. Moeck, P. Boulanger, A. Colomer-Pallas and L. Letellier, *Biochemistry*, 2001, **40**, 2606.
8   A.D. Ferguson, E. Hofmann, J.W. Coulton, K. Diederichs and W. Welte, *Science*, 1998, **282**, 2215.
9   K.P. Locher, B. Rees, R. Koebnik, A. Mitschler, L. Moulinier, J.P. Rosenbusch and D. Moras, *Cell*, 1998, **95**, 771.
10 S.K. Buchanan, B.S. Smith, L. Venkatramani, D. Xia, L. Esser, M. Palnitkar, R. Chakraborty, D. Van Der Helm and J. Deisenhofer, *Nat Struct Biol*, 1999, **6**, 56.
11 G. Carmel and J.W. Coulton, *J Bacteriol*, 1991, **173**, 4394.
12 H. Killmann, M. Braun, C. Herrmann and V. Braun, *J Bacteriol*, 2001, **183**, 3476.
13 D.C. Scott, Z. Cao, Z. Qi, M. Bauler, J.D. Igo, S.M. Newton and P.E. Klebba, *J Biol Chem*, 2001, **276**, 13025.
14 M. Bonhivers, L. Plancon, A. Ghazi, P. Boulanger, M. Le Maire, O. Lambert, J.L. Rigaud and L. Letellier, *Biochimie*, 1998, **80**, 363.
15 L. Plançon., M. Le Maire , M. Desmadril, M. Bonhivers, L. Letellier and P. Boulanger, submitted.
16 L. Plançon, M. Chami and L. Letellier, *J Biol Chem.* 1997, **272**, 16868.
17 O. Lambert, L. Plançon, J.L. Rigaud and L. Letellier, *Mol Microbiol*, 1998, **30**, 761.
18 O. Lambert, L. Letellier, W.M. Gelbart and J.L. Rigaud, *Proc Natl Acad Sci U S A*, 2000, **97**, 7248.
19 J. Böhm, O. Lambert, A.S.Fangakis, L. Letellier, W. Baumeister and J.L. Rigaud, *Current Biology*, 2001, **11**, 1168.
20 G. Guihard, P. Boulanger and L. Letellier, *J Biol Chem*, 1992, **267**, 3173.

MORPHOLOGICAL ASPECTS OF *IN CUBO* MEMBRANE PROTEIN CRYSTALLISATION

C. Sennoga[1], B. Hankamer[2], A. Heron[1], J. M. Seddon[1], J. Barber[2] and R. H. Templer[1]

[1]Department of Chemistry, Imperial College, London SW7 2AY, UK
[2]Wolfson Laboratories, Department of Biochemistry, Imperial College, London SW7 2AY, UK

1 INTRODUCTION

Comprehensive understanding of the molecular mechanisms underlying membrane protein function(s) is often obscured by the dimensions, at which such processes occur. Comprising as they do specific receptors, signal and energy transducers, channel-forming proteins, active transport pumps, and electron transport systems, membrane proteins play key roles in a variety of cellular processes. They are believed to comprise *ca.* 25-40% of all proteins in the cell.[1] In order to have the greatest possible impact on applications that rely on membrane protein functionality, for example in the pharmaceutical formulation of highly specific drugs, three-dimensional (3D) structural information at atomic resolution is required as a starting point. Despite their importance and diversity less than 40 membrane proteins structures have been determined to atomic or near atomic resolution. This is due, primarily to the rate-limiting step of producing 2D or 3D crystals of sufficient electron diffraction or X-ray crystallographic quality. One of the essential challenges of post-genomic biology therefore is to determine to high resolution the structure of membrane proteins.

Conventionally, membrane protein crystallisation has involved lipid-based 2-dimensional (2D) or detergent-based 3-dimensional (3D) methods[2]. In 2D crystallization both the purified and detergent solubilized membrane proteins are mixed with specially selected lipids. The resultant protein/lipid/detergent blend is then depleted of the solubilizing detergent in a controlled manner thereby inducing formation of lipid bilayers in which 2D membrane protein crystals are embedded. In contrast, detergent-based 3D crystallization systems are aimed at inducing crystal formation by the controlled addition of precipitants to pure detergent/protein complexes. In trials utilising either method, hydrophobic membrane proteins have proven to be particularly intractable to crystallization. Perhaps the reason for this stems from the fact that hydrophobic membrane proteins are characterised by hydrophilic surface domains that are insufficiently large to favour protein-protein crystal contacts, due to the steric hindrance of the bound detergent micelles. A novel method based on crystallization of the membrane protein in lipid-based, liquid crystalline bicontinuous cubic phases, *in cubo*[3] crystallisation, has the potential to overcome this and a number of related problems. Indeed one of the attractions of the cubic phase method is that in theory it combines almost all of the advantages of 2D crystallization with those of detergent based 3D crystallization methods. As its name suggests the process is based on the incorporation of the target protein into the complex 3D structure of liquid crystalline bicontinuous cubic phases (Figure 1). The power of this method is clearly demonstrated by the fact that it has yielded two of the highest resolution structures of any membrane protein published to date.[4]

1.1 *in cubo* Crystallisation Hypothesis

For brevity, we have schematically illustrated the proposed incorporation, stabilisation and crystallisation mechanism in Figure 1. In a marked departure from conventional methods, the pioneers of the *in cubo* crystallisation method argued that membrane proteins would more readily crystallise in lipid bilayers than in nonbilayer environments (2D and detergent 3D crystallisation) provided they were incorporated, stabilised and were capable of isotropic diffusion in an appropriate lipid phase. The systems they chose are the well studied inverse bicontinuous cubic phases[5-8] formed by the mono-glyceride, 1-*monooleoyl-rac-glycerol*, or monoolein (MO) in water (Figure 2), known to form a stable, structured matrix in which diffusion of both water and soluble membrane proteins could take place[9]. The premise underlying *in cubo* crystallisation is that labile membrane proteins once incorporated into continuous lipid bilayers are stabilised by the near native environment and diffuse freely along the bilayer in an analogous way to the lateral diffusion of lipids. Upon nucleation, this leads to well-ordered crystals.

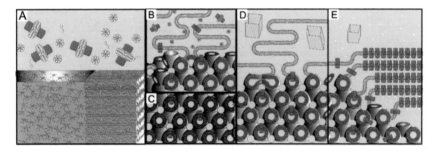

Figure 1 *Proposed mechanism of cubic phase crystallisation based on the work of Landau and co-workers. A) Dry monoolein (MO) is mixed with detergent solubilized protein. B) A hydration gradient is established through the monoolein wherein lamellar and cubic phases are proposed to coexist and the solubilized protein inserts into the near native environment of the lamellar phase bilayer. C) At equilibrium 60% (w/w) MO (see Figure 2), the system is purely cubic and the protein is then able to readily diffuse through it. D) Addition of salt or suitable crystallisation solution induces localised cubic to lamellar phase transition in the bulk cubic phase. E) Protein domains phase separate out into the lamellar phase and crystallise in the presence of suitable counter-ions.*

Whilst this method has facilitated macromolecular crystallization of 5 well ordered 3D crystals with different structural characteristics, namely: the light-driven proton pump bacteriorhodopsin (bR)[3,10], halorhodopsin (hR)[10], sensory rhodopsin II (sRII)[11,12] and the reaction centres of the purple bacteria *Rhodobacter sphaeroides* (RCsph) and *Rhodopseudomonas viridis* (RCvir) as well as the light-harvesting complex 2 from *Rhodopseudomonas acidophila* (LH2),[10] little is known about the underlying mechanism(s) of this method, although a working hypothesis has been proposed (see Figure 1)[13]. The role played by the cubic phase remains unclear although it exists both as the only phase prior to crystallization and as a major phase subsequent to protein crystal nucleation and early growth[14]. Early events subsequent to protein incorporation into the lipid bilayer might be important, although these remain largely unexplored. Clearly a number of physical and chemical parameters (i.e., protein concentration, temperature, solubilizing detergents, salts, pH and lipid type) need to be screened if a high crystallisation throughput is to be realised, in addition to extending our understanding of the crystallisation mechanism.

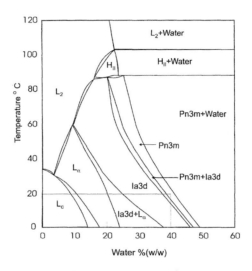

Figure 2 *The equilibrium temperature-composition phase diagram of the monoloein/water system indicating the composition and temperature used in the crystallisation of bR. The binary phase diagram of MO/H₂O was determined by Lutton[15], extended by Hyde[16] and later confirmed by Caffrey[17]. Thus at conditions above 25°C and 4–5 (w/w)% hydration, the MO/H₂O equilibrium phase diagram is characterised by a reversed micellar phase (L₂). Upon further hydration 8–22 (w/w)% or an increase in temperature, the L₂ phase undergoes a phase transition to a fluid lamellar phase (Lₐ). In the hydration region 25–40 (w/w)% H₂O, two bicontinuous cubic structures, the gyroid (Ia3d) and the double diamond (Pn3m), occur at low and high water content, respectively. At compositions with a water content above 40 (w/w)% the Pn3m phase co-exists with excess water while in the region 20–30 (w/w)% H₂O the two bicontinuous phases undergo phase transition to form a reversed hexagonal (H₁₁) phase at temperatures above 80°C.*

One of the essential factors relevant to the understanding of how and what the *in cubo* crystallisation mechanisms might involve, is how the protein is localised in the lipid matrix and what effect its presence has on the structure and stability of the lipid membrane. We have carried out studies with bacteriorhodopsin (bR) as the model system in order to address some of these questions, and our results are summarised here. The arrangement, distribution and transformations were investigated using polarizing microscopy (PM) and small angle X-ray diffraction (SAXD).

2 METHODS AND RESULTS

2.1 Preparation and Crystallisation of bR as the Model System

A major control for our work was the ability to produce bacteriorhodopsin (bR) crystals using the approach of Landau and Rosenbusch[3] as the "model" system. Thence, aqueous solutions of the purified detergent solubilized bR were mixed with dry monoolein (MO) in composition ratios based on the phase diagram of the pure lipid water system using the centrifugation approach of Rummel et al[13]. Purified bR was concentrated by centrifugal extrusion (10 mg/ml in 0.025 M NaP$_i$, pH 5.6) using an Amicon (PM-10) filter, and admixed with dry MO in the ratio 40:60%(w/w). In an eppendorf tube crystallisation admixtures were homogenised by temperature controlled (20°C) centrifugal mixing in a desktop microfuge (3000 g) for a period of 10 minutes. Subsequent mixing, was effected by rotating the eppendorf through 180° about its long axis, and repeating the mixing process with increments (1000rpm) to reach 13000rpm and giving a uniformly coloured purple gel. Crystallization was initiated by layering a solution of 1 M NaP$_i$ (pH 6) onto the purple gel. Control experiments were conducted in which either 1 M NaP$_i$ (pH 6) or 2-methyl-4-pentanediol (MPD) was used as the precipitant. The resultant admixture was sealed and incubated at 20°C in the dark. Crystallisation progress was monitored on duplicate samples by way of polarising microscopy. After a period of five weeks small purple crystals were observed to grow within the cubic phase (Figure. 3).

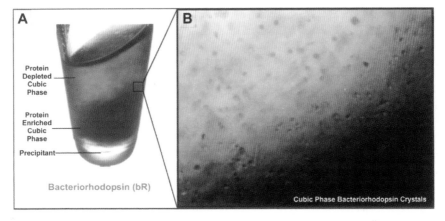

Figure 3 *Lipidic cubic-based crystallisation of bacteriorhodopsin. A) A Crystallisation admixture of 60% (w/w) MO lipid and 40% (w/w) aqueous protein suspension prepared by the centrifugation approach. All operations were conducted at 20°C. Bacteriorhodopsin solutions contained: n-octyl-β-D-glucopyranoside (OG) solubilized monomeric bR protein, Na/K P$_i$ (pH 4.6) salt/buffer. Additional NaP$_i$ 1 M (pH 4.6) was used as the precipitant. B) bR crystals were observed to grow within the cubic matrix, reaching their full size after five weeks incubation of the dispersion admixture.*

2.2 Multi-Well Batch Crystallisation

The centrifugation approach of mixing membrane proteins with MO has two main limitations. Firstly, the method is extremely time-consuming, as dry MO has to be weighed out in individual samples with the subsequent mixing step taking a further 2 hours. Secondly, weighing out the required amount of MO with any degree of accuracy renders the process difficult for trials having a volume of less than 15 μl, making the process expensive in terms of protein requirements. To address these problems we have built a syringe-mixing device, which we use in combination with an IMPAX® automatic micro-dispenser to screen a large number of crystallisation parameters. Figure 4 shows such a device with the green chlorophyll binding protein (LHC-II) loaded into the left syringe (Figure. 4A) with the appropriate amount of solid MO (white powder) in the right. A screw fit needle with a constriction connects the two syringes. During mixing, the protein sample is slowly injected through the needle connector into the syringe containing the MO (Figure 4B). The resultant gel is then gently pushed back and forth between the syringes until a homogeneously coloured gel (Figure 4C) is obtained. The resultant protein based admixture is then dispensed into wells of nonbirefringent multi-well sitting drop plates with different precipitants (Figure 4D). Using this approach we have reduced the mixing times from 2 hours to 10 minutes and achieved a 15-fold reduction in protein requirement due to bulk mixing. We are now able to routinely set up in excess of 300 crystallisation trials per day. By using nonbirefringent plates, we are able to routinely examine the crystallisation progress as a function of added precipitant on the cubic phase or physical conditions by way of polarising microscopy. The images are digitally captured both with and without cross-polarizers and then transferred to an in-house database. This approach has allowed us to screen a large number of crystallisation parameters.

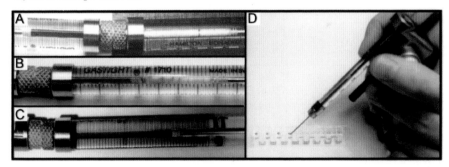

Figure 4 *Micro-batch mixing and set up. A) Solubilized and buffered chlorophyll binding protein i.e., Light Harvesting Complex II (LHC-II) loaded into syringe (left) and dry MO lipid loaded in right syringe. B) LHC-II is slowly injected into the syringe containing dry MO lipid. C) After gentle mixing (to and fro) through a small constriction in the syringe connector, a uniform green cubic phase gel is obtained. D) Using a dispensing device, 1μl batches of cubic phase gel are injected into ideally silica based nonbirefringent sitting drop plates. 6μl of a crystallisation solution is subsequently dispensed onto each 1μl gel batch. A further 100μl of solution is pipetted into the equilibration reservoir before sealing the wells with clear nonbirefringent tape.*

2.3 Bulk Morphological Changes that accompany bR Protein Crystallisation

Small angle X-ray diffraction (SAXD) was used to determine the phase changes that accompany the crystallisation of bR. In so doing we aimed to answer a number of questions: Is the transition from the cubic to the lamellar phase the key stage in crystal production? Is the protein size relative to the cubic phase unit cell dimensions important? And finally, could the cubic phase be accessed at low temperatures?

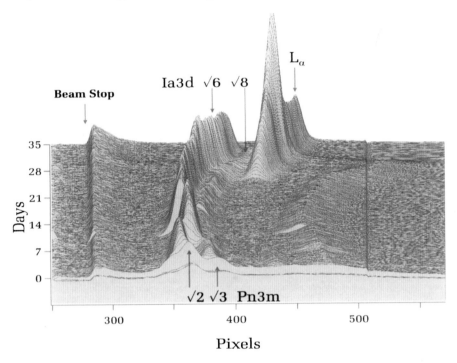

Figure 5 *The changes to the small-angle X-ray diffraction pattern with time that accompany detergent solubilized bR incorporation into the cubic bilayer, and subsequent crystallisation (continual image capture over a duration of six weeks). It is noteworthy that only a few crystals were observed with this run that quickly dissolved as the system transformed to a purely lamellar phase. The corresponding lattice parameters and mean curvature of the bilayer are given in Table 1.*

2.3.1 SAXD Sample Preparation. Purple membranes were isolated from *Halobacteria Salinaria S9* strain according to reference[18]. Partially denatured bacterio-opsin (bO) the de-retinalized form of bacteriorhodopsin (*ca.*2mg/ml) was purified in 2% sodium dodecyl sulphoxide (SDS) at a 5:1 SDS:protein (w/w)[19]. Purified bacterio-opsin (bO) (200μl) was then supplemented with solid dimyristoylphosphatidylcholine (DMPC) and OG together with 9M of all *trans* ethanolic retinal, to yield a protein suspension comprising OG:DMPC:retinal:protein in the ratio 20:5:1:1 all buffered with 50mM NaP_i pH 6.0. Under these conditions the retinal is essentially in excess as only 75% of the protein refolds. Protein refolding and thus concentration was determined spectrophotometrically by comparing the absorption peaks at 280nm (total protein, $\varepsilon280nm = 66,000$ M^{-1} cm^{-1}) and 555nm (refolded bR, ε555nm = 55,300 M^{-1} cm^{-1}). Residual SDS was eluted from reconstituted bR by sequential washing and centrifugation with an elution buffer of 1.2% OG, 50mM NaP_i pH 5.6 using an Amicon 10 microconcentrator to give a final protein concentration of 4.2mg/ml. Known aliquots of the resultant protein stock were mixed with dry MO in X-ray transparent capillaries (W. Müller, Berlin, Germany) using the centrifugation method of Rummel and coworkers[13]. Crystallisation was initiated by adding 1M NaP_i to yield a lipid-protein dispersion with 40% (v/w) protein dispersion medium. Sample homogeneity was monitored using polarizing optical microscopy to ensure the formation of a uniform cubic phase before the tube was sealed. All samples were optically isotropic, nonbirefringent and exhibited a uniform microstructure. Crystallisation progress was monitored by time resolved low-angle X-ray measurements (see §2.3.2). Control experiments were conducted in which one sample was used to monitor progress by polarizing microscopy at the 20°C isotherm or merely incubated in the dark thus allowing us to eliminate any possibility of radiation damage.

2.3.2 SAXD Measurements. X-ray diffraction measurements were conducted in a time-resolved mode using nickel-filtered Cu Kα lines (λ=1.542 Å) from an *Elliot GX20* rotating anode generator (Enraf-Nonius, Netherlands). The generator utilised a 100μm focus cup with a typical loading of 30kV × 25mA = 750Watts. X-ray focusing was achieved by using Frank's optics in a Kirkpatrick-Baez arrangement with 200 mm focal length. The X-ray focal beam spot was 160μm in height and 110μm in width (full width at half-maximum). A set of adjustable tungsten slits placed approximately 15 mm before the sample reduced parasitic scatter to levels where we were able to measure lattice spacings up to 300 Å. X-ray capillaries were placed in a copper sample holder with thermoelectric temperature control. The temperature was maintained at 20°C (±0.03)°C. The optics and the sample cell were both held under vacuum to minimize air scatter. The sample chamber included a variable length, evacuated flight path, which allowed the X-ray detector to be placed between 120 and 300 mm from the sample, thereby taking advantage of the long focal depth of the optics. The design was limited to wide-angle measurements down to a minimum spacing of 3.5 Å. Time-resolved diffracted intensity was recorded on a two-dimensional CCD-based image-intensified detector. The entire X-ray system was placed under computer control, allowing us to automate the acquisition and analysis of data. The acquired powder diffraction images were radially integrated and subsequently indexed using our TV4 fitting program. TV4, was designed by Professors E. Eikenberry and S. Gruner (Cornell, USA). Modified to suit our exacting requirements, TV4 has been tailored to operate our home-built detector and communicate with the off-board control system we use to operate our thermoelectric heaters. All diffraction measurements were performed at 20°C in the dark. Samples were allowed to equilibrate for at least 180 minutes before acquiring an X-ray pattern. Repeated runs on a single sample using this protocol gave results, which were reproducible to ±1.0 Å in the lattice parameter, and no observable radiation damage was recorded. Phase transition temperatures measured on a similar sample, using this protocol were reproduced to ±1 °C.

Our findings reveal that the phase transitions accompanying *in cubo* crystallization of bR are dominated by two cubic structures, one of which exists both prior and subsequent to crystal growth. The two structures with inverse bicontinuity are closely related to the *Schoen* G (gyroid) and *Schwarz* D (double diamond) infinite periodic minimal surfaces (IPMS), and occur at low and high aqueous content, respectively. The two cubic phases belong to space group Ia3d (Q^{230}) for the G and space group Pn3m (Q^{224}) for the D surface. The structure of the Pn3m cubic phase is characterised by two independent networks of aqueous tubes joined four-by-four at tetrahedral angles. Each tube forms a continuous channel system, which is surrounded by a lipid bilayer separating the aqueous phase from the hydrocarbon continuum of the supramolecular lipid organization (hence double diamond). The Pn3m cubic lattice is characterised by X-ray diffraction peaks (Figure 5) spaced in the ratio $\sqrt{2}:\sqrt{3}:\sqrt{4}:\sqrt{6}:\sqrt{8}:\sqrt{9}...$. In contrast the structure of the Ia3d cubic phase is characterised by two independent networks of aqueous tubes joined three-by-three. Adjacent networks of the aqueous channels are mirror images of each other. The Ia3d cubic lattice is characterised by X-ray diffraction peaks spaced in the ratio $\sqrt{6}:\sqrt{8}:\sqrt{14}:\sqrt{16}:\sqrt{20}:\sqrt{22}...$while lamellar ($L_\alpha$) phases are characterised by Bragg peaks with s values spaced in the ratio $1:2:3:4:5:6...$. The repeat d-spacing of lamellar structure are determined as a mean value of the spacing $d_n = n/s_{00n}$, against n, where n is the Bragg peak reflection order number.

Prior to salt addition, the observed Q^{224} phase is characterized by lattice parameters, a_Q of approximately 98Å. The Q^{224} lattice parameter was observed to initially increase to 114Å, before decreasing as the osmotic dehydration took effect upon adding the precipitant in solution form. After three days the bulk crystallization system was observed to undergo a phase transition from the Q^{224} to the Q^{230} phase, with a lattice parameter of 168Å. The Q^{230} lattice parameters were observed to decrease from 168Å down to 138Å before undergoing a gradual phase transition to the L_α phase (d=50Å). The sequence of morphological transformations was: $Q^{224} + salt \rightarrow Q^{224}/Q^{230} \rightarrow Q^{230} \rightarrow Q^{230}/L_\alpha \rightarrow L_\alpha$.

2.4 Detergent Effects and Crystallisation at Lower Temperatures

For the purposes of *in cubo* crystallization, detergent solubilized membrane proteins are mixed with dry MO, with the aim of incorporating them into the cubic phase. The resultant structural topology and stability thereof is dependent on both the detergent type and aqueous concentration present. As a function of their composition, therefore, detergent/water mixtures give rise to specific phase behaviour, which may include cubic phases of the normal type[20]. Because of their tendency to form interfaces curved away from water, high concentrations of protein solubilizing detergents are incompatible with stabilisation of the inverse cubic phases. Indeed, it has been reported that at high concentrations, n-octyl-β-D-glucopyranoside[21] or n-dodecyl-β-D-maltoside[22] when mixed with MO induces formation of lamellar phases. Since the stability of the cubic phase in the crystallisation system is dependent on both the detergent type and concentration present we have examined the temperature-dependence of glycoside-induced phase transitions on the MO cubic stability for a number of commonly used mild detergents, under equilibrium and metastable conditions at the relevant detergent concentration for crystallisation of 1 and 3 times the critical micelle concentration (CMC). Figure 6 shows the phase behaviour for the systems MO/n-dodecyl-β-D-maltoside/H₂O at the CMC (Figure 6A) and n-heptyl-β-D-thio-glucopyranoside at 3 × CMC (Figure 6B) as a function of temperature. In both cases, noncubic-to-cubic and cubic-to-cubic phase transitions are observed between - 15°C and 350°C in the heating direction (equilibrium). In the cooling direction (metastable) however, no cubic-to-noncubic phase transition is observed over the temperature range 35 to -5°C suggesting that through careful exploitation of the lipids' metastability, the cubic phase can be accessed at temperatures above 4°C in some MO/detergent blends. We have examined the MO/detergent metastability on a wide range of commonly used detergents by gradual cooling of the cubic/detergent blends to 4°C. The blends were held at 4°C for one month prior to re-examining the structural form of the MO/detergent mesophase. In the majority of systems studied the cubic structure was maintained. Depending on the detergent type and concentration, we find this phenomenon to be long lived to well over a month. Our full findings are described elsewhere[23].

It has been demonstrated that the crystallisation process can be carried out using detergent-free blends[24]. Indeed this might suggest that the influence of the detergent on the crystallisation process is a minor one. However, it is noteworthy that firstly, this approach is only feasible in cases where the membrane protein can be obtained in an enriched form. Secondly, detergents are used because they act as a conduit to maintaining the proteins integrity prior to incorporation into the cubic bilayer. To this end they have to be selected in such a way that the concentration used in no way destabilises but maintains both, the integrity of the protein and topology of the cubic matrix. Finally, based on these findings we believe, that low temperature crystallisation can be accessed by exploiting the lipid's metastabilty. A low temperature crystallisation strategy (for example at 4°C) has the potential to improve the crystal quality of temperature sensitive proteins.

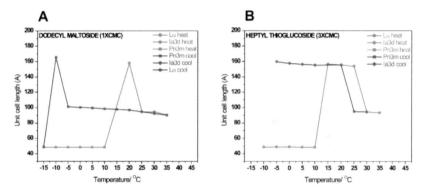

Figure 6 *Temperature dependent phase behaviour at 60%(w/w) MO:40%(w/w) aqueous detergent A) 1 × CMC, n-dodecyl-β-D-maltopyranoside and, B) 3 × CMC, n-heptyl-β-D-thioglucopyranoside mild detergents. The red and blue data set indicate phase changes induced by sample heating (equilibrium) and cooling (metastable), respectively.*

2.5 Spectroscopic Modifications of PSII and LCH-II Macromolecules

We have characterized the events following the incorporation of Light Harvesting Complex (LHC-II) and Photosystem II (PSII) membrane proteins into the bicontinuous cubic phases of monoolein. We find that the phase behaviour of the crystallisation matrix is well correlated to that of the model system, but find that the spectral properties of both protein candidates to be modified by the incorporation process (data not shown). In both cases the structure of the cubic mesophase was not modified by the incorporation of either protein. The incorporation of the PSII core complex for example, into the MO cubic phase was accompanied by a 5 nm blue shift in the chlorophyll absorption bands. On the assumption that the curvature of the cubic phase might be responsible for the observed spectroscopic modifications, we have incorporated both proteins in lamellar phases. In both cases, the spectral modifications observed were similar to those observed in cubic mesophases. Given that the same modifications were observed both in cubic and lamellar phases, this points to the possibility that the spectral modifications originate from a dimensional mismatch between protein moieties and the MO bilayer, and not from the strain exerted by the bilayer curvature on the protein. The 5 nm blue shift in PSII core complexes is known to be associated with a decoupling of the integral manganese-cluster from the core complex. Based on previous studies, we know that PSII is relatively labile and thus readily undergoes spectral modification owing to loss of the manganese-cluster from the complex. We believe the loss of the manganese-cluster upon incorporation in the cubic phase has its origins in the said mismatch and thus exposure of the membranes hydrophobic subunits to the aqueous cores of the cubic matrix.

Based on correlations of the estimated lateral (trans-membrane) hydrophobic dimensions between the target protein and the estimated parameters of the *in cubo* mesophases (see §2.6), we argue, that the observed macromolecular spectral modifications have their origins in the bilayer/protein dimensional mismatch (see § 2.7). We argue further that the mismatch will be obviated by using lipid blends with a maximal swelling of the cubic mesophase, doped with additional lipids that promote reverse hexagonal and lamellar structures[23]. A suitable blend for the PSII core complex might involve doping with synthetic phytanyl-chained glycolipids, which have shown successful functional reconstitution of PSII core complex[26]. We are exploring the use of this blend and will report its potential in another publication.

2.6 Parametric Estimation of *in cubo* Mesophases

The structure of lipid aggregates depend on the bending energy of the lipidic monolayers within the aggregates. The bending energy therefore leads to the formation of lipid mesophases with different radii of curvature and geometry. Here we use observed SAXD data for aqueous bR hydrated MO and apply an established formalism to estimate the mean curvature of the phases observed. A prerequisite of using this approach is that the location and indeed existence of the pivotal surface has to be established as a starting point[27]. Before defining the pivotal surface we shall first restate the structural nature of bicontinous cubic mesophases. As outlined earlier inverse bicontinuous cubic phases are best envisaged as consisting of an amphiphilic bilayer draped onto an infinite periodic minimal surface (IPMS) with the minimal surface occupying the average location of the bilayer mid-plane. In applying this formalism[27], all molecularly defined surfaces, for example the polar/nonpolar interface and the pivotal surface are defined as occupying parallel surfaces to the minimal surface. Thus we define the pivotal surface as a particular cross-sectional area, A_n of a molecule within a curved bilayer which lies parallel to the underlying infinite periodc minimal surface (IPMS) and remains invariant to isothermal bending i.e., the mass of the sample on either side of the pivotal surface remains contant during molayer bending. Therefore any change in molecular shape due to isothermal change is accompanied by a change in the distance, ξ, from the IPMS to this surface.

To derive the relationships relating the lattice parameter to the water volume fraction we have to define the geometry of the "pivotal surface". The cross-sectional area per lipid at this surface, and the molecular volume between this surface and the ends of the hydrophobic chains are given by A_n and v_n, respectively. The *"stoichiometric"* lipid geometry therefore is characterised by these two parameters, and by the total molecular volume of the lipid, v. The location of the pivotal surface can be determined by applying the rules of differential geometry[27] and for our purposes it is sufficient to use values determined earlier, which place the pivotal surface at or about the C_3 position on the hydrocarbon chain [27, 28]. For MO we have

$$\langle v \rangle = 612 \text{Å}^3; \langle v_n \rangle = 465 \text{Å}^3; \langle A_n \rangle = 33 \text{Å}^2 \tag{1}$$

Our X-ray data for the crystallization process are summarized in Table 1. The location of the pivotal surface, ξ, was established using the experimentally determined lattice parameter, a_Q, and the aqueous volume fraction, ϕ_{aq} via equation (1). Here σ, is the ratio of the IPMS area to the (unit cell volume)$^{2/3}$, and χ is a topology index of the IPMS known as the Euler-Poincaré characteristic for a given cubic phase. The constants σ and χ are 1.919 and –2 for the Q^{224} phase, and 3.091 and –8 for the Q^{230} phase.

$$\frac{\langle v_n \rangle}{\langle v \rangle}(1 - \phi_{aq}) = 2\sigma\left(\frac{\xi}{a_Q}\right) + \frac{4}{3}\pi\chi\left(\frac{\xi}{a_Q}\right)^3 \tag{2}$$

The magnitude of the surface averaged mean curvature defined to the pivotal surface, $|\langle H_p \rangle|$, was determined for the Q^{224} and Q^{230} cubic mesophases of the crystallisation matrix based on the lattice parameter, a_Q, and the pivotal surface, ξ at the limiting hydration in each phase according to the relationship[29]

$$\langle H_p \rangle = \frac{2\pi\chi\xi}{\sigma a_Q^2} \tag{3}$$

The average radius of cubic aqueous network defined at the polar/apolar interface was determined via equation (4)

$$\langle R_w \rangle = \frac{1}{\sqrt{\langle -K \rangle}} - l \tag{4}$$

where, the surface average Gaussian curvature, $\langle K \rangle = 2\pi\chi / \sigma a_Q^2$ and l, the length of the lipid amphiphile is determined via the lipid volume fraction, ϕ_l according to equation (5)[30]

$$\phi_l = 2\sigma\left(\frac{l}{a_Q}\right) + \frac{4}{3}\pi\chi\left(\frac{l}{a_Q}\right)^3 \tag{5}$$

The diameter of the aqueous channel within Q^{244} and Q^{230} cubic mesophase is then determined by using the expression in (6)

$$\langle \varnothing \rangle = 2\left(\left[-\sqrt{\frac{\sigma}{2\pi\chi}}\right]a_Q - l\right) \tag{6}$$

Table 1 *Selected Structure and Lattice Parameters from bR crystallisation mixtures prepared as described earlier in X-ray transparent tubes. Column 2 and 3 show recorded structural transformations, cubic lattice parameters, and lamellar repeat spacing in the bulk mesophase following crystallization induction at 20°C. Column 7 shows the corresponding variation in mean curvature due to osmotic pressure (see §2.6 Parametric Estimation of in cubo Mesophases) exerted by the precipitant salt. Column 6 shows the variation in lipid length (l), and thus trans-membrane bilayer thickness (2l) while column 9 shows the local variation in the average diameter of the cubic aqueous network. Finally, the time (-1) refers to the duration of time prior to salt addition. The dimension in parentheses refer to the water separation in the lamellar phase.*

Time Duration	Space Group	Unit Cell Lattice a	Lipid volume fraction ϕ_{aq}	Pivotal ξ_n	Lipid Length l	Mean Curvature $|\langle H_p \rangle|$	$|\langle R_w \rangle|$	Cubic Channel Diameter $|\langle \varnothing \rangle|$
(days)		[Å]			[Å]	$\text{Å}^{-1} \times 10^{-3}$	[Å]	[Å]
–1	Pn3m	97	0.399	11.9354	16.17	– 9.74	21.75	43.51
1	Pn3m	104	0.404	12.6826	17.17	– 7.68	23.49	46.99
3.5	Pn3m	92	0.337	12.5899	17.21	– 8.31	18.76	37.52
7	—	—	—	—	—	—	—	—
21	Ia3d	138	0.289	12.6332	17.36	–10.79	16.86	33.73
35	L$_\alpha$	42	0.158	—	17.68	—	—	(6.64)

The crystallization dependence of the mean curvature, $\langle H_p \rangle$, determined at the MO pivotal surface is presented in column 7 of Table 1. The trend shows that the curvature of the lipid monolayer (defined at the pivotal surface) in the bicontinuous crystallization matrix at 20°C is initially constant but as expected decreases as the aqueous additives are added. Once the sample achieves structural homogeneity (1 day) as judged by the diffraction, the mean curvature reaches a maximum. Subsequently the system undergoes a cubic-to-cubic (Q^{224} to Q^{230}) phase transition (1 week). After three weeks the system undergoes a cubic to noncubic (Q^{230} to L$_\alpha$) phase transition (Figure 5). That the curvature is decreasing indicates that the additives have an osmotic effect on the cubic matrix, dehydrating the cubic phase and ultimately leading to a lamellar phase transition. Thus we find that the crystallization process occurs via an increase in the bilayer membrane spanning thickness and a decrease in the aqueous space available to the protein. One would expect that protein-lipid and protein–protein interactions might be least favorable under conditions of high curvature compared to those of zero curvature. Part of the free energy driving force that leads to the dehydration of the cubic carrier phase might therefore be related to this. Phase diagrams for potential lipid systems for crystallizing membrane proteins may therefore require that high water composition cubic phases border on to a lower water composition lamellar phase.

2.7 Effects of Protein/lipid Hydrophobic Mismatch on Crystallization

The dimensional matching between the hydrophobic length of an integral membrane protein and the hydrophobic thickness of the target membrane lipid bilayer is an important factor in protein-lipid interaction[28]. In addition to the well-documented protein sorting, hydrophobic mismatch plays important roles in other membrane processes like reconstitution. Given that membrane protein reconstitution forms a key step in the *in cubo* membrane protein crystallization method, the effects of hydrophobic matching or mismatch between an embedded membrane protein and the supporting lipid bilayer might be central to the early stages of protein stabilization and as a consequence crystallization in the cubic matrix. The lateral hydrophobic dimension in the case of bacteriorhodopsin is estimated to be *ca.* 30-31 Å[31] a value that falls within the range of that determined for the MO cubic-based bilayer. In Table 1 we have that the pivotal surface has a width of approximately 12.6 Å. This is the width to the C_3 carbon on MO, so if we add on 2 Å for the C_1 and C carbons, the hydrophobic thickness of the bilayer will be 29.2 Å. This is just somewhat thinner than the estimate of the hydrophobic thickness of bacteriorhodopsin.

The exact unitary dimensions of a bR monomer are unknown, however, given its relative molecular size and a comparison of this to the other proteins investigated here, it can be argued that bR is an unrepresentative system given that many proteins are characterized by much larger dimensions. Based on our cubic parametric estimates (Table 1) it would appear that bR monomers are reasonably readily accommodated both within the MO bilayer and the cubic aqueous channels, which take diameters of the order 31-42 Å. The two other proteins studied, the light-harvesting complex (LHC-II) and photosystem II (PS II) are characterised by dimensional length scales of the order (*ca.* 33.9 × 38.29 × 53.41 Å) and (*ca.* 220 × 150 × 95 Å), respectively. Based on correlations between our determined estimates of the membrane-spanning parts of bR, LCH-II or PSII and the hydrophobic thickness of MO we are drawn to conclude that while the hydrophobic thickness of bR is reasonably well matched to that of MO, the reverse is true in the case of the LCH-II and PSII. It might therefore be reasonable to attribute the observed spectral modifications in both the LCH-II and PSII protein complexes to hydrophobic mismatch. Figure 7 B-C illustrates the two main forms of hydrophobic mismatch. In both LCH-II and PSII the case illustrated in (Figure 7C) applies.

In planar mesophases (lamellar phases), hydrophobic mismatch between the lipid alkyl chain region and the protein hydrophobic region may lead to bilayer deformations that carry a free energy penalty. The same will hold true in non-planar structures (such as the bicontinuous cubics). Noting that the hydrophobic thickness in the cubic phases of MO are thinner than that of the lamellar phase it is conceivable that proteins with a narrow hydrophobic region might drive the system to stabilise a bicontinuous cubic structure. In cases where the protein has a thicker hydrophobic domain the latter is not possible and where the mismatch is extreme it is quite possible that the protein will simply not be stable within the MO matrix. Under such circumstances it would be quite probable that one would observe the aggregation of protein into amorphous forms. In the case of PSII, we believe that something of this sort may be occurring, leading to the loss of Mg in the PSII subunit and the observed spectroscopic modifications. The observed blue shift is diagnostic of PSII denaturation. This and related issues will be addressed in a future publication.

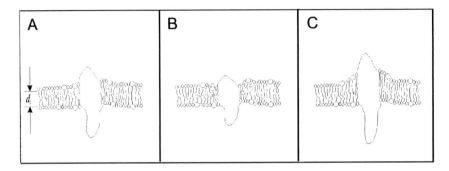

Figure 7 *A schematic illustration of the protein/lipid hydrophobic matching. A) The hydrophobic thickness of the protein and that of the lipid membrane (d$_l$) are well matched. In this case, the interfacial and chain pressure of the two opposing monolayers are well matched and as a consequence the net moment is zero. To this end, both opposing monolayers have no desire to bend. B) The protein-lipid hydrophobic thickness is mismatched, i.e., the protein hydrophobic dimension is shorter than that of the host membrane lipid bilayer. C) Here the protein hydrophobic dimension is much larger than the dimensional length of the membrane lipid alkyl chain.*

3 CONCLUSION

In experiments based on the pioneering work of Landau and Rosenbusch we have used a combination of optical microscopy and time resolved SAXD to identify the structural phase transitions that accompany the crystallisation of bR as a model system. This work was aimed at facilitating continuous monitoring of the membrane lipid structural properties during crystallisation, in such a way that crystallisation conditions, and their parameters could be optimised to produce protein crystals with a high structural resolution. To our knowledge this is the first time that it has been possible to follow the crystallisation of membrane proteins in this way. In our efforts to screen a whole series of crystallisation parameters, an increase in the number of trials undertaken has made automation a necessity. We have therefore implemented Multi-Well batch crystallisation methods. Using this approach we have screened a number of *in cubo* crystallisation conditions. New insights into the necessary crystallisation conditions has been achieved, although more research is required before a detailed analysis is complete, and thus providing a full understanding of the mechanism of *in cubo* membrane protein crystallisation. In addition we have characterized the events following Light Harvesting Complex (LHC-II) and Photosystem II (PSII) membrane protein incorporation into bicontinuous cubic phases of MO. Our experiments show that whilst, the spectral properties in both proteins were modified by the incorporation process (data not shown) the structure of the cubic phase matrix was stabilised by the incorporation of either protein, as long as the protein to cubic ratio was kept to that required by the standard crystallisation protocol.

On the assumption that the curvature of the cubic phase might be responsible for the observed spectroscopic modifications, we have incorporated both proteins in lamellar phases. The resultant spectral analyses prove that the modifications are similar to those observed in cubic phases. These observations point to the possibility that the spectral modifications originate from a dimensional mismatch between protein moieties and the MO bilayer, and not from the strain exerted by the bilayer curvature on the protein. Based on lipid packing calculations, we have shown that the observed macromolecular spectral modifications might indeed originate from the bilayer/protein hydrophobic mismatch. We argue further that the mismatch can be obviated by using lipid blends with a maximal swelling of the cubic mesophase, doped with additional lipids that promote reverse hexagonal and lamellar structures.

C.S acknowledges support in form of a studentship from the EPSRC and Unilever Research, Port Sunlight.

References

1. D.T. Jones, *FEBS Lett.* 1998, **423**, 281.
2. H. Michel and D. Oesterhelt, *Proc. Natl. Acad. Sci. USA,* 1980, **77**, 1283.
3. E. M. Landau and J.P. Rosenbusch, *Proc. Natl. Acad. Sci. USA,* 1996, **93**, 14532.
4. H. Luecke, B. Schobert, H.-T. Richter, J-P. Cartailler and J. K. Lanyi, *J. Mol. Biol.* 1999, 291, 899
5. G. Lindblom and L. Rilfors, *Biochim. Biophys. Acta.* 1989, **988**, 221.
6. R.H. Templer, *Curr. Opin. Coll. Interf. Sci.* 1998, **3**, 255.
7. J.M. Seddon and R.H. Templer, *Phil. Trans. R. Soc. Lond. A* 1993, **344**, 377
8. V. Luzzati, *Curr. Opin. Struct. Biol.,* 1997, **7**, 661;
9. S. Cribier, A. Gulik, P. Fellman, R. Vargas, P. Devaux and V. Luzzati, *J. Mol. Biol.* 1993, **229**, 517.
10. M.L. Chiu, P. Nollert, M.C. Loewen, H. Belrhali, E. Pebay-Peyroula, J.P. Rosenbusch and E.M. Landau, *Acta Cryst. D,* 2000, **56**, 781.
11. A. Royant, P. Nollert, K. Edman, R. Neutze, E.M. Landau, E. Pebay-Peyroula, J. Navarro, *Proc. Natl. Acad. Sci. USA,* 2001, **98**, 10131.
12. H. Luecke, B. Schobert, J. K. Lanyi, E. N. Spudich and J. L. Spudich, *Science,* 2001, **293**, 1499.
13. G. Rummel, A. Hardmeyer, C. Widmer, M.L. Chiu, P. Nollert, K.P. Locher, I.I. Pedruzzi, E.M. Landau and J.P. Rosenbusch, *J. Struct. Biol.,* 1998, **121**, 82.
14. P. Nollert, H. Qiu, M. Caffrey, J.P. Rosenbusch, and E.M. Landau, *FEBS Lett.* 2001, **504**, 179.
15. E.S. Lutton, *J. Am. Oil Chem. Soc.* 1965, **42**, 1068
16. S. Hyde, S. Andersson, B. Ericsson and K. Larsson, *Z Kristallogr.* 1984, **168**, 213
17. H. Qui and M. Caffrey, *Biomaterials,* 2000, **21**, 223
18. D. Oesterhelt, and W. Stockenius, *Methods Enzymol* 1974, **31**, 667
19. M. S. Braiman, et al. *J. Biol. Chem.* 1987, **262**, 9271
20. P. Sakya, J.M. Seddon and R.H. Templer, *J. Phys. II.* 1994, **4**, 1311
21. B. Angelov, M. Ollivon and A. Angelova, *Langmuir,* 1999, **15**, 8225
22. X. Ai and M. Caffrey, *Biophys. J,* 2000, **79**, 394
23. C. Sennoga, B. Hankamer, J.M. Seddon and R.H. Templer, (in preparation)
24. P. Nollert, A. Royant, E. Pebay-Peyroula and E.M. Landau *FEBS Lett.* 1999, **457**, 205
25. C. Sennoga, Ph.D. thesis, Imperial College, London, (in preparation)

26. T. Baba, H. Minamikawa, M. Hato, A. Motoki, M. Hirano, D. Zhou and K. Kawasaki *Biochem. Biophys. Res. Commun.,* 1999, **265**, 734
27. R. H. Templer, Langmuir, 1995, **11**, 334
28. B. J. Khoo, Ph.D. thesis, Imperial College, London , 1996
29. R. H. Templer, J. M. Seddon and N. A. Warrender Biophys. Chem. 1994, **49**, 1
30. D.C. Turner, Z.-G. Wang, S.M. Gruner, D.A. Mannock, R.N. McElhaney, *J. Phys. II France* 1992, **2**, 2039
31. B. Piknová, E. Pérochon and J.-F, Tocanne, *Eur. J. Biochem.,* 1993, **218**, 385

MOBILITY OF PROTEINS AND LIPIDS IN THE PHOTOSYNTHETIC MEMBRANES OF CYANOBACTERIA

C.W. Mullineaux and M. Sarcina

Department of Biology, University College London, Darwin Building, Gower St. ,London WC1E 6BT, UK

1 INTRODUCTION

In recent years there has been tremendous progress in understanding the structures of photosynthetic light-harvesting antennae and reaction centres. However, we know much less about the dynamics of these complexes *in vivo*. How do light-harvesting complexes and reaction centres interact in the intact photosynthetic membrane? Are the interactions permanent or transient, and how are they affected by regulatory mechanisms? We have been probing these questions using a variant of Fluorescence Recovery after Photobleaching (FRAP) which allows us to observe the diffusion of fluorescent pigment protein complexes in photosynthetic membranes *in vivo*.[1] Our currently preferred model organism is the cyanobacterium *Synechococcus sp* PCC7942. In common with some other cyanobacteria, *Synechococcus* 7942 has elongated cells with the thylakoid membranes arranged as regular concentric cylinders aligned along the long axis of the cell. The cells may be further elongated by growth in the presence of cell division inhibitors, without any detectable side-effects in terms of photosynthetic function or membrane structure.[2] Cyanobacterial thylakoid membranes have a rather uniform composition, with no significant lateral heterogeneity. Their regular geometry is in contrast to the photosynthetic membranes of virtually all other photosynthetic organisms, which tend to exhibit lateral heterogeneity and/or intricate and irregular fine-scale membrane structure. These properties make *Synechococcus* 7942 ideal for FRAP measurements. In addition, *Synechococcus* 7942 is well-characterised and transformable, and numerous mutants are available.

Since FRAP is an optical technique it has limited spatial resolution. Thus, we require a regular membrane geometry for quantitative measurements. We use a one-dimensional FRAP technique which exploits the cylindrical geometry of the *Synechococcus* thylakoids (Fig. 1). As in other cyanobacteria, the principal light-harvesting complexes in *Synechococcus* are phycobilisomes, large and highly-ordered protein aggregates coupled to the cytoplasmic surface of the thylakoid membrane.[3] The Photosystem I and Photosystem II reaction centres are integral membrane protein complexes. Energy transfer measurements indicate that phycobilisomes can interact with both kinds of reaction centre.[4] We have used FRAP to measure the diffusion rates of the phycobilisomes and Photosystem II reaction centres. In *Synechococcus*, as in all the other cyanobacteria that we have examined, we find the Photosystem II is essentially immobile. However, phycobilisomes diffuse rapidly on the surface of the thylakoid membrane. This shows that the interaction between phycobilisomes and reaction centres is transient and unstable. We report the use of FRAP measurements on mutants to further explore the nature of the interaction between phycobilisomes and thylakoid membrane components, and we discuss the possible physiological role(s) of phycobilisome mobility.

Figure 1

Geometry of a one-dimensional FRAP measurement (adapted from Ref. 1).
A. A cell aligned in the y-direction is selected. A highly-focused confocal laser spot is scanned rapidly across the cell in the x-direction, bleaching the pigments in a line across the cell.
B. The laser power is reduced to prevent further bleaching, and the spot is scanned in the XY plane to record a series of two-dimensional fluorescence images of the cell.
C. The images are integrated in the x-direction to produce plots of fluorescence intensity versus position along the long axis of the cell. A series of plots at different times after the bleach shows the diffusion of the fluorescent complex.

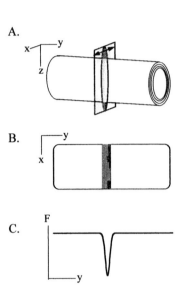

2 MATERIALS AND METHODS

Synechococcus sp. PCC7942 was grown in BG11 medium [5] supplemented with 10 mM NaHCO$_3$ and appropriate antibiotics for mutants. Liquid cultures were grown in an orbital shaking incubator at 30 °C with white illumination at about 10 μE. m^{-2} s^{-1}. For use in FRAP measurements the cells were elongated by treatment with thiobendazole. This resulted in increased mean cell length without any detectable alteration in photosynthetic function.[2] FRAP experiments were carried out at CLRC Daresbury Laboratory (Warrington, Cheshire, UK) using the scanning confocal microscope Syclops with a 633 nm Helium-Neon laser or a 442 nm Helium-Cadmium laser. Fluorescence was selected using a Schott RG665 red glass filter, transmitting light above about 665nm. Under these conditions excitation with 442 nm light allows observation of fluorescence predominantly Photosystem II, and excitation with 633 nm light allows observation of fluorescence predominantly from phycobilisome cores[1]. Cells were spread on 1.5% agar containing growth medium, covered with a glass cover slip and placed on a temperature-controlled stage under the microscope objective lens. A 40 x oil immersion lens (numerical aperture 1.3) was used with 20 μm pinholes to create a confocal spot with FWHM dimensions of about 0.9 μm in the Z-direction and 0.3 μm in the XY plane. The confocal spot was scanned for about 1 second in the X-direction to create the bleach. The confocal spot was then scanned in the XY plane to record a sequence of images of the cell at 3 s intervals. Images were analysed and diffusion coefficients calculated as described by Mullineaux *et al.*[1]

3 RESULTS AND DISCUSSION

3.1 Mobility of phycobilisomes and Photosystem II in *Synechococcus* 7942

Figure 2 is a FRAP image sequence showing the mobility of phycobilisomes. We found that the phycobilisomes diffused rapidly. At 30 °C, the average diffusion coefficient for phycobilisomes was $(3.1 \pm 1.0) \times 10^{-10}$ cm^2. s^{-1}. However, in the case of Photosystem II we could detect no diffusion on the timescale of the measurement (not shown). As in the other cyanobacteria we have examined, it appears that the association between phycobilisomes and reaction centres is transient and unstable.

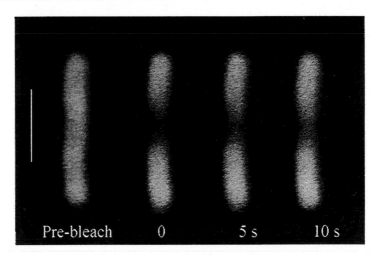

Figure 2
FRAP image sequence showing phycobilisome fluorescence. The scale bar is 3 microns. Note that the bleached line spreads and becomes shallower with time, indicating diffusion of the phycobilisomes

3.2 Effect of phycobilisome size

We have explored the effect of phycobilisome size by measuring the diffusion coefficient in a mutant lacking the phycobilisome rod elements. The mutant, R2HECAT, lacks genes coding for the α-and β-subunits of phycocyanin and rod linker polypeptides. However the phycobilisome cores are still assembled and functional.[6] The phycobilisome cores have a molecular mass of 1200-1300 kDa and dimensions of about 22 x 11 x 12 nm. The intact phycobilisomes of wild-type cells are hemidiscoidal structures with a typical molecular mass of about 6000 kDa and a longest diameter typically about 60 nm.[7] At 30 °C, the mean diffusion coefficient for the phycobilisome cores in R2HECAT was $(7.1 \pm 0.8) \times 10^{-10}$ cm^2 s^{-1}. This compares to a mean diffusion coefficient of $(3.1 \pm 1.0) \times 10^{-10}$ cm^2. s^{-1} in the wild-type. Thus phycobilisome diffusion at growth temperature is faster by a factor of 2.3 ±

0.7 in R2HECAT. This suggests that cytosolic crowding[8] plays a role in limiting the rate of diffusion of the phycobilisomes.

3.3 Diffusion of intact phycobilisomes or detached rod elements?

We have interpreted our FRAP results in terms of the movement of intact phycobilisomes, since we excite the phycobilisomes with short wavelength light predominantly absorbed by phycocyanin in the phycobilisome rods, and observe long-wavelength fluorescence predominantly from the phycobilisome cores.[1] However, spectral overlap makes it hard to completely exclude an alternative possibility, that the phycobilisome cores are immobile and the diffusion we see is of rod elements that may not be stably coupled to the phycobilisome cores *in vivo*. Our studies with the R2HECAT mutant (see above) shed further light on this problem. We find that the phycobilisomes are mobile in this mutant. Since the rod elements are lacking, the cores must be moving. Thus the diffusion we observe in the wild-type is most probably of intact, fully assembled phycobilisomes.

3.4 How do phycobilisomes interact with the membrane?

Phycobilisomes are assembled and are membrane-associated even in the absence of Photosystem II and Photosystem I reaction centres.[9] Thus, when phycobilisomes diffuse, we imagine them decoupling from a reaction centre, but remaining attached to the membrane surface. The phycobilisome will then diffuse freely on the membrane surface before coupling to another reaction centre. However, the nature of the interaction with the membrane is unclear. The ApcE protein of the phycobilisome core is implicated. Proposals for the association of ApcE with the membrane have included an integral membrane domain[10] or a covalently attached acyl group.[11] We have explored this problem by measuring the diffusion coefficient for phycobilisomes at a range of temperatures. We found that cooling below the phase transition temperature of the membrane had no significant effect on the mobility of phycobilisomes. Under the same conditions, the diffusion coefficient of a lipid-soluble fluorescent marker was reduced by a factor of six (data not shown). This strongly suggests that there is no integral membrane component in the phycobilisome. Instead, we propose that phycobilisomes interact with lipid head-groups at the membrane surface. A precedent is spectrin, a component of the erythrocyte cytoskeleton. Spectrin is proposed to interact with the membrane via multiple weak interactions with lipid head-groups.[12] As with phycobilisomes, spectrin can diffuse rapidly on the membrane surface, and the diffusion coefficient is not strongly affected by cooling to the phase transition temperature of the membrane.[12]

3.5 Effect of lipid desaturation - role of lipids in controlling phycobilisome-reaction centre interaction?

Mutants in which the thylakoid membrane lipid composition is altered provide a further opportunity to explore the interaction between phycobilisomes and membranes. We have used *des*A[+], a transformant of *Synechococcus* 7942 which contains *des*A, the Δ12 fatty acid desaturase gene from *Synechocystis* 6803.[13] *Des*A[+] cells have a much higher proportion of unsaturated fatty acids than the wild-type. As would be expected, the thylakoid membranes are more fluid in *des*A[+] than in the wild-type (unpublished data). Unexpectedly, we found that phycobilisome diffusion was far slower in *des*A[+] than in the wild-type. At 30 °C, the mean phycobilisome diffusion coefficient in *des*A$^+$ was $(2.5 \pm 1.2) \times 10^{-12}$ cm^2 s^{-1}, slower than in the wild-type by a factor of 120 ± 70. The most likely explanation is that the interaction with the reaction centres is stabilised in *des*A$^+$. We know that Photosystem II is immobile (Fig. 3). Therefore, if the binding of phycobilisomes to Photosystem II is stabilised, the diffusion coefficient for phycobilisomes will be reduced. How could lipid desaturation alter phycobilisome-reaction centre interaction? Maybe specific lipids, or the general lipid environment of membrane, play a crucial role in mediating phycobilisome-reaction centre interaction. Alternatively, it could be an indirect effect. Phycobilisome-reaction coupling may be influenced by the redox state of electron transport cofactors (see below) and this may differ in wild-type and *des*A$^+$ cells under our measuring conditions.

3.6 Physiological role(s) of phycobilisome mobility?
Phycobilisome mobility is characteristic of all the cyanobacteria that we have examined. What physiological role(s) could it play? Three possible explanations are explored below:

a. Phycobilisome mobility is required for regulation of light-harvesting through state transitions.
The physiological adaptation mechanism known as state transitions involves the redistribution of phycobilisomes between Photosystem II and Photosystem I.[14] This presumably requires movement of the phycobilisomes. However, state transitions occur on a timescale of a few seconds to about a minute. At the diffusion rates we observe, we can estimate that a phycobilisome could diffuse from Photosystem II to Photosystem I in about 15 milliseconds. Thus it is likely that the rate at which state transitions occur is controlled by the signal transduction pathway, rather than by the diffusion of the complexes. We could predict that state transitions could still occur if the diffusion rate of phycobilisomes were hundreds of times slower. In fact we find that state transitions occur normally in the *des*A$^+$ transformant, in which phycobilisome diffusion is about 120 times slower than in the wild-type (see above).

b. Phycobilisome mobility is required for synthesis and turnover of thylakoid membrane components
Phycobilisomes are large complexes, which normally occupy much of the cytoplasmic surface of the thylakoid membrane.[15] It could be argued that phycobilisome mobility is necessary to allow access of ribosomes, proteases, and regulatory enzymes to the membrane surface, in order to allow synthesis, turnover and regulation of thylakoid membrane components. One prediction of this idea would be that the turnover of the D1 polypeptide should be slower in the *des*A$^+$ transformant, in which phycobilisome mobility is greatly reduced (see above). However, it appears that D1 turnover is actually faster in *des*A$^+$ than in the wild-type.[16]

c. Phycobilisome mobility increases the efficiency of light-harvesting
Phycobilisomes are mobile on the same timescale as the secondary electron transport reactions. Could phycobilisomes decouple from photochemically "closed" reaction centres and re-associate with open reaction centres, thus minimising the wasteful transfer of excitons to closed reaction centres? In this model, phycobilisome mobility would be a way to allow a limited pool of phycobilisomes to act as efficient light-harvesting antennae for a much larger pool of reaction centres. Experiments to test this idea are in progress.

Acknowledgments

We thank Mark Tobin (CLRC Daresbury Laboratory, Warrington, UK) for his assistance with the FRAP measurements, and Abosede Felix and Jessamy Findlater for electron microscopy. We also thank Norio Murata (National Institute for Basic Biology, Okazaki, Japan) and Petter Gustafsson (University of Umeå, Sweden) for providing mutants. Supported by a BBSRC research grant to CWM.

References

1. C.W. Mullineaux, M.J. Tobin and G.R. Jones, *Nature*, 1997, **390**, 421.
2. M. Sarcina and C.W. Mullineaux, *FEMS Microbiol. Lett.*, 2000, **191**, 25.
3. R. MacColl *J. Struct. Biol.*, 1998, **124**, 311.
4. C.W. Mullineaux, *Aust J. Plant Physiol.*, 1999, **26**, 671.
5. R.W. Castenholz in *Methods in Enzymology*, eds L. Packer and A.N. Glazer, Academic Press, San Diego, 1988, Vol **167**, 68.
6. R.P. Bhalerao, T. Gillbro and P. Gustafsson *Photosynth. Res.*, 1995, **45**, 61.
7. A.N. Glazer, *Biochim. Biophys. Acta*, 1984, **768**, 29.
8. R.J. Ellis, *Curr. Opin. Struct. Biol.*, 2001, **11**, 114.
9. J. Yu, Q. Wu, H. Mao, N. Zhao and W.F.J. Vermaas, *IUBMB Life*, 1999, **48**, 625.
10. T. Redlinger and E. Gantt, *Proc. Nat. Acad. Sci. USA*, 1982, **79**, 5542.
11. D. Bald, J. Kruip and M. Rögner, *Photosynth. Res.*, 1996, **49**, 103.
12. P.J. O'Toole, C. Wolfe, S. Ladha and R.J. Cherry, *Biochim. Biophys. Acta*, 1999, **1419**, 64.
13. Z. Gombos, E. Kanervo, N. Tszvetkova, T. Sakamoto, E.-M. Aro and N. Murata *Plant Physiol.*, 1997, **115**, 551.
14. J.J. van Thor, C.W. Mullineaux, H.C.P. Matthijs and K.J. Hellingwerf, *Botanica Acta*, 1998, **111**, 430.
15. L. Mustardy, F.X. Cunningham and E. Gantt, *Proc. Nat. Acad. Sci. USA*, 1992, **89**, 10021.
16. K. Sippola, E. Kanervo, N. Murata and E.-M. Aro, *Eur. J. Biochem.*, 1998, **251**, 641.

PARTITIONING AND THERMODYNAMICS OF CHLORDIAZEPOXIDE IN n-OCTANOL/BUFFER AND LIPOSOME SYSTEM

Catarina Rodrigues[a], Paula Gameiro[a*], Salette Reis[b], J.L.F.C. Lima[b] and Baltazar de Castro[a]

[a] CEQUP/Departamento de Química, Faculdade de Ciências, Universidade do Porto, 4169-007 Porto, Portugal
[b] CEQUP/Laboratório de Química-Física, Faculdade de Farmácia, Universidade do Porto, 4050-047 Porto, Portugal

1 INTRODUCTION

Chlordiazepoxide is a benzodiazepine and belongs to a class of neuroactive drugs that enjoy a wide administration because of their muscle relaxant, hypnotic anticonvulsivant and ansiolytic characteristics. The ability of a drug to elicit a pharmacological action in a biological system requires, in most cases, an interaction between the drug and the biological membrane. The permeability of these membranes is determined by physical factors including the partition coefficient, diffusion coefficient, membrane thickness and interfacial barriers. [1]

Thermodynamic studies have been employed to gain an insight into drug transport and relative biological activity and are also used to predict drug incorporation into phospholipid vehicles like liposomes, which are considered to be useful drug carriers.[2]

Recently, interest has focused on determination of the partition coefficients of drugs between lipid bilayer vesicles and the aqueous phase in order to investigate the behaviour of drugs towards biomembranes. Partition coefficients are normally determined in an isotropic two-phase solvent system like water/octanol but as membranes possess an ordered molecular arrangement, with structural characteristics varying from a liquid crystalline state to a rigid gel and are able to exert electrostatic influences, liposomes are believed to be a better model membrane to predicted drug/membrane partition coefficients. This is because they mimic better the hydrophobic part and the outer polar and negatively charged surface of the phospholipids of natural membranes.[3]

The present study compares the partitioning behaviour of chlordiazepoxide in n-octanol/buffer and buffered DMPC and DPPC multilamellar liposome using a thermodynamic approach and, from this, evaluates the interaction of chlordiazepoxide with lipid bilayers in comparison with the n-octanol/buffer system.[4]

2 MATERIAL AND METHODS

Dimyristoylphosphatidylcholine (DMPC) and Dipalmitoylphosphatidylcholine (DPPC) were obtained from Avanti Polar Lipids Inc. Chlordiazepoxide was a gift from Hoffman-La Roche. *n*-Octanol was from Fluka and chloroform from Merck (pro-analysi). All lipid suspensions were prepared with aqueous 10 mM Hepes buffer solution (I=0.1 M NaCl; pH 7.4).

2.1 Distribution studies in *n*-octanol/buffer systems

Convenient volumes of aqueous phase (5ml Hepes buffer) containing the appropriate concentration of drug (23.6 µM) and *n*-octanol (0.1 mL) were equilibrated for 4h at constant temperature (±0.1°C) in a shaking water-bath. Concentrations of chlordiazepoxide in the aqueous phase were determined by UV spectrophotometry in an UNICAM UV-300 spectrophotometer equipped with a constant-temperature cell holder at $\lambda_{máx}$ (261nm). The distribution of chlordiazepoxide was obtained from the average of duplicate determinations at each temperature over the range 27-50 °C.

2.2 Distribution studies in liposome systems

Liposomes of DMPC and DPPC were prepared by evaporation to dryness of a lipid solution in chloroform (2mg/mL) under a stream of argon and then left under vacuum overnight to remove all traces of the organic solvent. The resultant dried lipid films were dispersed with 5mL Hepes buffer containing 23.3µM chlordiazepoxide and the mixtures were vortexed above the phase transition temperature to produce multilamellar liposomes (MLV). The distribution of the drug was determined using UV analysis over the range 4-50°C. Determinations were made in duplicate and the results averaged.

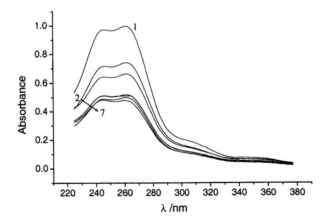

Figure 1   *Absorption spectra of chlordiazepoxide (23.3µM) in Hepes buffer solution (1) and in DMPC liposome suspension after incubation at different temperatures: (2) 4°C, (3) 7°C, (4) 23°C, (5) 25°C, (6) 30°C, (7) 40°C.*

3 CALCULATION OF PARTITION COEFFICIENTS

The molar partition coefficients, K_p, were calculated from the distribution results by:

$$K_p = \frac{(C_t - C_w)W_1}{C_w.W_2} \tag{1}$$

where C_t is the total initial concentration of chlordiazepoxide in the aqueous buffer phase before equilibration, C_w is the final aqueous phase concentration of chlordiazepoxide, W_1 is the molar concentration of water and W_2 is the molar concentration of lipid or *n*-octanol.

The partition coefficient is related to the standard transfer Gibbs energy, $\Delta G^0{}_{w \to L}$, that is the change in Gibbs energy when one mole of solute is transferred from water to lipid at infinite dilution, by:

$$\Delta G^0{}_{w \to L} = -2.303RT \log K_p \tag{2}$$

Considering the standard transfer enthalpy and standard transfer entropy constants in the range of temperature used in this work it is possible to calculate their values by a linear plot of the partition coefficients *vs* temperature, $\log K_p$ *vs* T^{-1}.

$$\log K_p = \frac{\Delta H^0{}_{w \to L}}{2.303RT} + \frac{\Delta S^0{}_{w \to L}}{2.303R} \tag{3}$$

The linear plot of equation 3 enables calculation of $\Delta H^0{}_{w \to L}$ from the slope and $\Delta S^0{}_{w \to L}$ from the intercept. $\Delta H^0{}_{w \to L}$ and $\Delta S^0{}_{w \to L}$ have the physical meaning of the change in enthalpy and in entropy, respectively, when one mole of solute is transferred from water to lipid at infinite dilution.

However, the values of the change in entropy of partitioning, $\Delta S^0{}_{w \to L}$, were more conveniently obtained from

$$\Delta S^0{}_{w \to L} = \frac{\Delta H^0{}_{w \to L} - \Delta G^0{}_{w \to L}}{T} \tag{4}$$

Table 1 *Effect of temperature on partition coefficients of chlordiazepoxide in DMPC and DPPC liposome suspensions.*

T (K)	Below Tc			T (K)	AboveTc		
	$\log K_{p\,(w/o)}$	$\log K_p$ (DMPC)	$\log K_p$ (DPPC)		$\log K_{p\,(w/o)}$	$\log K_p$ (DMPC)	$\log K_p$ (DPPC)
277		3.83		297		4.29	
280		3.99		303		4.30	
295	3.20	4.28		313	3.33	4.33	3.84
303	3.28		3.53	318			4.02
307			3.80	322	3.40	3.83	4.13

a) $K_{p\,(w/o)}$ is partition coefficient in water/octanol

4   RESULTS AND DISCUSSION

Partition coefficients were higher in liposomes than in *n*-octanol and for the liposomes they are higher in DMPC than in DPPC but in both systems they increase with temperature below and above T_c (Table 1). From figure 2 it is also evident that the change of K_p with temperature is similar for water/octanol and for DMPC above the T_c, this observation can be explained by the less-ordered structure, of DMPC, above phase transition.

As liposomes mimic better the hydrophobic part and the outer polar and negatively charged surface of the phospholipids of natural membranes, the K_p values obtained in these systems characterise better the drug/membrane interaction.

The values of enthalpy and entropy obtained for the three systems are positive but are lower for DMPC liposomes. This latter result shows that the partitioning of chlordiazepoxide depends on the rigidity of lipid bilayers and once more it can be concluded that liposomes constitute a more selective partitioning system than does the *n*-octanol.

The observed positive entropy changes arise from the loss in water structure surrounding the drug molecules on transfer to the bilayer. The partitioning disrupts the ordering of the bilayer and the magnitude of the effect is dependent on the structure of the membrane phase (Table 2).

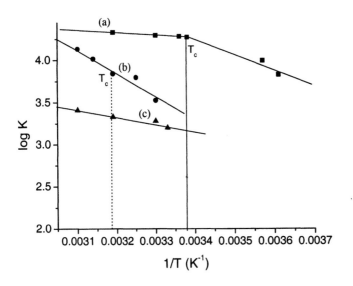

Figure 2   *Log K_p values vs T^{-1} for chlordiazepoxide:a-DMPC, b-DPPC and c- n-octanol.*

Analysing the enthalpy of partitioning of chlordiazepoxide in the three systems it is evident that the least endottermic process occurs in DMPC above the phase transition which shows that less energy is required to insert a drug molecule into its fluid bilayer state.

All the $\Delta G_{w \to L}$ values are negative, with positive $\Delta S°$ and $\Delta H°$, which show that in liposomes and *n*-octanol/water systems partitioning is, as expected, entropy dominated.[5] The higher K_p values obtained for the DPPC liposome system are clearly a consequence of the high standard transfer entropy obtained when the drug is transferred from water to DPPC liposomes. The even higher K_p values obtained for DMPC liposomes are not only a consequence of the high standard transfer entropy but also of the small standard transfer enthalpy compared with the other two systems.

4.1 Concluding Remarks

A comparison of the K_p values determined in liposomes and determined in water/octanol media reveals that liposomes are a different and (better) membrane model than an isotropic two-phase solvent system, such as water-octanol. They allow for a better prediction of drug-membrane partition coefficients, since they mimic better the hydrophobic part and the outer, polar and negatively charged, surface of the phospholipids in natural membranes.

The analysis of the thermodynamic functions shows that partition is a process where entropy dominates, but as is usual when there is a system with more favourable enthalpy and entropy, the Gibbs energy of transfer (and consequently the partition coefficient) is greater. This is observed for DMPC liposome system.

Table 2 *Thermodynamics of partitioning of chlordiazepoxide in the n-octanol or phospholipid/Hepes buffer systems at pH 7.4.*

T = 40°C	$logK_p$	$\Delta G°_{w \to L}$ (KJmol^{-1})	$\Delta H°_{w \to L}$ (KJmol^{-1})	$\Delta S°_{w \to L}$ (Jmol^{-1})
n-octanol	3.33	-19.95	15.03	111.74
DMPC	4.30	-26.01	3.54	93.70
DPPC	3.84	-23.17	52.98	243.31

References

1 A. R. James and S. D. Stanley, *Biochimica et Biophysica Acta*, 1980, **598**,392.

2 T. Tomita, M. Watanabe, T. Takahashi, K. Kumai,T. Tadakuma and T. Yasuda, *Biochimica et Biophysica Acta,* 1989, **978**,185.

3 A. A. Omran, K. Kitamura, S. Takegami, A. Y. El-Sayed and M. Abdel-Mottaleb, *Journal of Pharmaceutical and Biomedical Analysis*, 2001, **25**, 319.

4 G.V.Betageri and S. R. Dipali, *J.Pharm. Pharmacol.*,1993, **45**, 931.

5 M. Arrowsmith, J. Hadgraft and I.W. Kellaway, *Biochimica et Biophysica Acta*, 1983, **750**, 149.

DISTRIBUTION OF VITAMIN E IN MODEL MEMBRANES

P.J. Quinn

Division of Life Sciences, King's College, London SE1 9NN, U.K.

1 INTRODUCTION

Vitamin E is found in all mammalian cell membranes where it represents a minor component of the polar lipid fraction. Although it is classified as a fat-soluble vitamin it possesses a hydroxyl group attached to the ring structure that confers a weakly amphipathic character on the molecule. This physical property is believed to influence the way vitamin E interpolates into the membrane and hence how it performs its biochemical functions.

Despite the relatively low concentration of vitamin E compared to other membrane lipids it is believed to play an important part in preserving the integrity of membranes. Foremost amongst its putative functions is its ability to protect polyunsaturated lipids of the lipid bilayer matrix of membranes against free radical oxidation.[1] Another important function in membranes is the formation of complexes between vitamin E and products of membrane lipid hydrolysis such as lysophospholipids and free fatty acids.[2] The complexes thus formed tend to stabilize membranes and prevent the detergent-like actions of lipid hydrolytic products on the membrane.

One of the key factors in understanding how vitamin E fulfils its functions in membranes is how it is localized in the lipid bilayer matrix and what effect its presence has on the structure and stability of membranes. Studies with model membrane systems containing vitamin E are described which address these questions.

2 METHOD AND RESULTS

2.1 Preparation and Characterisation of Model Membranes

The arrangement and distribution of vitamin E in model membranes has been investigated using aqueous dispersions of mixtures of vitamin E with defined phospholipids. The principle method used to characterise these model membranes was synchrotron X-ray diffraction in which X-ray scattering intensity in the small-angle (SAXS) and wide-angle (WAXS) scattering regions were recorded simultaneously during heating and cooling scans.[3] The dependence of mesophase structural parameters observed in dispersions containing varying proportions of vitamin E was used to determine the mode of interaction of vitamin E with the phospholipid and its distribution in the particular mesophase. Differential scanning calorimetry was used to assess the effect of vitamin E on the thermotropic phase behaviour of the phospholipid.

2.2 Effect of vitamin E on phospholipid phase behaviour

Examination of the thermal properties of mixed aqueous dispersions α-tocopherol (vitamin E) with distearoyl-, dipalmitoyl-, dimyristoyl- and dilauroyl-derivatives of phosphatidylcholine (DSPC, DPPC, DMPC,DLPC) by differential scanning calorimetry showed that increasing proportions of α-tocopherol cause a progressive broadening of the gel to liquid-crystalline phase transition and a decrease of enthalpy change of the transition.[4-9] Similar results were obtained by Fourier transform infrared and Raman spectroscopy.[10] Proportions of α-tocopherol greater than about 20mol% resulted in almost complete loss of transition enthalpy.

A consistent observation in the thermal studies of mixtures of α-tocopherol with saturated phosphatidylcholines is that the presence of α-tocopherol appears to eliminate the pretransition enthalpy. This does not mean that the ripple structure itself is eliminated. Evidence from synchrotron X-ray diffraction and freeze-fracture electron microscopy have shown that different ripple structures are induced in mixed aqueous dispersions of α-tocopherol and PC.[11] Figure 1 shows the SAXS patterns of codispersions of DSPC containing up to 10mol% α-tocopherol at 25°C. It can be seen

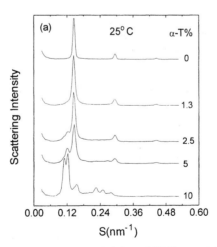

Figure 1 *.SAXS intensity patterns recorded from DSPC containing different amounts of α-tocopherol.*

that the number and intensity of the additional peaks increase with increasing proportions of α-tocopherol, suggesting that these peaks originate from α-tocopherol enriched domains. Together with freeze-fracture electron microscopy it was concluded that one effect of α-tocopherol on phosphatidylcholine was to form disordered ripples of large periodicity (about 50-150nm). When the proportion of α-tocopherol is increased the ripples of large periodicity are replaced by ordered ripples of a periodicity of 16nm, which produce at least 6 reflections in the SAXS region (see Fig. 1). Comparison of enthalpy changes and SAXS intensity values of DPPC containing varying proportions of vitamin E indicate that a complex of stoichiometry 1 vitamin E:10 phospholipid molecules forms in both gel and fluid phases in equilibrium with domains of pure phospholipid.[9]

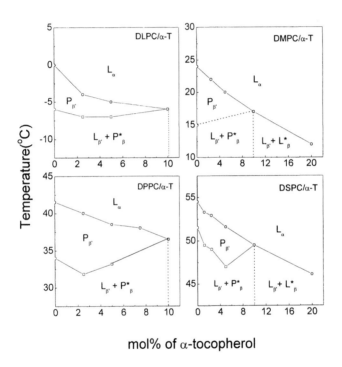

mol% of α-tocopherol

Figure 2   *Partial phase diagrams of aqueous dispersions of saturated phosphatidylcholines and α-tocopherol over temperature ranges about the gel to liquid-crystal phase boundary. L_α, lamellar liquid-crystal; $L_{\beta'}$, lamellar gel; $P_{\beta'}$, ripple phase; *, phases enriched in α-tocopherol.*

Partial phase diagrams of vitamin E and saturated phosphatidylcholines dispersed in excess water have been constructed from calorimetric and X-ray scattering data. These are presented in Figure 2.

Unlike bilayer-forming phospholipids, the effect of α–tocopherol on the phase behaviour of nonbilayer forming lipids appears to be different. X-ray diffraction experiments have given strong evidence of phase separation in codispersion of α-tocopherol and phosphatidylethanolamines.[12-14] There are three interesting aspects of the α-tocopherol enriched domain formed in the mixture of α-tocopherol and phosphatidylethanolamine. Firstly, they form a lamellar crystal phase during low temperature equilibrium. Secondly, they form an H_{II} phase at temperatures much lower than that of the L_α to H_{II} transition of the pure PE, even below the main transition of the pure PE. Finally, they can form Pn3m cubic phases at higher temperatures. Increasing the hydrocarbon chain length favours the formation of the lamellar crystal phase, but not the formation of the Pn3m cubic phase. Moreover, the cubic phase was only observed in mixtures of PE containing α-tocopherol between 5 and 20mol%. In the mixture containing 20mol% α-tocopherol, however, only the H_{II} phase was observed, which is independent of the hydrocarbon chain length of the phospholipid.

2.2   Specific Interactions Between Vitamin E and Phospholipids

The effect of α-tocopherol on the phase behaviour of mixed diacylphosphatidylcholines to determine whether any preferential interactions take place has been reported. Ortiz *et al* [8] have studied the effect of α-tocopherol on the equimolar mixtures of DPPC/DSPC and DMPC/DSPC. The equimolar DPPC/DSPC mixture only showed a single endotherm, which was modified by the presence of α-tocopherol in a similar manner to the individual PCs. The equimolar DMPC/DSPC mixture, however, showed monotectic behaviour. With increasing proportions of α-tocopherol the highest temperature endotherm was broadened and shifted towards lower temperatures, while the lowest temperature endotherm was weakened and disappeared in the mixture containing 20mol% α-tocopherol. This was interpreted such that α-tocopherol preferentially affects the lower melting component which corresponds to DMPC. Increasing proportions of α-tocopherol in an equimolar mixture of 18:0/18:1- and 18:0/22:6-PC also results in a two component transition enthalpy.[15] The lower temperature transition is progressively broadened as the α-tocopherol content of the mixture increases but the higher temperature component is virtually unaffected by the presence of up to 10mol% α-tocopherol. This was interpreted as a phase separation of α-tocopherol within the mixed phospholipid dispersion with a preference for an association with the more unsaturated molecular species of phospholipid.

The inverted hexagonal phase induced by α-tocopherol in PEs can be used as a marker for (PE+α-tocopherol) domain in PC/PE mixtures.[16,17] X-ray diffraction study of mixtures of α-tocopherol with DOPC and DOPE showed that the effect of α-tocopherol on unsaturated phospholipids follows a similar pattern to that observed for their saturated counterparts. However, in the X-ray diffraction study of mixed aqueous dispersions of α-tocopherol with DOPE/DOPC (1:1) and DOPE/DMPC (1:1), it was found that lamellar gel and liquid-crystalline phases dominated the phase structure. This is consistent with the fact that α-tocopherol does not preferentially interact with PE. The changes in the SAXS patterns are similar to those of mixtures of α-tocopherol and phosphatidylcholines or of equimolar mixtures of PE/PC.[16] This again suggests that α-tocopherol preferentially interacts with the PC in the mixture or distributes randomly in domains containing of both PE and PC regardless of the saturation and the length of hydrocarbon chains of the two phospholipids.

The effect of α-tocopherol on equimolar mixtures of saturated PE and PC has been investigated using DSC[8] and X-ray diffraction.[17] The DSC results were interpreted that α-tocopherol preferentially partitioned in the most fluid phase irrespective of whether this was PE or PC. However, the conclusions drawn from the X-ray data appears to be different.[17] The change in the WAXS intensity of the sharp peak at 0.43nm to temperature was plotted in Figure 3a of the equimolar mixtures of DMPC and DPPE containing up to 20mol% α-tocopherol. The curves show maxima corresponding to the midpoint of the phase transition of the DMPC (Peak1) and DPPE (Peak2) components of the mixture. The ratio in height, Peak2:Peak1, is found to increase with increasing α-tocopherol in the mixture. This is consistent with a preferential partition of α-tocopherol into DMPC so as to reduce the contribution of change in the intensity of the wide-angle

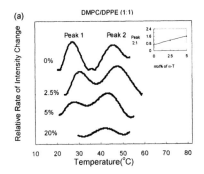

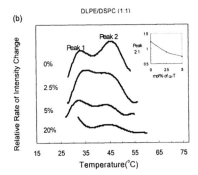

Figure 3   *Rate of change of normalised X-ray scattering intensity, I, of the WAXS peak (dᵢ/dᵢ) recorded from heating scans at 2°/min of codispersions of DMPC/DPPE (a) and DLPE/DSPC (b) containing indicated mol% α-tocopherol plotted as a function of temperature. Inset shows a plot of the relative heights, Peak2:Peak1, plotted as a function of α-tocopherol in the mixture.*

reflection due to DMPC relative to DPPE. A similar analysis of the WAXS intensity data of mixtures of DLPE and DSPC containing different proportions of α-tocopherol are presented in Figure 3b. The inset to the figure shows the relationship between the relative heights of Peak 2 and 1 and the mol% α-tocopherol in the mixture. This shows that as the proportion of α-tocopherol in the mixture increases the contribution to the change in scattering intensity of the WAXS peak from DSPC decreases relative to that from DLPE. This can be interpreted as a preferential partitioning of α-tocopherol into the high melting point phospholipid component of the mixture, which is again phosphatidylcholine, DSPC.

3   CONCLUSIONS

Present evidence indicates that, although it is found in relatively minor proportions in biological membranes, α-tocopherol segregates in membranes and forms complexes with specific lipid constituents, for example, phosphatidylcholines. These interactions have the effect of stabilizing the lipid bilayer matrix since α-tocopherol is prevented from destabilizing non-bilayer forming lipids. The wider significance of these preferential interactions, especially in regard to the putative antioxidant function of α-tocopherol, remains to be established. Moreover, where the distribution of phospholipid classes across membranes is asymmetric the resultant asymmetric distribution of vitamin E may also have implications in the function of the vitamin.

References

1   E.B. Burlakova, S.A. Krashakov N.G. Khrapova, Membr. Cell Biol. 1998, **12**, 173.
2   V.E. Kagan, Annal N.Y. Acad. Sc., 1989, **570**,121.
3   P.J. Quinn, J. Appl. Cryst., 1997, **30**, 733.
4   B. DeKruijff, P.W.M., Van Dijck, R.A. Demel, A. Schuijff, F. Brants, and L.L.M. van Deenen, Biochim. Biophys. Acta, 1974, **356**,1.
5   J.B. Massey, H.S. She and H.J. Pownall, Biochem. Biophys. Res. Commun. 1982, **106**, 842.
6   E.J. McMurchie. and G.H. McIntosh, J. Nutr. Sci. Vitaminol. 1982, **32**, 551.
7   J. Villalaín, F.J. Aranda, and J.C. Gómez-Fernández, Eur. J. Biochem. 1986, **158**, 141.
8   A. Ortiz, F.J. Aranda, J.C. Gómez-Fernández, Biochim. Biophys. Acta, 1987, **898**, 214.
9   P.J. Quinn, Eur. J. Biochem. 1995, **233**, 916.
10   T. Lefevre, and M. Picquart, Biospectroscopy, 1996, **2**, 391.
11   X. Wang K. Semmler, W. Richter and P.J. Quinn, Arch. Biochem. Biophys., 2000, **377**, 304.
12   X. Wang and P.J. Quinn, Prog. Lipid Res., 1999, **38**, 309.
13   X. Wang and P.J. Quinn, Biophys. Chem., 1999, **80**, 93.
14   X. Wang, H. Takahashi, I. Hatta and P.J. Quinn, Biochim. Biophys. Acta, 1999, **1418**, 335.
15   W. Stillwell, T. Dallman, A.C. Dumaual, F.T. Crump and L.J. Jenski, Biochemistry, 1996, **35**, 13353.
16   X. Wang and P.J. Quinn, Biochim. Biophys. Acta, 2000, **1509**, 361.
17   X. Wang and P.J. Quinn, Eur. J. Biochem., 2000, **267**, 1.

THEORY ON OPENING-UP OF LIPOSOMAL MEMBRANES BY ADSORPTION OF TALIN

Y. Suezaki

Physics Laboratory, Department of General Education, Saga Medical School, Saga 8498501 Japan. E-mail: suezaki@post.saga-med.ac.jp

1   INTRODUCTION

The lipid bilayer membrane is a good model system of biological cell membranes and has been studied extensively. Lipid membranes in aqueous solution form various morphologies such as spherical vesicles, tubules, lamellar structures and others depending on the materials, preparations, and other factors. By mixing with another kind of lipid or additive molecules, a wide variety of further morphologies of membrane are found.

Saitoh and others showed that spherical lipid vesicles opened up to cup-like vesicles and other vesicles with more complicated topologies by adding protein[1]. The protein, talin, was adsorbed by the orifices of the cup-like vesicles. Figure 1 shows the concentration dependence of the change of the shape of cup-like vesicles. In Figure 1, an increase of concentration of talin from A to G, makes a cup-like vesicle open up more and more, and a reduction of the concentration, G to L, recovers the original spherical vesicle reversibly. The reversibility as a function of concentration shows that these vesicles are thermodynamically stable at least for the time scale of the observation of the system. Other vesicles with more than single orifices were observed at higher concentrations. Sheet-like vesicles as shown in H in Figure 1 were also observed.

Above a threshold concentration cup-like vesicles appeared, and cup-like vesicles and sheet-like vesicles coexist above a certain concentration.

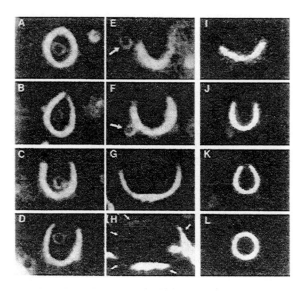

Figure 1 *A sequence of morphological change of vesicles at talin concentrations from 0 to 2 µM (A to H); After talin dilution (H to L) the spherical vesicle is recovered*[1]

Later, we analyzed the adsorption isotherm of talin between aqueous solution and the vesicle membrane by taking account of the bending energy of the membrane and the line tension induced by talin at the periphery of cup-like vesicles[2]. The shape of cup-like vesicles was determined by the balance of the energy gain for the adsorption of talin to the periphery of the cup-like vesicles and the change of the bending energy of the lipid membrane. Observed coexistence of cups and sheet-like vesicles was reproduced. Vesicles with two orifices were also analyzed and theoretically reproduced. In this paper, the above theoretical analysis will be improved and further analysis will be done. First, in the next section the effect of Gaussian bending modulus will be discussed. Next, the theory will be improved to be more consistent with observed facts than the previous paper. Analysis of concentration dependence in the neighborhood of the threshold concentration will show that very small orifices begin to form just above the threshold concentration, while previous theory predicted a finite size of cup-like vesicle at that concentration. Lastly, in the previous work[2], we modeled the cup-like vesicle as a partial sphere cut by a plane. However, this model does not satisfy the mechanical force balance and mimics realistic shapes only in an approximate way. In the section to the following, we will show the result of Monte Carlo simulation to determine the shape of cup-like vesicles and vesicles with two orifices. The resulting shapes resemble those observed in experiment.

2   THEORETICAL   ANALYSIS

2.1   Bending energy of membrane

Here, we briefly review our previous theory and replace it with an improved version. According to Helfrich's model for the bending energy of membranes[3], the general forms of the bending energy, that of a sphere, and that of a cup, E_{bend} , E_{sphere} and E_{cup} , respectively, are given as follows

$$E_{bend} = \frac{\kappa}{2} \iint (c_x + c_y - c_0)^2 \, dxdy + \kappa' \iint c_x c_y dxdy \qquad (1)$$

$$E_{sphere} = \frac{\kappa}{2} \int (c_x + c_y - c_0)^2 \, dS = 2\pi\kappa (2 - c_0 R_0)^2 + 4\pi\kappa' \qquad (2)$$

$$E_{cup} = \frac{\kappa}{2} \int (c_x + c_y - c_0)^2 \, dS = 2\pi\kappa (2x - c_0 R_0)^2 + 2\pi\kappa' (1 + \cos\theta) \qquad (3)$$

where κ, κ',c_x, c_y and c_0 are the bending modulus of the membrane, Gaussian bending modulus, two principal curvatures and the spontaneous curvature, respectively. Also, R_0 and $x(=R_0/R$, R is the radius of curvature of cup-like vesicle$)$ and θ are the radius of the initial spherical vesicle, the relative curvature of the cup with respect to sphere, and the angle between the normal to the axis of the cup and the tangent of the cup periphery as shown in Figure 2. Then the bending energies of a spherical vesicle and cup-like vesicle follow the forms of Eqs. (2) and (3). The spontaneous curvature of the bilayer membrane is usually assumed to be zero. However, this idea is based on the assumption that the membrane system is in equilibrium. Vesicles were not necessarily in equilibrium but were made by mechanical agitation (sonication) or were under some specific boundary conditions or initial condition of the system. As the turnover time of a lipid molecule from one monolayer to the other is very long, artificially made vesicles could initially possess non-zero spontaneous curvatures. The last terms of Eqs. (1), (2), and (3) had not been considered in the previous paper.

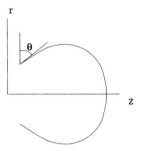

Figure 2 *Schematic drawing of a cup-like vesicle. The cup is axi-symmetric with the z-axis.*

Now, we will discuss the problem of the Gaussian bending modulus. For a thin elastic plate, the value of the Gaussian bending modulus, κ' should be the same order of magnitude as κ. For liquid membranes, however, Peliti and Nelson showed that the Gaussian bending modulus should be zero from the shear free condition of a liquid membrane[4]. This result is reasonable because the Gaussian bending modulus itself is the modulus for shear deformation in the membrane surface. For real membranes, there could exist frustrated stresses along the perpendicular direction to the surface. In this case, a finite value of the Gaussian bending modulus could exist. To simulate the frustration, the author analyzed the surface elasticity model devised by Petrov and others[5]. By inserting an intermediate surface to their dumbbell surface elasticity model, we could reproduce a frustrated liquid membrane[6]. By analyzing this model correctly, we could show that the Gaussian bending modulus can take both positive and negative values depending on the frustration of the membrane inside, while that of the elastic plate should always be negative. However, the value of the Gaussian bending modulus is two orders of magnitude smaller than that of the usual bending modulus[6].

Some experimental workers have tried to evaluate the Gaussian bending modulus indirectly[7,8]. However, there has been no direct evidence of the quantitative observation of a Gaussian bending modulus so far. For lipid membranes, the magnitude of the Gaussian bending modulus is expected to be small. This is due to the fact that the frustration of a membrane is not large because the cross sectional area of the head group is not much different from that of two alkyl chains. Actually, cell fusion and fission often occur in biological cell systems. This might be due to the fact that the energy gain or loss during the fusion or fission is small. To explain, for instance, bi-continuous structures seen in microemulsions or lipid vesicles, the role of Gaussian bending energy term has been considered[9]. It might be possible where delicate and small energy balance works in the conformational change of the system. For our further analysis in this report, therefore, we will neglect the term of Gaussian bending modulus.

2.2 Statistical mechanics of adsorption of talin

The actual shape of the cup is a little different from our model, but we employ the partial sphere for cup as a first approximation. To evaluate the adsorption equilibrium of talin, we define the total number, X, and the adsorbed number, N, of talin to orifices of the vesicles. The partition function Z of the total system is written as $Z=Z_{ad}(N)Z_{bulk}(X-N)$ and respective partition functions are represented as

$$Z_{ad}(N)=\frac{1}{N!}(pf)_{ad}^{N}\binom{n_0}{n}\exp\left[-\frac{1}{kT}\left(-N\varepsilon_0+n\left(E_{cup}-E_{sphere}\right)\right)\right] \qquad (4)$$

$$Z_{bulk}(X-N)=\frac{1}{(X-N)!}(pf)_{bulk}^{X-N} \qquad (5)$$

where, $(pf)_{bulk}$ and $(pf)_{ad}$ are entropic parts of the molecular partition functions of dissolved talin in the aqueous solution and an adsorbed talin, respectively. We assumed, ε_0, as an affinity free energy of talin from water to the orifice of the vesicle. By definition, the free energy, F, of the total system, becomes $F=-kT \ln Z$. In Eq. (4), the factor n_0 is the total number of original spherical vesicles and n is the number of cup-like vesicles. By minimizing F with respect to N and n, we obtain the thermal equilibrium and the mechanical force balance. The equations, $\partial F/\partial N$ and $\partial F/\partial n$ determine the chemical potential balance of talin and a number, n, of cups as follows

$$\frac{\partial F}{\partial N}=\mu_{ad}-\mu_{bulk}=0 \qquad (6)$$

where

$$\mu_{bulk}=kT\ln\left(\frac{X-N}{V}\right),\qquad \mu_{ad}=-\varepsilon_0-kT\ln\left(ed^3\right)+8\pi n\kappa\left(2x-c_0R_0\right)\frac{\partial x}{\partial N} \qquad (7)$$

and

$$1 + R_0 c_0 - x^2 - \frac{R_0 c_0}{x} = \frac{kT}{8\pi\kappa} \ln\left(\frac{n_0}{n} - 1\right) \tag{8}$$

From Eq. (7), we derived a threshold concentration of talin. Eq. (8) determines the number n, in other words, it shows the mechanical force balance of cup-like vesicles. By our model calculation the observed shape change of vesicles was well reproduced, and the observed coexistence of cup and sheet-like vesicle[1] was also reproduced[2]. In that case, however, we should assume a finite spontaneous curvature of membranes to solve a non-trivial solution. Also, we expected a finite size for the orifices of cups at a threshold concentration. Experimentally, however, very small orifices appeared at just above threshold concentration[1]. In this report, we improve and reanalyze our previous theory that is not consistent with observed data.

By analyzing the concentration dependence of Eq. (8) carefully, we obtained an improved result. A factor n_0/n in Eq. (8) becomes as follows

$$\frac{n_0}{n} = \frac{s}{2R_0 d} \frac{C_L}{C_X - C_{th}} \sqrt{1 - x^2} \tag{9}$$

where C_L, C_X, and C_{th} are concentrations of lipid and talin and threshold concentration, respectively. By carefully evaluating the right-hand side of Eq. (8) by use of Eq. (9), the relative curvature, x, is determined as a cross section as shown schematically in Figure 3 for concentration C_X that is close to the threshold concentration C_{th}. Figure 3 is different from Figure 2 in Ref. [2].

The minimized free energy per molecule, f, scaled by kT becomes

$$f = \frac{F_{min} - F_0}{XkT} = 1 - \ln\left(\frac{C_X}{C_{th}}\right) - \frac{C_{th}}{C_X} \tag{10}$$

The value of Eq. (10) is always negative and thus the formation of cup-like vesicles has been shown to be stable[3]. Here, stable formation of cup-like vesicle, in reality, is metastable due to the fact that spherical vesicles themselves are metastable and realized only in the finite time scale of experiments. Observations[1] were made around several tens of minutes and many colloidal systems suffer from change by the aging effect as is well known.

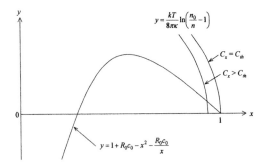

Figure 3   *Schematic drawing of left and right hand sides of Eq. (7) as function of x*

The energy profile of Eq. (10) is a sigmoidal curve with an inflection point. Although we have not shown it here, the energy calculation of sheet-like vesicles with plane ellipses showed that a common tangent exists for both energy profiles of cup-like vesicles and plane elliptical vesicles[2]. This fact shows that there exists coexistence of cup-like vesicles and sheet-like vesicles. This coexistence was also observed in experiment made by Saitoh and others[1].

3. THE MONTE CARLO SIMULATION

3.1   The methodology of the Monte Carlo simulation

The model for cup-like vesicles in the previous section could explain the observed adsorption isotherm qualitatively. However, the partial sphere model does not satisfy the local mechanical force balance within the vesicle. Actually, observed shapes of various vesicles are not a partial sphere, but deformed significantly from a spherical shape[1]. To reproduce realistic shapes of vesicles, we performed Monte Carlo simulation at equilibrium. We assume that the deformation proceeds by keeping axial symmetry around the z-axis as shown in Figure 4.   Then the vesicle shape is described by the line contour θ_i, where θ_i is the angle made by the contour's normal and the z-axis. In the Monte Carlo simulation, we discretize the vesicle contour into joint segments. The i-th joint is described by the distance r_i from the z-axis and the coordinate z_i as shown in Figure 4. Two consecutive joints i and $i+1$ are connected by a segment of length s_i.

Local self-avoidance can be achieved by confining $a/2 < s_i < 2a$, where a is a unit of length. The procedure of the Monte Carlo simulation is similar to that made by Morikawa and others[10].

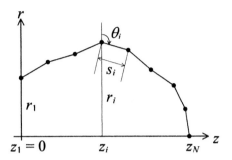

Figure 4 *Schema of the joint-segment model of a cup-like vesicle. The θ_i is the angle variable between the segment i and r-axis, and s_i is the segment length between the joint i and i+1.*

The model energy, E, of the Monte Carlo simulation is written down as

$$E = \frac{\kappa}{2} \int (C_1 + C_2 - C_0)^2 \, dA + 2\pi r(1)\varepsilon \; (+2\pi r(N)\varepsilon) \tag{11}$$

where ε is the line tension energy of the orifice of the cup-like vesicle. The factors $r(1)$ and $r(N)$ are the radii of the orifice of the cup and that of the other orifice when the vesicle has two orifices. The factor, N is the number of joints, which changes during the course of the simulation. In the Monte Carlo simulation, the first term of Eq. (11) is replaced by

$$E_B = \frac{\kappa}{2} \sum_{i=1}^{N} (C_{1i} + C_{2i} - C_0)^2 \Delta A_i \tag{12}$$

where

$$C_{1i} = \frac{2\sin(\theta_i - \theta_{i-1})}{\sqrt{s_i^2 + s_{i-1}^2 + 2s_i s_{i-1} \cos(\theta_i - \theta_{i-1})}} \tag{13}$$

$$C_{2i} = \frac{\sin(\overline{\theta}_i)}{r_i} \tag{14}$$

$$\Delta A_i = \frac{\pi}{4}[(r_{i-1} + 3r_i)s_{i-1} + (3r_i + r_{i+1})s_i] \tag{15}$$

In Eq. (14), θ_i is the defined as

$$\overline{\theta}_i = \frac{r_{i+1} - r_{i-1}}{z_{i+1} - z_{i-1}} \tag{16}$$

The line tension energy in Eq. (11) is represented as

$$E_L = 2\pi r_1 \varepsilon \qquad (+2\pi r_N \varepsilon \text{ when two orifices}) \tag{17}$$

The total energy E is written as

$$E = E_b + E_L + \gamma(A - A_0)^2 \tag{18}$$

where A and A_0 are the total area and the initial total area, and γ is the Lagrange multiplier to conserve the surface area of the vesicle.

The Monte Carlo simulation consists of the following sequences of procedures. A joint i is chosen randomly with a probability to its associated local area of ΔA_i, and we try to shift its coordinates randomly in the (z, r) plane according to

$$\begin{cases} z_i \to z_i + \Delta z_i \\ r_i \to r_i + \Delta r_i \end{cases} \tag{19}$$

We calculate energy change ΔE associated with this position change and accept or reject the trial according to the standard Metropolis algorithm. If ΔE is negative the trial is automatically accepted, and if ΔE is positive it is accepted with a probability proportional to $\exp(-\Delta E /kT)$ at temperature T. In the following subsection, we will show the results of the Monte Carlo simulations of cup-like vesicles and vesicles with two orifices for each line tension.

3.2 Numerical result of the simulation

The preliminary result of the Monte Carlo simulation for the cup-like vesicle and vesicle with two orifices are given in the following. The initial number, N was 100 and 200. The radius of the spontaneous curvature was assumed to be twice of that of the original spherical vesicle as a preliminary trial of the analysis. In future work, we will calculate the system with many conditions. In Figure 5a and 5b, we show the shapes of cup-like vesicles with line tension energy $\varepsilon/\kappa=0.1$ and 0.7 respectively. The shapes of the cups are different from the partial sphere and well reproduce the observed shapes. In Figure 6a and 6b, the shapes of vesicles with two orifices of the line tension energy $\varepsilon/\kappa=0.1$ and 0.7, respectively, are shown in Figures 5 and 6.

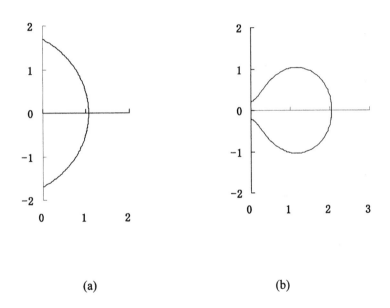

(a)   (b)

Figure 5 *Shapes of cup-like vesicle obtained by MC simulation. a; $\varepsilon/\kappa=.1$, b; $\varepsilon/\kappa=0.7$*

Similar vesicles with shapes of Figure 6a have also been observed[1]. The shapes of cups are different from a partial sphere especially when the line tension is large and the shapes by this simulation well reproduce the observed shapes of vesicles[1]. Although we do not show the data of the local curvatures of respective vesicles, the sum of c_1 and c_2 is increasing function when the contour proceeds from the orifice to the bottom of the cup.

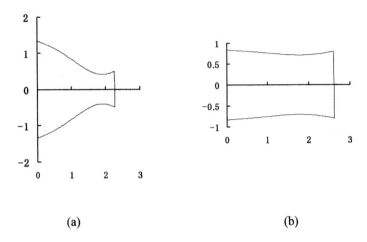

(a)   (b)

Figure 6   *Shapes of vesicles with two orifices obtained by MC simulation. a; ε/κ=.1, b; ε/κ=.7*

And the sum has a maximum value at about the center in the case of vesicles with two orifices. This is consistent with the fact that a freely sustained beam at both ends possesses maximum curvature at the center while it is bent. In Table 1, the values of energies E_b and E_L of respective vesicles obtained by the Monte Carlo simulation are given.

Table 1   *Values of energies for cup-like vesicles and vesicles with two orifices*

Type of vesicle	ε/κ	E_b/κ	E_L/κ	(E_b+E_L)/κ
Cup-like vesicle	0.1	0.015	1.067	1.082
	0.7	0.870	5.856	6.726
Vesicles with two orifices	0.1	0.026	1.152	1.178
	0.7	0.332	6.857	7.189

From Table 1, The total energies of vesicles with two orifices are larger than those of cup-like vesicles, which is reasonable. For both vesicles, the ratio, E_L/E_b, is larger for vesicles with $\varepsilon/\kappa=0.1$ than those of $\varepsilon/\kappa=0.7$. This fact shows that the line tension and the bending of membrane are mechanically similar to the series connection of two springs with different spring constants. The weaker spring stores more energy than the stronger when it is pulled. More detailed analysis will be done in future work. The obtained shapes of vesicles with two orifices are not symmetrical although the symmetrical shape is expected from topological considerations. In this case, the symmetric point might be a saddle point or a summit of the energy profile. Actually, the simulation by assuming an axially symmetric shape cannot lower the total energy more than those given in Figure 6. At this point, further study is left for future work.

4 CONCLUSION AND DISCUSSION

In the previous sections, we reviewed the previous theoretical paper on the formation of cup-like vesicles, and a more reasonable discussion on the energy term of Gaussian bending modulus was made. The previous theoretical analysis of the formation of cup-like vesicles just above the threshold concentration was reanalyzed in more detail, and the improved calculation showed that the calculated shape and the observed shape of cup-like vesicles at just above that concentration came closer to the observed one than the previous analysis[2]. To realize local force balance in the cup-like vesicle and vesicles with two orifices, we performed the Monte Carlo simulation by assuming the line tension energy and bending energy as well. Starting from the partial sphere, the Monte Carlo simulation by random fluctuation of z and r resulted in the realistic vesicle shapes and they approximately reproduced observed shapes. The obtained energies of bending and line tension possess mechanically reasonable meaning.

Although the author believes in the physical nature of Gaussian bending modulus as pointed out in the theoretical section, as far as biological membranes or their model membranes are concerned, many people think that the essential nature remains unclear. To resolve this point on the Gaussian bending modulus, it is useful to refine the experimental accuracy of observing cup-like vesicles and vesicles with more than single orifices. Furthermore, it is important to theoretically analyze the problem in more detail by using our model including Gaussian bending energy. We are continuing the Monte Carlo simulation more systematically to resolve the problem stated above. In this case, the effect of changing the spontaneous curvature of the membrane is now being taking into account.

Acknowledgment

The author would like to thank to Drs. Umeda and Morikawa for their useful discussion on the problem. He also thanks Drs. Takefu and Horimoto and Mr. Ichinose for their technical support of the numerical calculation. Mr. B. Andersen kindly assisted the author with the English presentation. Financial support came from the Ministry of Education and Science, Japan (grant #10640372).

References

1 A. Saitoh, K. Takiguchi, Y. Tanaka, and H. Hotani, _Proc. Natl. Acad. Sci. USA_ 1998, **95**, 1026.

2 Y. Suezaki, H. Ichinose, K. Takiguchi and H. Hotani, _Biohys. Chem._ 1999, **80**, 119.

3 W. Helfrich, _Z. Naturforsch._ 1973, **28c**, 693.

4 Peliti, L and Nelson, D., in Physics of Amphiphilic Layers,1987,106, Springer-Verlag, Berlin.

5 A.G. Petrov and A.J. Derzhanski, _Phys. Coll._, 1976, **37**, C3.

6 Y. Suezaki and H. Ichinose, _J. Phys. I France_, 1995, **5**, 1469.

7 A. Fogden, S.T. Hyde and G.J. Lundberg, _Chem. Soc. Faraday Trans._ 1991, **87** 949.

8 B. Farago and D. Richter, _Phys. Rev. Lett._, 1990, **65**, 3348.

9 S. Leibler, _Statistical mechanics of membrane and surfaces,_ 1988, **5** (World Science, Singapore), 46.

10 R. Morikawa, Y. Saito and H. Hyuga, _J. Phys. Soc. Japan_, 1999, **68**, 1760.

DIFFERENTIAL SCANNING CALORIMETRY AND X-RAY DIFFRACTION STUDIES OF GLYCOLIPID MEMBRANES

O. Ces[1], J. M. Seddon[1], R. H. Templer[1], D. A. Mannock[2] and R. N. McElhaney[2]

[1]Department of Chemistry, Imperial College, London SW7 2AY, UK
[2]Department of Biochemistry, University of Alberta, Edmonton, Canada T6G 2H7

1 INTRODUCTION

Using X-ray diffraction and differential scanning calorimetry (DSC) we have studied the phase behaviour of 1,2-di-0-tetradecyl-3-O-(β-glucopyranosyl)-*sn*-glycerol (di-14:0-β-D-GlcDAG) and 1,2-di-0-tetradecyl-3-O-(β-galactopyranosyl)-*sn*-glycerol (di-14:0-β-D-GalDAG) as a function of temperature and water content. In this paper we present an introduction into the observed phase behaviour of these synthetic dialkyl glycolipids. The X-ray diffraction data has been used to calculate the lattice repeat vector, d and the limiting hydrations of the various phases. The behaviour of the two systems upon heating is similar as they both adopt a fluid lamellar phase L_α and an inverse hexagonal phase H_{II}. Upon cooling however, striking differences are seen with di-14:0-β-D-GlcDAG forming a metastable L_β gel phase below the chain-melting transition in contrast to di-14:0-β-D-GalDAG which forms only crystalline lamellar phases on the timescale of our experiments. Bicontinouos cubic phases are not observed in either system. We compare the findings from these glycolipids with those from our previous studies of the phospholipid didodecyl phosphatidylethanolamine (di-12:0-PE).

2 BACKGROUND

Glycolipids, biological amphiphiles whose polar headgroups contain one (monosaccharide) or more (oligosaccharide) cyclic sugar groups, are vitally important components of biological membranes, being involved in many regulatory and recognition processes. There is increasing interest in their structural roles in membranes, since they are the major membrane lipids of all plants (mainly galactose-based) and many micro-organisms (mainly glucose-based)[1]. The reason why such species utilise glycolipids rather than phospholipids may be to do with conserving phosphorus, which is usually a precious commodity in the environment. Frequently, monosaccharide-based lipids are the major glycolipid component of such membranes. The reasons why some organisms utilise galactose-based lipids, and some glucose-based lipids, as the main structural component of their membranes is poorly understood. It is therefore important to establish whether there are any differences in the lyotropic phase behaviour of these two classes of glycolipid.

If one compares the two sugars, D-glucose and D-galactose, they have rather similar solubilities in water of 45.1 wt% and 40.6 wt% at 20°C, respectively[2]. They also have similar hydration numbers (8.4 and 8.7), but slightly different partial molar volumes at infinite dilution (111.7 and 110.2 cm^3 mol^{-1}). These differences are due to the change from an equatorial (glucose) to an axial (galactose) configuration of the hydroxyl group on carbon C(4) of the sugar ring. This changes the arrangement of hydrophilic sites around the sugar group, and will lead to differences in the pattern of both headgroup-headgroup and headgroup-water hydrogen bonding. A further striking effect is the large change in the distance from the hydroxyl oxygen atom on C(4) to the hydroxyl oxygen on C(2), which is reduced from approximately 4.8 Å for glucose to approximately 4.4 Å for galactose. There is evidence from compressibility studies that galactose is less well accommodated within the structure of water, and consequently has a stronger perturbing effect[3].

When these two sugars are attached to diacyl (or dialkyl) glycerol, these differences may be expected to lead to differences in the phase behaviour for the two glycolipids. A well-established and important finding is that fully-hydrated monoglycosyl diacylglycerols and dialkylglycerols, with either glucose-based or galactose-based headgroups, have a strong tendency to adopt inverse, non-lamellar phases such as the inverse hexagonal H_{II} and bicontinuous cubic phases. This is found both in naturally-occurring glycolipids extracted from biological membranes, and in purely synthetic glycolipid systems[4]. Non-lamellar structures may exist within cells, both animal (smooth endoplasmic reticulum) and plant (etiolated chloroplasts)[5]. The former contain primarily phospholipids, whereas the latter contain glycolipids. Furthermore, non-lamellar phases are topologically and geometrically related to defects such as fusion channels between bilayers, and thus are also highly relevant to dynamic processes in membranes.

The phase behaviour of monosaccharide-based glycolipids is strikingly similar to that of the phosphatidylethanolamines (PE), one of the major classes of phospholipid in animal cell membranes. For synthetic dialkyl (saturated, ether-linked) PE's it was found that the effect of reducing the chainlength was first to raise the fluid bilayer – inverse hexagonal ($L_\alpha - H_{II}$) phase transiton temperature, and then when the chainlength was reduced below C_{14}, to induce inverse bicontinuous cubic phases to form, between the L_α and H_{II} phases[6,7,8,9].

This behaviour can be understood in terms of the chain packing frustration which exists in the H_{II} phase, and which becomes progressively more severe as the chainlength is shortened. This frustration can be partially relieved by the bilayer deforming to a saddle surface, with negative gaussian curvature ($K < 0$), allowing each monolayer to bend towards the water, but without creating voids within the hydrocarbon chain region.

For a symmetric bilayer, the two monolayers should curve in opposite directions but by equal amounts, leading to the mid-plane having zero mean curvature ($H = 0$). Thus the mid-plane should correspond to a minimal surface, which can be extended to form an infinite periodic minimal surface (IPMS) with cubic symmetry. The three simplest of these are the P, D and G minimal surfaces, which are related to each other by a Bonnet transformation, a mathematical transformation which preserves the mean and gaussian curvature at each point on the surfaces. Draping a lipid bilayer onto these three IPMS leads to the formation of the bicontinuous cubic phases of spacegroups, Im3m, Pn3m and Ia3d, respectively[10]. It should be noted that the interfacial area is actually maximal on the minimal surface at the bilayer midplane, and diminishes on moving towards the polar headgroup regions. The system is still frustrated, because it is impossible to satisfy both uniform interfacial mean curvature and uniform bilayer thickness. However, it has been shown that the frustration in an inverse bicontinuous cubic phase is smaller than in the H_{II} phase[11].

It was subsequently found that the above-mentioned effects of chainlength also applied to glycolipids, which was to be expected since the underlying physical mechanisms are rather general in nature, and not specific to phospholipid systems. In fact the glycolipids have a stronger tendency to form inverse phases than the phospholipid PE system, the $L_\alpha - H_{II}$ phase transition in excess water occurring at 55 - 56 °C and 62 - 63 °C for di-14:0-β-D-GlcDAG[12, 13] and di-14:0-β-GalDAG[14], but at 96 °C for the di-14:0-PE[6]. Changing the chain linkage from ether to ester is sufficient, for the di-14:0 glycolipids, to cause the appearance of inverse bicontinuous cubic phases above the L_α phase, but at higher temperatures of 72 °C for the ester-linked Glc[15] and 81 °C for the ester-linked Gal[16], compared to the $L_\alpha - H_{II}$ transition temperatures of the corresponding dialkyl compounds.

Many membrane lipids contain chiral centres, and so it is natural to ask whether lipid membranes exhibit chiral recognition, or whether the molecular chirality manifests itself in any detectable way, in terms of the membrane structure or interactions. Very little evidence of chiral discrimination has so far been found in phospholipid bilayers, where the molecules are usually enantiomeric, with a single chiral centre at carbon atom C2 in the glycerol backbone. For glycolipids, chirality may be expected to play a greater role than in phospholipids, since the 1,2-*sn* and 2,3-*sn* glycolipid diastereomers are not mirror images of one another and so can have different physical properties. Indeed, studies in excess water of ether-linked di-12:0-β-D-GlcDAG found significant differences in phase behaviour between the 1,2-*sn* and 2,3-*sn* isomers[17,18]. For the longer chainlength β-D-glucosyl compounds, however, the effects of chirality were very small. For galactose-based lipids, on the other hand, the effects of chirality are still quite marked for the 14 carbon chainlength[14,19], and are very striking for the di-12:0 and di-13:0 compounds[20].

We have been interested in comparing the detailed phase behaviour of glucose- and galactose-based glycolipids as models for glycolipid membranes. We have chosen to study the dialkyl compounds because the ether linkages (of the saturated hydrocarbon chains to the glycerol group) confer enhanced chemical stability compared to the ester-linked compounds. The particular homologues we chose for detailed study are di-14:0-

β-D-GlcDAG and di-14:0-β-D-GalDAG which are diastereomers having the natural 1,2-sn glycerol configuration.

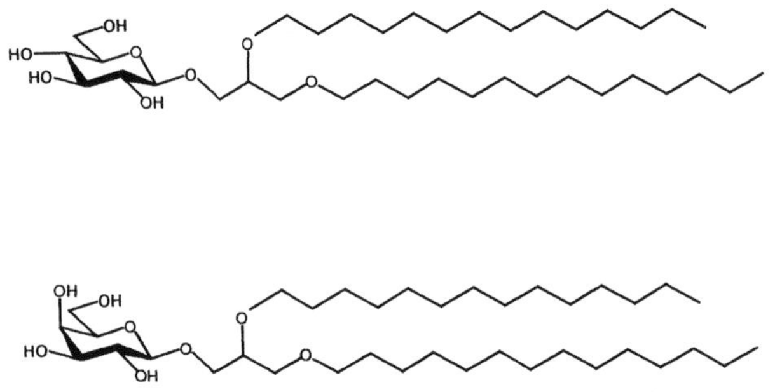

Figure 1 *Chemical structures of the synthetic dialkyl glycolipids di-14:0-β-D-GlcDAG and di-14:0-β-D-GalDAG.*

This chainlength of C_{14} is just long enough to suppress the formation of any cubic phases. For the corresponding PE phospholipid system, we have previously found that the effect of hydrostatic pressure is to induce the formation of inverse bicontinuous cubic phases (of spacegroups Pn3m, and Im3m) between the fluid lamellar L_α and H_{II} phases for pressures in excess of 600 bar[21]. We plan to extend these high-pressure structural studies to the di-14:0-β-D-GlcDAG and di-14:0-β-D-GalDAG glycolipid systems, to determine whether this pressure-induced formation of inverse bicontinuous cubic phases, is a universal effect for systems which exhibit such phases at atmospheric pressure at slightly shorter chainlengths. However, the first stage is to establish the binary temperature-composition phase diagrams for these two compounds at atmospheric pressure, and some of our preliminary findings are reported here.

3 EXPERIMENTAL

The lipids 1,2-di-0-tetradecyl-3-O-(β-glucopyranosyl)-sn-glycerol (di-14:0-β-D-GlcDAG) and 1,2-di-0-tetradecyl-3-O-(β-galactopyranosyl)-sn-glycerol (di-14:0-β-D-GalDAG) were synthesized according to literature methods which have been previously described[22]. The compounds were pure as determined by TLC, using a solvent system of $CHCL_3/CH_3OH$ (9:1 v/v). Differential scanning calorimetry measurements were performed on a TA Instruments DSC 2910 equipped with a data station or a Perkin-Elmer DSC7 calorimeter. Heating and cooling scans were typically repeated three times to ensure homogeneity of the samples and were recorded with heating-/-cooling rates of 0.5 and 1 °C per minute. Fixed hydration samples were made by weighing out exact amounts of lipid and HPLC grade water into stainless steel DSC pans which were hermetically sealed. The uncertainty in the concentration $\Delta c/c$ is estimated to be 1 wt%.

Low and wide angle X-ray patterns were obtained using two different systems, one based on a high intensity point source and the other around a line source. In the former case the X-rays were produced by a Philips PW 2213/20 generator operated at 40 kV and 30 mA with a fine focus tube with a copper target. The diffraction patterns were obtained using a Guinier camera (Huber Diffraktionstechnik) which was equipped with a quartz crystal monochromator set to isolate the $CuK_{\alpha 1}$ ($\lambda=1.5405$ Å) radiation. This was operated under vacuum to reduce air scatter with a fixed sample-detector distance of 114.6 mm. The samples were held in position by an electrically heated copper block which was able to control the sample temperature to an accuracy of $\pm$ 0.5 °C by employing an electronic controller. The temperature probe was embedded as close as possible to the sample (within 1 cm). The electronic controller allowed the temperature of the sample to be varied linearly with time whilst the X-ray film holder was scanned vertically thereby producing a continuous diffraction pattern. Kodak Scientific Imaging Film was used and developed according to standard techniques. The scanned films were then analysed using software written within the IDL 5.4 (Research Systems) data processing package. The estimated accuracy of the X-ray spacings is ± 0.5 Å (low-angle) and ± 0.05 Å (wide-angle).

The X-ray point source was provided by an Elliot GX20 rotating anode X-ray generator operating at 30 kV and 25 mA, with a 100 μm focus cap, and was focused by Franks optics to a point of dimensions 160×110 μm^2. The diffraction patterns were captured by a custom-built electronic two-dimensional CCD detector. Photometrically accurate images could be captured in a matter of minutes and then analysed and indexed from a single computer terminal. The estimated accuracy of the layer spacing is ± 0.5 Å.

4 RESULTS AND DISCUSSION

To determine the phase behaviour of both compounds significant numbers of fixed hydration samples were prepared with the percentage of water ranging from 0 wt% to 80 wt% at intervals of 1-2 wt%. DSC scans were measured for each of these samples. Figure 2 shows representative scans for the dry samples and samples containing excess water.

DSC measurements conducted upon a dry sample of the di-14:0-β-D-GlcDAG compound show a single highly energetic phase transition at 54.5 °C upon heating. In the case of the excess water sample two peaks are instead observed corresponding to a highly energetic phase transition at 53.2 °C and a low-enthalpy transition at 59.5 °C. The cooling scans indicate that these transitions are indeed reversible.

The behaviour of the di-14:0-β-D-GalDAG on the other hand is somewhat different. On heating, samples at all water contents exhibit a single phase transition which progressively drops in temperature with increasing hydration from 78.9 °C in the case of the dry sample to a limiting value of 70.2 °C. Cooling scans revealed that this behaviour is not reversible indicating that the Gal compound forms metastable phases. The single exothermic phase transition observed in the cooling scans of the dry sample at 64 °C is replaced in samples containing excess water by a poorly exothermic phase transition at 61.5 °C, and a highly exothermic phase transition at 54.2 °C.

The thermodynamic data are summarised in Table 1.

Sample	Heating Sequence	Phase Transition Temperature (°C)	Enthalpy ΔH (kJ mol^{-1})	Entropy ΔS (Jmol^{-1}K^{-1})
Glc (Dry)	Heating	54.5	43.0	131.3
	Cooling	53.2	42.6	130.6
Glc (Excess Water)	Heating	53.2	27.6	84.6
		59.5	6.9	20.8
	Cooling	55.7	7.3	22.2
		49.7	27.8	86.2
Gal (Dry)	Heating	78.9	80.9	229.9
	Cooling	64.0	63.3	187.8
Gal (Excess Water)	Heating	70.2	86.6	252.3
	Cooling	61.5	6.2	18.5
		54.2	81.2	248.2

Table 1 *Thermodynamic data (transition temperatures, enthalpies and entropies) relating to the DSC scans shown in figure 2.*

(a)

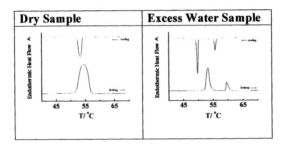

(b)

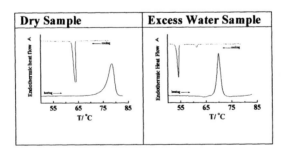

Figure 2 *DSC scans of (a) di-14:0-β-D-GlcDAG and (b) di-14:0-β-D-GalDAG.*

In order to understand the DSC results, the two glycolipids were studied by X-ray diffraction. In addition to establishing the structures and symmetries of the phases, these measurements also yielded the lattice parameters of each phase. Fig. 3 shows representative low-angle X-ray patterns from the various lyotropic phases observed for these two glycolipids.

The stable low temperature phase of the di-14:0-β-D-GlcDAG compound is a lamellar crystalline phase (L_{C1}). Four orders of diffraction are visible in the low-angle region, their diffracted orders being in the ratio 1:2:3:4. The interlamellar repeat distance d of the L_{C1} phase is 52.5Å and two strong reflections are seen in the wide-region at 4.6 and 3.90 Å which demonstrate a highly ordered arrangement of the hydrocarbon chains. The structural characteristics of this phase do not change upon heating. After heating above the chain-melting transition the low-temperature phase becomes an untilted $L_β$ lamellar gel phase. The two peaks at 4.6 and 3.90 Å are replaced by a single, quite sharp symmetrical wide-angle peak at 4.15Å along with an increase in the layer spacing which rises to 55.7 Å in excess water. At low hydrations, the chain-melting transition occurs directly to the inverse hexagonal H_{II} phase, whereas in excess water the transition at 53.2 °C is to the fluid lamellar $L_α$ phase (d = 51.8 Å), followed by a further transition at 59.5 °C to the H_{II} phase (lattice parameter a = 67.8 Å).

The behaviour of the di-14:0-β-D-GalDAG compound is rather different. At low temperatures, depending on the sample history, it can form either the L_{C1} phase, with a layer spacing of 54.0 Å and strong wide-angle peaks at 4.54, 3.87 and 3.50 Å, or a second tilted lamellar crystalline phase L_{C2}, with a smaller layer spacing of 47.2 Å and a very different wide-angle pattern, containing numerous sharp peaks (Fig.3 (e)). It has not yet been possible to identify the packing modes from these wide-angle powder patterns, but it has been suggested that monoclinic chain subcells may be involved[14]. On the other hand, Hinz and co-workers suggest that the wide-angle patterns indicate hybrid orthorhombic / triclinic subcell packing modes[23]. What does seem clear from both our data and these earlier results is that the packing within the L_{C1} phase is essentially the same for both the Glc and the Gal compounds. The wide-angle diffraction pattern we observe from the L_{C2} phase appears to be different from any previously reported, possibly indicating the formation of a new lamellar crystalline phase, with a different subcell packing, not previously seen. At all hydrations, the lamellar crystal phase melts directly to the H_{II} phase upon heating, generally from the L_{C2} phase at low hydration (at 78.9 °C), but from the L_{C1} phase in excess water (at 70.2 °C). The lattice parameter of the H_{II} phase in excess water is 62.3 Å.

On the timescale of our experiments, we see no evidence for gel phase formation in the di-14:0-β-D-GalDAG compound. However, previous X-ray experiments did observe a transient $L_β$ lamellar gel phase, which reverted rapidly (less than one hour) to a lamellar crystalline phase[14,23]. The rate of reversion from the gel phase to the more ordered crystalline lamellar phases is known to be faster for β-D-Gal than for the β-D-Glc lipids, and also to depend on the chirality of the glycerol backbone[24]. Generally speaking, the rate is fastest for the 1,2-*sn* compounds, intermediate for the racemate, and slowest for the 2,3-*sn* compounds. The explanation for these differences is still unclear, but must lie in the different patterns of headgroup-headgroup and headgroup-water hydrogen bonding.

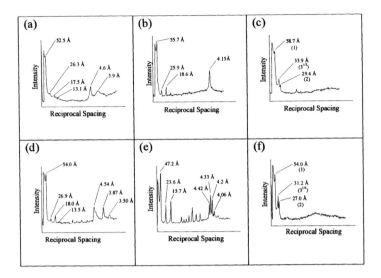

Figure 3 *X-ray powder diffraction patterns from the various phases of di-14:0-β-D-GlcDAG (a: L_{C1}; b: L_β; c: H_{II}), and di-14:0-β-D-GalDAG (d: L_{C1}; e: L_{C2}; f: H_{II}) in excess water.*

For hydration levels in excess of 20 wt% water, we observe a fluid lamellar L_α phase, for di-14:0-β-D-GalDAG upon cooling, at 61.5 °C, followed by a transition to the lamellar crystal L_{C1} phase at 54.2 °C. We infer that not only the L_β gel phase, but also the fluid lamellar L_α phase is metastable for this lipid, although this has not as yet been directly observed.

By plotting the lattice parameters as a function of water content (data not shown), we have determined the limiting hydration of each phase. We find the following values for the L_β gel, L_α fluid bilayer, and H_{II} phases (these are nearly independent of temperature within each phase):

System	L_β	L_α	H_{II}
di-14:0-β-D-GlcDAG	7	13	21
di-14:0-β-D-GalDAG	-	11	19
di-12:0-PE[s]	6	12	16

[s] Reference [7]

Table 2 *The limiting hydrations (waters per lipid molecule) of the L_β gel, fluid lamellar L_α, and inverse hexagonal H_{II} phases.*

The limiting hydrations in the gel phase are similar to the hydration numbers of 8.4 and 8.7 for free glucose and free galactose in solution[3]. Our results are in quite good agreement with the hydration values for dialkyl glycolipids inferred indirectly by Hinz and co-workers[23], and demonstrate that the limiting hydrations of the three phases for the glycolipid systems are rather similar to those established earlier for the same phases of the phospholipid system didodecyl phosphatidylethanolamine (di-12:0-PE)[7]. It has been established that, as in the case of PE, the lamellar crystalline phases of the glycolipids are essentially anhydrous (less than one water per lipid[23]).

The binary temperature-composition phase diagrams of these two synthetic glycolipid systems will be published at a later date.

Acknowledgments

This work was supported by grant GR/L48065 from the EPSRC (UK) and by a grant from the Canadian Institute of Health Research.

References

1 I. Ishizuka and T. Yamakawa, *"Glycoglycerolipids"*, in Glycolipids, ed H. Wiegandt (Elsevier, Amsterdam), 1985, 101.
2 H. Stephen, *Solubilities of Inorganic and Organic Compounds*, Pergamon Press, 1963.
3 S. A. Galema and H. Hoiland, *J. Phys. Chem*, 1991, **95**, 5321.
4 R. N. A. H. Lewis, D. A. Mannock and R. N. McElhaney, in *Lipid Polymorphism and Membrane Properties,* ed. R. M. Epand, 1997, 25.
5 Y. R. Deng and T. Landh, *Zool Stud*, 1995, **34 (1)**, 175.
6 J. M. Seddon, G. Cevc and D. Marsh, *Biochemistry*, 1983, **22**, 1280.
7 J. M. Seddon, G. Cevc, R. D. Kaye and D. Marsh, *Biochemistry,* 1984, **23**, 2634.
8 J. M. Seddon, J. L. Hogan, N. A. Warrender and E. Pebay-Peyroula, *Prog. Colloid Polym. Sci.,* 1990, **81**, 189.
9 R. N. A. H. Lewis, D. A. Mannock, R. N. McElhaney, D. C. Turner and S. M. Gruner, *Biochemistry,* 1989, **28**, 541.
10 J. M. Seddon and R. H. Templer, *Phil. Trans. Ro. Soc. A.*, 1993, **344**, 377.
11 D. M. Anderson, S. M. Gruner and S. Leibler, *Proc. Nat. Acad. Sci. U. S. A.*, 1988, **85**, 5364.
12 R. Tenchova, B. Tenchov, H.-J. Hinz and P. J. Quinn, *Liquid Crystals*, 1996, **20**, 469.
13 H.-J. Hinz, H. Kuttenreich, R. Meyer, M. Renner and R. Freund, *Biochemistry*, 1991, **30**, 5125.
14 D. A. Mannock, R. N. McElhaney, P. E. Harper and S. M. Gruner, *Biophys. J.,* 1994, **66**, 734.
15 D. A. Mannock, R. N. A. H. Lewis, A. Sen and R. N. McElhaney, *Biochemistry,*1988, **27**, 6852.
16 D. A. Mannock, P. E. Harper, S. M. Gruner and R. N. McElhaney, *Chem. Phys. Lipids,* 2001, **111**, 139.

17 D. A. Mannock, R. N. A. H. Lewis, R. N. McElhaney, M. Akiyama, H. Yamada, D. C. Turner and S. M. Gruner, *Biophys. J.*, 1992, **63**, 1355.
18 D. C. Turner, Z.-G. Wang, S. M. Gruner, D. A. Mannock and R. N. McElhaney, *J. Phys. II France*, 1992, **2**, 2039.
19 H. Kuttenreich, H.-J. Hinz, R. D. Koynova and B. G. Tenchov, *Chem. Phys. Lipids*, 1993, **66**, 55.
20 N. Zeb, PhD Thesis, *Imperial College, London*, 1998.
21 P. M, Duesing, J. M. Seddon, R. H. Templer and D. A. Mannock, *Langmuir*, 1997, **13 (10)**, 2655.
22 D. A. Mannock, M. Kreichbaum, P. E. Harper, S. M Gruner and R. N. Mcelhaney, *Synthesis and Thermotropic Characterization of an Homologous Series of Racemic β-D-Galactosyl Dialkylglycerols*. Manuscript in preparation, 2002.
23 M. Köberl, H.-J. Hinz and G. Rapp, *Chem. Phys. Lipids*, 1998, **91**, 13.
24 D. A. Mannock, M. Akiyama, R. N. A. H. Lewis and R. N. McElhaney, *Biochim. Biophys. Acta*, 2000, **1509**, 203.

Subject Index

Ab initio methods, 23
Activation free energy, 153
Acylated protein, 61
β_2-Adrenergic receptors, 86
Albmycin, 216
Alcohol dehydrogenase, 26
AMBER-95, 79
Amide, 22, 141
Amphipathic helix, 169, 191, 197
Amyloidosis, 103, 115, 116
Amyloids, 94, 103, 104
Apoptosis, 166
Artificial lipid membranes, 3
Atomic force microscopy (AFM), 31, 33, 50–54
ATP, 64

B cell receptors, 61
B850, 119
Bacteriophage, 147
Bacteriorhodopsin, 26, 208, 211, 222, 224, 226
Barrier crossing, 147
Bath modes, 123
Benzodiazepine, 243
Bicontinuous cubic phase, 177, 178
Bilayer, 138, 200, 206, 215, 230, 251, 254
Binding force measurements, 35
Bioimaging, 61
Bioluminescence resonance energy transfer (BRET), 86
Brownian motion, 65, 155
Butanedione monoxime, 62

Calcium-sensing receptors, 85
Carbonic anhydrase, 26
Cardiomyocytes, 41
CASPT2, 23
CASSCF2, 23, 27
Catalase, 78, 81, 83
CD spectroscopy, 15, 18, 20, 21, 26, 140
CD4, 73
CD45, 59

CD54, 63, 65
CD86, 65
CHARMM, 7
Chitin, 53, 56
Chlordiazepoxide, 243, 244, 245
Chlorophyll, 229
Chloroplasts, 147
Cholera toxin, 61, 62
Cholesterol, 60, 67, 170
Chiral thin film, 139
Chromatin, 165, 166
Chymotrypsin, 26, 170
Chymotrypsinogen, 26
Classical molecular interaction potential (CMIP), 82
CNDO/S, 27
Colicin M, 216
Collective excitation dynamics, 118
Concanavalin A, 26
Confocal microscopy, 31, 59
Confocal spot, 238
Constant mean curvature, 189
Cork domain, 216, 217
Crossing time, 155
Cryo-electron microscopy, 218
Crystallisation, 221, 225, 227, 228
C-type lectin, 58
Curvature elastic energy, 178
Cyanobacteria, 237
Cyclohexane, 181
Cytidylyltransferase (CCT), 163
Cytochalasin D, 62
Cytochrome *c*, 4, 15, 26
Cytoskeleton, 35, 40, 41, 42, 43, 60, 62, 64, 65, 240
Cytotoxic peptide, 191
Cytotoxic T cells, 65

Dendritic cells, 62, 65
Density functional theory, 23
Detergent effects, 228
Diamide, 25
Differential scanning calorimetry, 218, 249, 267, 270, 272, 273

Dilauroylphosphatidylcholine, 177
Dioleoylphosphatidylcholine, 202
Direct immunofluorescent microscopy, 166
DNA, 147, 149, 158, 218, 219
DsbB, 209

Ectodomain shedding, 64, 65
Elastase, 26
Elastic stress, 169
Electro-magnetic properties, 136
Electron tomography, 219
Electrostatic free energy, 101
Electrostatic interactions, 22
Electrostatic stability, 94
Ellipsommetry, 141
Endoplasmic reticulum, 147, 165
Entropy analysis, 86, 89, 90
Erabutoxin, 26
Erythrocytes, 193
Escherichia coli, 199, 215
Evolutionary trace (ET) method, 86, 87
Excess water measurements, 182
Exciton, 118, 129, 132
Exosomes, 64, 65

F-actin, 41, 43, 62
Familial amyloidotic polyneuropathy (FAP), 94
Fascin, 62
Feedback regulation, 172, 173
FhuA, 216
Fibrilogenesis, 115
Flavodoxin, 26
Flow linear dichroism, 3, 18
Fluid lamellar phase, 274
Fluorescence imaging, 58
Fluorescence lifetime imaging (FLIM), 58, 66–68
Fluorescence recovery, 237
Fluorescence resonance energy transfer (FRET), 61, 86
Fluorescence spectroscopy, 208
Formamide, 23
FRAP, 238–240
Frenkel excitons, 119
Friction image, 52
FTIR spectra, 141

GABA$_B$ receptor, 85, 87, 89, 90

Gangliosides, 61
Gaussian bending modulus, 257, 265
Gaussian curvature, 231
Gene therapy, 4
Glutamate receptors, 85
Glutathione reductase, 26
Glycolipids, 267–269, 275
Glycoprotein, 44
Glycosylation, 50, 57
GPI anchor, 61
G-protein coupled receptors (GPCRs), 85–89
Gramicidin, 4, 12, 15
Green fluorescent protein (GFP), 59, 63–67
GROMACS, 73, 95, 105
GTPγS, 88

Halorhodopsin, 222
Harmonic oscillators, 159
α-Helix, 11, 20, 21, 140, 168, 191, 194, 196, 208, 209, 213
Heme, 81
Hemoglobin, 26
Hemolytic activity, 191, 193
HIV-1, 72
Huang-Rhys factor, 128, 130–132, 134
Hydrogen bonds, 81, 217
Hydrogen peroxide, 78, 79, 81–83

ICAM-1, 59, 62, 63
Immune synapse, 58–60, 62, 63, 65, 66
Immunoregulatory tyrosine activating motifs (ITAMs), 61, 62
Influenza virus, 72
Integrin, 38, 39
Inverse hexagonal phase, 177, 251, 274
Inverse lyotropic mesophases, 177
Ion channel, 42, 72
IR spectroscopy, 140, 141,
Iron transporter, 215

Kinases, 61
Kink mechanism, 147, 149, 150, 154, 156
Kramers problem, 147, 148

Lactate dehydrogenase, 26
Lamellar phase, 226, 232
Langevin equation, 119, 124, 126, 159
Langmuir films, 139

Laplace's law, 204
Laser scanning confocal microscopy, 59, 63
Lateral force microscopy, 51
Lattice parameter, 183
Lauric acid, 177
Lck, 61
LD cell, 7
LD spectroscopy, 3, 15, 18
LFA-1, 59, 63
LH1, 119
Light emitting diodes, 136
Light harvesting complexes, 222, 225, 229, 233, 234
Light harvesting systems, 118
Light microscopy, 32
Lipid bilayer, 18, 72, 221
Lipid rafts, 59, 60–62
Lipid vesicles, 203
Lipids, 167
Liposome, 3, 4, 18, 218, 243–245, 254
Lymphocytes, 65
Lysophosphatidylcholine, 169, 172
Lysozyme, 26, 103, 104

Macrophages, 173
Mahalanobis distance, 106
MD simulation, 73, 114
Mean curvature, 232
Mechanosensation, 199
Mechanosensitive channel (MscL), 199, 206
Membrane bending energy, 256
Membrane curvature, 169
Membrane dynamics, 118
Membrane pore, 147
Membrane proteins, 208, 217, 221
Membrane synthesis, 163
Membrane-tension-gated channel, 199
Mesophase, 184, 230, 231
Metropolis algorithm, 262
MHC, 58, 59, 63–65
Mitochondria, 147
Molecular aggregates, 118
Molecular dynamics, 20, 78–80
Monolayer, 139, 141, 177, 191, 193–196
Monoloein, 222, 223
Monosaccharides, 267
Monte Carlo simulation, 260, 261, 262, 263

Multi-reference configuration interaction (MRCI), 23, 27
Myosin, 62
Myristoylated protein, 61

Natural killer (NK) cells, 58, 62–65
NB protein, 73–76
NhaA, 209
Nonadiabaticity, 124

Octadecene, 181
Octylglucosides, 61
Oligosaccharides, 267
Opioid receptors, 85
Opsin receptors, 86
Oregon Green, 165
Osmotic stress, 179, 181, 186
Osteoblasts, 42
Osteoclasts, 35, 39

Papain, 26
Parametric estimation, 230
Patch-clamp, 202
Pauli repulsion, 24
Periplasm, 215
Peroxisome, 79
Phage receptors, 215
Phase behaviour, 180
Phosphatidic acid, 171, 172
Phosphatidylcholine, 163, 164, 168, 172
Phosphatidylethanolamine, 163, 200, 202
Phospholipase, 172
Phospholipid, 60, 243, 251, 275
Phosphorylation, 61
Photosynthetic membranes, 237
Photosystem I, 240, 241
Photosystem II, 229, 237, 240, 241
Phycobilisomes, 237–239, 241
Plasma membrane, 65
Plastocyanin, 26
Poisson-Boltzmann equation, 96, 97
Polarization modulation infra-red reflection absorption spectroscopy (PM-IRRAS), 192
Polarizing microscopy, 223
Polaron, 118, 128, 129, 132
Polyalanine, 136, 140, 142, 143
Polyethylene glycol, 35
Polypropylene, 138
Pore, 156

Porin, 26, 215
Prealbuim, 94
Prealbumin, 26
pre-PsbW, 4, 15
Protease inhibitors, 65
Protein folding, 208
Protein translocation, 147
Protein-lipid interactions, 199
Proteoliposomes, 219
Pyrene, 10

Ras, 89
Reaction centre, 222, 237, 240
Receptor activity modifying protein
    (RAMP), 90, 91
Renaturation, 213
Retinol, 94
Rhodanese, 26
Rhodobacter sphaeroides, 222
Rhodopseudomonas acidophila, 223
Rhodopseudomonas viridis, 222
Rhodopsin, 86, 87
Ribonuclease, 26
Rosenfeld equation, 21
Rouse model, 148, 149, 152, 157
Rydberg states, 23

Saccharomyces cerevisiae, 50, 51, 57
Scanning Electron Microscopy, 53
Scanning probe microscopy, 31
Schizosaccharomyces pombe, 51
SDS micelles, 17
Secondary structure, 80
Senile systematic amyliodosis (SSA), 94
Shear distorted liposomes, 4
β-Sheet, 11, 20, 191, 196, 197, 216
Siderophores, 215
Single photon counting, 66
Site-directed mutagenesis, 81
Small angle X-ray diffraction, 223, 226
Solar cells, 136
Solvent accessible surface area (SASA),
    108, 114
Soret band, 17
Spectrin, 240

Sphingolipids, 60
Sphingomyelin, 163, 172
Spiroplasma melifferum, 192, 193
Stark effect, 137
Strickler-Berg formula, 66–68
Subtilisin, 26
Superoxide dismutase, 26
Supramolecular activating clusters
    (SMAC), 59
Surface hopping, 125, 126
Surface roughness, 56
Synchrotron radiation, 20
Synechococcus sp., 237–239

T cell receptors, 59, 61, 62
Talin, 254, 255, 258, 259
TDDFT, 23
Texas Red, 165
Thermolysin, 26
Thylakoid membrane, 17, 237, 238, 240,
    241
TIP3P, 79, 82
Ton complex, 219
Transferrin receptors, 61
Transition polarisation, 11
Translocation, 157
Transmembrane helix, 208
Transmembrane segment (TM), 73, 73,
    75
Transthyretin (TTR), 94, 95
Triose phosphate isomerase, 26
Triton X-100, 61
Trypsin, 26
Tubulin, 65
Tyrosine kinase, 42

Vitamin E, 248, 250, 251, 252
Vpu protein, 73-76

Xplor, 73
X-ray diffraction, 181, 216, 227, 249,
    267, 271, 273
X-ray scattering, 250

Zinc, 64